I0888757

Submarine Cables:
The Handbook of Law and Policy

Submarine Cables

The Handbook of Law and Policy

Edited By

Douglas R. Burnett
Robert C. Beckman
Tara M. Davenport

MARTINUS
NIJHOFF
PUBLISHERS

LEIDEN • BOSTON
2014

Library of Congress Cataloging-in-Publication Data

Submarine cables : the handbook of law and policy / edited by Douglas R. Burnett, Robert C. Beckman, Tara M. Davenport.
pages cm
Includes index.
ISBN 978-90-04-26032-0 (hardback : alk. paper) — ISBN 978-90-04-26033-7 (e-book)
1. Cables, Submarine—Law and legislation. I. Burnett, Douglas R. II. Beckman, Robert C. III. Davenport, Tara.

K4317.S83 2014
384'.042—dc23

2013028490

This publication has been typeset in the multilingual "Brill" typeface. With over 5,100 characters covering Latin, IPA, Greek, and Cyrillic, this typeface is especially suitable for use in the humanities. For more information, please see www.brill.com/brill-typeface.

ISBN 978-90-04-26032-0 (hardback)
ISBN 978-90-04-26033-7 (e-book)

Copyright 2014 by Koninklijke Brill NV, Leiden, The Netherlands.
Koninklijke Brill NV incorporates the imprints Brill, Global Oriental, Hotei Publishing, IDC Publishers and Martinus Nijhoff Publishers.

All rights reserved. No part of this publication may be reproduced, translated, stored in a retrieval system, or transmitted in any form or by any means, electronic, mechanical, photocopying, recording or otherwise, without prior written permission from the publisher.

Authorization to photocopy items for internal or personal use is granted by Koninklijke Brill NV provided that the appropriate fees are paid directly to The Copyright Clearance Center, 222 Rosewood Drive, Suite 910, Danvers, MA 01923, USA.
Fees are subject to change.

This book is printed on acid-free paper.

CONTENTS

PART II

INTERNATIONAL LAW ON SUBMARINE CABLES

PART III

CABLE OPERATIONS—LAW AND PRACTICE

PART IV

PROTECTING CABLESHIPS AND SUBMARINE CABLES

SPONSORING INSTITUTES

Centre for International Law, National University of Singapore

The Centre for International Law (CIL) is a university-wide research centre established in 2009 at the National University of Singapore (NUS) in response to the growing need for international law expertise and capacity building in the Asia-Pacific region. CIL focuses on multidisciplinary research and collaborates very closely with the NUS Faculty of Law as well as other high calibre organizations and institutions to further its research and capacity-building objectives.

CIL focuses its activities on three core areas that are critical to the Southeast Asia region, these being Ocean Law and Policy, ASEAN Law and Policy, and Trade and Investment Law and Policy. As part of its activities in Ocean Law and Policy, CIL has undertaken work on piracy and international maritime crimes, the South China Sea disputes, biodiversity and environmental issues. This present publication, *Submarine Cables: The Handbook of Law and Policy,* is part of an extensive CIL research project on submarine cables, which has included two regional Workshops organized in collaboration with the International Cable Protection Committee (ICPC). CIL research and other relevant materials on submarine cables are available on our website, www.cil.nus.edu.sg.

International Cable Protection Committee (ICPC)

The ICPC is the premier international submarine cable authority providing leadership and guidance on issues related to submarine cable security and reliability. Founded in 1958, the ICPC membership spans over 63 nations and presently includes the owners and operators of over 97 per cent of the world's international submarine cable systems and the 18 submarine power cable owners. Since 2010 governments have been eligible to join and many have elected to do so. Membership is also open to submarine cable system suppliers and installers, marine survey companies, cableship owners and operators, international banks, and others with interest in critical submarine cable infrastructure.

The ICPC issues *Recommendations*, available to the public upon request, on aspects of submarine cable laying, repair, surveying, and protection. The ICPC works to promote education and compliance with the United Nations Convention on the Law of the Sea (UNCLOS) and customary international law impacting submarine cables among its members, States, international organizations, and other seabed users. More information is available from the ICPC website www.iscpc.org.

FOREWORD

Dean Veverka, Chairman, International Cable Protection Committee (ICPC)

The submarine cable industry has flourished in the world's oceans since 1850. Progressing from telegraph and telephony to high-speed data fiber optic cables and power cables, these submarine cables are increasingly recognized as critical international infrastructure by more and more nations. While the technical success of the industry represents steady evolution and innovation by countless people in companies worldwide, the role of international law in the success of the business is not well understood by many in governments involved with diplomacy and ocean policy decisions. This Handbook is welcomed by the industry as the first comprehensive book on the topic of submarine cable law and policy. My hope is that it will allow industry and governments to work better together in providing the world with ever improving international communications, power, energy, scientific knowledge and security based on submarine cables in the ocean environment.

I note with pride that the co-authors of this Handbook represent a diverse and seasoned group of leaders in the various sectors that comprise the cable industry. But what makes this Handbook so valuable is the partnering of these industry experts with recognized experts in international law of the sea. Too often governments make policy decisions about undersea cables without the knowledge and experience that is available to them from the cable industry. On the other hand, companies can act with imperfect knowledge of their rights and obligations under international law. By combining legal scholarship and sound industry experience in a readable volume, the Handbook is well on its way to becoming the 'go to' reference for both business and government, the essential purpose of the Handbook.

FOREWORD

Professor Tommy Koh, Chairman of the Governing Board of the Centre for International Law, (CIL) at the National University of Singapore (NUS)

In December 2009, the newly established Centre for International Law (CIL) at the National University of Singapore (NUS) organized its inaugural 'Workshop on Submarine Cables and Law of the Sea' in collaboration with the International Cable Protection Committee (ICPC).

The Workshop, one of the first of its kind in the region, brought together experts from the cable industry, law of the sea experts and government representatives from the region. Its objective was to examine the practice of industry and governments on submarine cables in light of the legal regime set out in the 1982 United Nations Convention on the Law of the Sea (UNCLOS).

The discussions at the Workshop revealed that governments did not fully appreciate the importance of submarine cables and that there was a lack of communication between governments and the submarine cable industry. It was acknowledged that this contributed to the adoption of international and national policies which were often detrimental to the integrity of the world's international telecommunications systems.

The challenges confronting the submarine cable industry prompted CIL and the ICPC to continue to collaborate in order to enhance discussion and understanding of the importance of submarine cables. Since the 2009 Workshop, CIL and the ICPC have worked closely together on a variety of projects to raise awareness and foster dialogue on this critical communications infrastructure. This Handbook, which marks the culmination of the joint efforts of CIL and the ICPC, is timely and significant for several reasons.

First, the Handbook provides a one-stop shop of essential information pertaining to the international governance of submarine cables. It is extensive in its scope and comprehensively covers a wide range of issues relating to submarine cables. It includes essential information on the development and uses of submarine cables, the submarine cable industry, the international legal regime governing submarine cables, the issues relating to cable operations and the protection of cables, as well as new uses of submarine cables.

Second, the majority of chapters are authored by both an international lawyer and an expert from the submarine cable industry. The result is a unique combination of legal and technical knowledge which allows the contributors to formulate effective policy recommendations on specific issues relating to submarine cables. Accordingly, the Handbook will be an invaluable source of knowledge to a large audience including academics, the submarine cable industry, government officials and policy-makers.

Third, the Handbook is the first of its kind available in the market. Despite the world's increasing reliance on submarine cables for a myriad of activities including the internet and telecommunications, there is very little contemporary literature on submarine cables. The Handbook will fill this void and hence make an important contribution to the discussion of possible solutions for the issues faced by both governments and the cable industry, in the governance of submarine cables.

ACKNOWLEDGEMENTS

The Editors would first like to thank their respective organizations, the International Cable Protection Committee (ICPC) and the Centre for International Law (CIL) at the National University of Singapore (NUS) for their unwavering support and encouragement for this project.

Second, the Editors would like to thank all the authors who, despite their busy schedules, have given up a great deal of time and resources to provide their expertise and have been instrumental in making this project successful. They have also helped to source the various images which have contributed to the quality of the Handbook. Special thanks to Professor Lionel Carter and Stewart Ash for their extensive assistance in preparing material for the Handbook.

Third, much appreciation goes to Kevin Summers from Submarine Telecoms Forum for his efforts in providing us with the images of the global submarine network, as well as Dr. Kevin Tan for his invaluable editorial advice and assistance. Thank you also to the staff of Squire Sanders (US) LLP and to Peter Gibson for his last minute problem solving skills.

Last, but not least, the Editors would like to express their heartfelt thanks to Monique Page, CIL Research Associate, for her tireless efforts in overseeing the publication of the Handbook, from liaising with the authors and publisher, to editing, drafting and re-drafting and ensuring that deadlines were met. The Handbook would not have been possible without her contribution.

Robert Beckman, Douglas Burnett and Tara Davenport

LIST OF CONTRIBUTORS

Stewart Ash, is an independent consultant. He specializes in assisting oil and gas companies with the design and implementation of submarine systems for offshore facilities. Stewart graduated in 1970, joining STC Submarine Systems (a British telecommunications company) as a transmission equipment design engineer. He subsequently worked in various capacities on surveys, terrestrial/marine installation and system commissioning and in submarine system supply. He moved to Cable & Wireless Marine (CWM) to lead the development of low cost installation solutions for the emerging repeaterless system market. In 1999, CWM was taken over by Global Crossing and Stewart was appointed General Manager Engineering Services, responsible for cable engineering, jointing technology, fault investigation and the protection of Corporate Intellectual Property. In this role he was chairman of the Universal Jointing Consortium until 2005. Stewart also writes a bi-monthly article on the history of the industry for SubTel Forum and in 2000 he co-wrote and edited a book on the first 150 years of the submarine cable industry, *From Elektron to 'e' Commerce.*

Professor Robert Beckman, is Director of the Centre for International Law (CIL), a university-wide research centre at the National University of Singapore (NUS). In addition to serving as Director of CIL, he also heads its Ocean Law and Policy programme. Professor Beckman received his J.D. from the University of Wisconsin and his LL.M. from Harvard Law School. He is an Associate Professor at the NUS Faculty of Law, where he has taught for more than 30 years. He currently teaches Ocean Law and Policy in Asia, Public International Law and International Regulation of Shipping. He also lectures at the summer programme for the Rhodes Academy of Oceans Law and Policy, Greece. Professor Beckman is an expert in law of the sea issues in Southeast Asia, including piracy and maritime security. He served for several years as a regional resource person in the workshops on Managing Potential Conflicts in the South China Sea. He is an Adjunct Senior Fellow in the Maritime Security Programme at the S. Rajaratnam School of International Studies (RSIS), Nanyang Technological University (NTU).

Douglas R. Burnett, International Law Advisor for the International Cable Protection Committee (since 1999), and Maritime Partner in the New York office of Squire, Sanders (U.S.) LLP, an international law firm with 39 offices in 19 countries. His practice focuses on international law, submarine cables, maritime and shipping, involving litigation and arbitration. Douglas is a graduate of the U.S. Naval Academy and University of Denver Law School and is a retired captain in the U.S. Navy. He has argued before the U.S. Supreme Court and testified as an industry expert on submarine cables before the 2007 Senate Foreign Relations Committee for the UNCLOS hearings. Douglas has worked on submarine cable cases for over 30 years. He has frequently instructed at the Rhodes Academy of Oceans Law and Policy, Greece.

Professor Lionel Carter, Marine Environmental Advisor for the International Cable Protection Committee (since 2003), and Professor of Marine Geology, Antarctic Research Centre, Victoria University, Wellington, New Zealand. Lionel trained in geology and oceanography at the universities of Auckland, New Zealand and British Columbia, Canada and has undertaken research in the North Pacific, North Atlantic and Southern oceans, as well as off New Zealand. This research led to publication of over 140 peer-reviewed papers. Lionel helped set up major international projects in New Zealand, including Ocean Drilling Program Leg 181 and the MARGINS "Source to Sink" initiative, which is designed to determine the processes that shape Earth's surface from the mountains to the abyssal ocean. His latest project relates to the Antarctic Drilling Programme, which seeks to identify the impacts of the Ross Ice Shelf on global climate and oceans. Lionel has expertise gained in marine geology/oceanography which is applied to ocean engineering projects, in particular submarine telecommunication and power cables.

Tara M. Davenport, Research Fellow, Centre for International Law, Singapore. Tara holds a Bachelor of Laws from the London School of Economics and a Masters of Law in Maritime Law from the National University of Singapore. She is a qualified lawyer in Singapore and has spent a large part of her career working as a lawyer in one of Singapore's top shipping law firms. As a Research Fellow Tara currently undertakes research in the area of Ocean Law and Policy, with a particular emphasis on maritime crimes, submarine cables, joint development, and the South China Sea. She is also an Assistant Editor of the Asian Journal of International Law. Tara attended the 2010 Rhodes Academy of Oceans Law and Policy and was the winner of the inaugural Rhodes Academy Submarine Cables Award sponsored by International Cable Protection Committee in 2010 for her paper "Submarine Cables: Problems in Law and Practice". She is a co-tutor with Professor Robert Beckman at the National University of Singapore for Ocean Law and Policy and was also a tutor at the Maritime Delimitation Workshop organized by the International Boundaries Research Unit, Durham and CIL in Singapore in September 2011. Tara is the recipient of Singapore's 2013 Fulbright Scholarship and will be commencing her LL.M. at Yale University in August 2013.

Malcolm Eccles, CEO and Managing Director of the Basslink Group based in Australia and Director / Executive Committee member of the International Cable Protection Committee (ICPC). Malcolm is the CEO and Managing Director of the Basslink Group of companies. The Basslink Group comprises a power business, telecoms business and engineering consultancy business. Malcolm is a non-executive Director of City Gas Pte Ltd in Singapore, City Gas is the largest gas retailer in Singapore with over 660,000 customers. He is also a non-executive Director of Gippsland Water in Australia, Gippsland Water is the second largest regional water authority in the state of Victoria. Malcolm has been an Executive Committee member and Director of the ICPC since 2007. Malcolm has a Master of Science degree in Electrical Engineering and Management Studies, a Post Graduate Diploma in Project Management, a Post Graduate Diploma in Strategic Management and is a Chartered Electrical Engineer. He is a member of the IET (UK) and a Senior Member of the IEEE (US). Malcolm has been actively involved with submarine power cables and HVDC transmission systems since 2002.

Graham Evans, Graham has more than 34 years experience as a marine geoscientist and is Business Development Director for the EGS Survey Group of companies worldwide having specific responsibilities for the Group's submarine telecommunications cable business; and is an Executive Director of EGS Survey Pty Ltd, Australia and EGS Americas Inc. Graham's involvement in the submarine cable industry began in 1990 after being encouraged to adapt geoscience procedures he developed for dredging applications to submarine cable burial assessment. Graham joined EGS in 1978 leaving in 1990 to join as one of the first three employees of what became Fugro Survey playing a key role developing that company's submarine cable business before rejoining EGS in 1996. Graham is a member of the SubOptic Executive Committee, serving as Vice Chair of the SubOptic 2010 Program Committee, and represents EGS in the ICPC and is a member of the ICPC Executive Committee. Graham is a regular speaker at international submarine telecommunications events around the world. Graham holds a Bachelor of Science in Geology and Bachelor of Arts in Earth and Environmental Sciences.

Captain Keith Ford-Ramsden, is an independent consultant providing technical and project management services for the survey, installation and maintenance of telecoms, power and renewable energy submarine cables on a global basis through his company, KFR Marine Ltd, since 2007. Keith spent the final 15 of his 23 years in the British Merchant Navy in the oil and gas and telecommunications industries before transferring ashore in 1996 to carry out a variety of roles with Cable & Wireless Marine Ltd (UK), Global Marine Systems Ltd (Singapore) and SB Submarine Systems (China) involved with the planning, installation and maintenance of submarine telecommunications cables. Keith moved to FLAG Telecom as Marine Maintenance Manager in 2001 and served on the Executive Committee of the International Cable Protection Committee until 2007.

Joska Ferencz, Technical Services Manager for the Basslink HVDC Interconnector, Chairman of the Oceania Submarine Cable Association, a member of the International Cable Protection Committee, CIGRE B1 Insulated Cable Australia Panel and a Technologist Member of the Institute of Engineers Australia. Joska holds a Bachelor in Technology in Electrical and Electronic Engineering and a Masters of Business Administration in Management and Logistics from the Australian Maritime College.

Mick Green, Head of Subsea Centre of Excellence, British Telecom; Mick joined the Submarine Cable Systems unit of British Telecom in 1980 with a degree in Physics and has over 31 years experience in the submarine cable industry. Mick has held positions in engineering, project management, operations and maintenance with responsibility for many major submarine cable projects including UK-Belgium 5, the first international optical system and TAT-12/13, the first to use optical amplifiers and ring switching. Mick headed the Subsea Centre of Excellence within BT, responsible for strategy, planning, implementation and operation of all BT's subsea cable interests. He is author of several industry papers and currently the Vice-Chairman of the International Cable Protection Committee.

Wayne Nielsen, Founder of WFN Strategies, a company that provides project development and engineering of remote communications for telecoms, defence, and oil and gas clients, and includes transoceanic submarine cables systems. Wayne is also founder and publisher of Submarine Telecoms Forum magazine. Wayne has 25 years of telecoms experience, and has developed and managed international projects in the Americas, Far East/Pacific Rim, Europe and the Middle East.

Monique Page, Research Associate, Centre for International Law, Singapore. Monique has a Masters of Laws in International and Comparative Law from the National University of Singapore, a first class honours degree in English Literature, a Bachelor of Laws, a Bachelor of Arts, and a Post Graduate Diploma in Art History and Classical Studies. Monique was formerly a practicing solicitor in Australia and also a Legal Editor for Butterworths Legal Publishing in New Zealand. As a Research Associate at CIL Monique currently works in the areas of Ocean Law and Policy, International Regulation of Shipping, and Treaty Law and Practice.

Captain J. Ashley Roach, JAGC, U.S. Navy (retired) was attorney adviser in the Office of the Legal Adviser, U.S. Department of State, from 1988 until he retired at the end of January 2009. He was responsible for law of the sea matters. He has taught, advised and published extensively on national maritime claims and other law of the sea issues, including piracy and armed robbery at sea. He has negotiated, and participated in the negotiation of, numerous international agreements involving law of the sea issues. He received his LL.M. (highest honors in public international law and comparative law) from the George Washington University School of Law in 1971 and his J.D. from the University of Pennsylvania Law School in 1963.

Professor Alfred Soons, studied law at Utrecht University, the Netherlands, followed by postgraduate studies in international law at the University of Washington (Seattle, USA) and Cambridge University (UK). He obtained a Ph.D. degree at Utrecht University in 1982 with a thesis on the international legal regime of marine scientific research. After having served from 1976 in various legal and policy positions as a civil servant at the Netherlands Ministry of Transport, Water Management and Public Works he became professor of public international law and director of the Netherlands Institute for the Law of the Sea (NILOS) at Utrecht University in 1987. He is, *inter alia*, a former president of the Netherlands Society of International Law, Director of Studies of the International Law Association (ILA), member and chairman of the Jury for the Hague Prize for International Law, and member and chairman of the Standing Advisory Committee on Public International Law of the Netherlands Ministry of Foreign Affairs. Currently he serves as chairman of the Scientific Advisory Council of the Netherlands Defense Academy, member of the Advisory Body of Experts on the Law of the Sea of the Intergovernmental Oceanographic Commission (IOC/ABE-LOS), and co-director of the Rhodes Academy of Oceans Law and Policy. As counsel and arbitrator he has been involved in international litigation at the International Court of Justice and arbitral tribunals.

Robert Wargo, serves on the Executive Committee of the International Cable Protection Committee. He is also is the President of the North American Submarine Cable Association—a trade organization similar to the ICPC with a focus on the unique challenges of laying and maintaining undersea cables in the US, Canada and the Caribbean. Bob is currently the Marine Liaison Manager with the AT&T Undersea Cable Systems Operation and Maintenance organization. In this position he handles a wide variety of issues related to the installation and maintenance of undersea cables, from planning through retirement and removal. Bob has been involved in the submarine cable industry at AT&T since 1990. Prior to his employment with AT&T, Bob was employed by Rutgers University on projects related to beach grasses, vegetative buffers adjacent to waterways, surf clam population estimates and oyster diseases. Bob holds a Bachelor of Science degree in Marine Science from Stockton State College.

TABLE OF MULTILATERAL CONVENTIONS

1988

1982

TABLE OF CASES

LIST OF FIGURES, IMAGES AND MAPS

ABBREVIATIONS*

AC	Alternating Current
ACMA	Atlantic Cable Maintenance Agreement
AIS	Automatic Identification System
APEC	Asia-Pacific Economic Cooperation
ARPA	Automatic Radar Plotting Aids
ATOC	Acoustic Thermometry of Ocean Climate
BAS	Burial Assessment Survey
BMH	Beach Manhole
BOEM	Bureau of Ocean Energy Management
BU	Branching Unit
C&MA	Construction and Maintenance Agreement
CANTAT-1	Canadian Trans-Atlantic Telephone Cable-1
CHIPS	United States Clearing House Interbank Payment System
CIGRE	International Council on Large Electric Systems
CIL	Centre for International Law, National University of Singapore
CLCS	Commission on the Limits of the Continental Shelf
CLS	Cable Landing Station
CMA	Cable Maintenance Agreement
COLREGS	Convention on the International Regulations for Preventing Collisions at Sea
COTDR	Coherent Optical Time Domain Reflectometers
CPT	Cone Penetration Test
CTD	Conductivity/Temperature/Depth
DA	Double Armor
DART	Deep-ocean Assessment and Reporting of Tsunamis

* Only commonly used abbreviations from the Handbook are listed. Names of submarine cable companies are not included.

DC	Direct Current
DEFRA	Department of Environment, Food and Rural Affairs (United Kingdom)
DGPS	Differential Global Positioning Systems
DMA	Defense Mapping Agency (United States)
DPS	Dynamic Positioning Systems
DTS	Desktop Study (Cable Route Survey)
EC	European Commission
EDFA	Erbium Doped Fiber Amplifier
EEZ	Exclusive Economic Zone
EIA	Environmental Impact Assessment
EIAO	Environmental Impact Assessment Ordinance
EIG	Europe India Gateway
EIS	Environmental Impact Study
EMF	Electromagnetic Field
EU	European Union
Gbit	Gigabit per second
GIS	Geographic Information System
GPS	Global Positioning System
HDD	Horizontally Directional Drill
HV	High Voltage
HVDC	High Voltage Direct Current
Hz	Hertz
ICAO	International Civil Aviation Organization
ICPC	International Cable Protection Committee
IEEE	Institute of Electrical and Electronics Engineers
ILC	International Law Commission
IMO	International Maritime Organization
IOC	International Oceanographic Commission
IODP	Integrated Ocean Drilling Program
IRU	Indefeasible Right of Use
ISA	International Seabed Authority
ISPS	International Ship and Port Facility Security Code
ITT	Invitation to Tender
ITU	International Telecommunication Union
K bit/z	Kilobit
kHz	Kilohertz
km	Kilometer
LCE	Linear Cable Engine
LRAD	Long Range Acoustic Device
LW	Lightweight
LWP	Lightweight Protected
LWS	Lightweight Screened
m	Meter

MA	Maintenance Authority
MAI	Marine Archeological Investigation
MARS	Monterey Accelerated Research System
MBARI	Monterey Bay Aquarium Research Institute
Mbit/s	Megabit per second
MHz	Megahertz
MOU	Memorandum of Understanding
MPA	Marine Protected Area
MCPT	Mini Cone Penetration Test
MSP	Marine Spatial Planning
MSR	Marine Scientific Research
NASCA	North American Submarine Cables Association
NEPTUNE	North-East Pacific Time-series Undersea Networked Experiments
nm	Nautical Miles (NM) (1 nm = 1.852 km)
NMS	National Marine Sanctuary
NOAA	National Oceanic and Atmospheric Administration
NOC	Network Operations Center
O&M	Operation and Maintenance
OADM	Optical Add Drop Multiplexer
OOI	Ocean Observatories Initiative
OSPAR	Oslo/Paris Convention for the Protection of the Marine Environment of the North-East Atlantic
P&I Insurance	Protection and Indemnity Insurance
PFE	Power Feed Equipment
PLB	Post Lay Burial
PLGR	Pre-Lay Grapnel Run
PLSE	Pre-Laid Shore End
POPS	Points of Presence
PSO	Power Safety Officer
PSSA	Particularly Sensitive Sea Areas
RCPC	Regional Cable Protection Committees
RLO	Restoration Liaison Officer
ROV	Remotely Operated Vehicle
RPL	Route Position List
RSN	Regional Scale Node
SA	Single Armor
SHOM	Service Hydrographique et Océanographique de la Marine (France)
SLD	Straight Line Diagram
SOA	State Oceanic Agency (China)
SOSUS	Sound Surveillance System
SPA	Single Protection Armor
SPV	Special Purpose Vehicle

SSP	Ship Security Plan
STEWS	Seismic Tsunami Early Warning System
SWIFT	Society for Worldwide Interbank Financial Telecommunications
TAT	Trans-Atlantic Telephone Cable
Tbit	Terabits per second (1 terabit = 1000 gigabits)
TDR	Time Division Reflectometer
UHF	Ultra High Frequency
UJC	Universal Jointing Consortium
UN	United Nations
UNCLOS	United Nations Convention on the Law of the Sea
UNEP	United Nations Environment Programme
UNESCO	United Nations Educational, Scientific and Cultural Organization
UNODC	United Nations Office on Drugs and Crime
VCLT	Vienna Convention on the Law of Treaties
VHF	Very High Frequency
VMS	Vessel Monitoring System
WAC	Wholly Assigned Capacity
WDM	Wave Division Multiplexing
WMO	World Meteorological Organization
XLPE	Cross-linked Polyethylene

CHART OF MARITIME ZONES FOR ALL OCEAN SPACES WITH TREATY ARTICLE REFERENCES

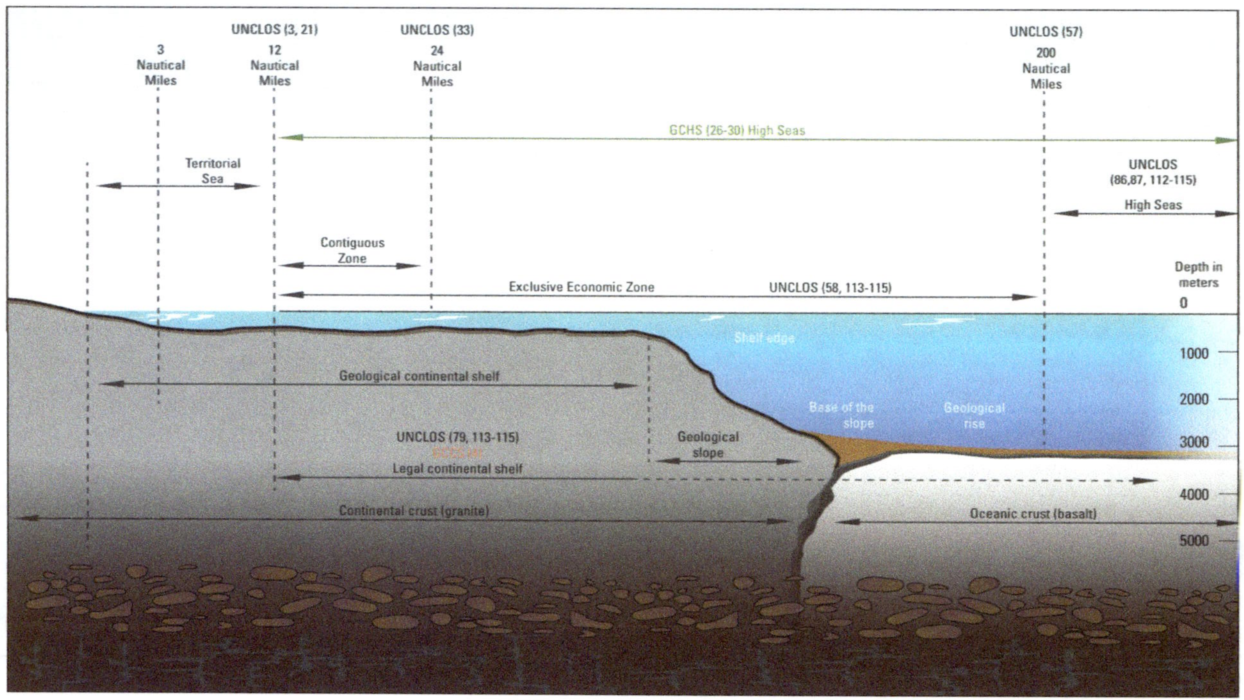

Chart 1 Maritime Zones (Courtesy of D. Burnett and Squire Sanders (US) LLP).

MAPS OF SUBMARINE CABLE SYSTEMS BY REGION

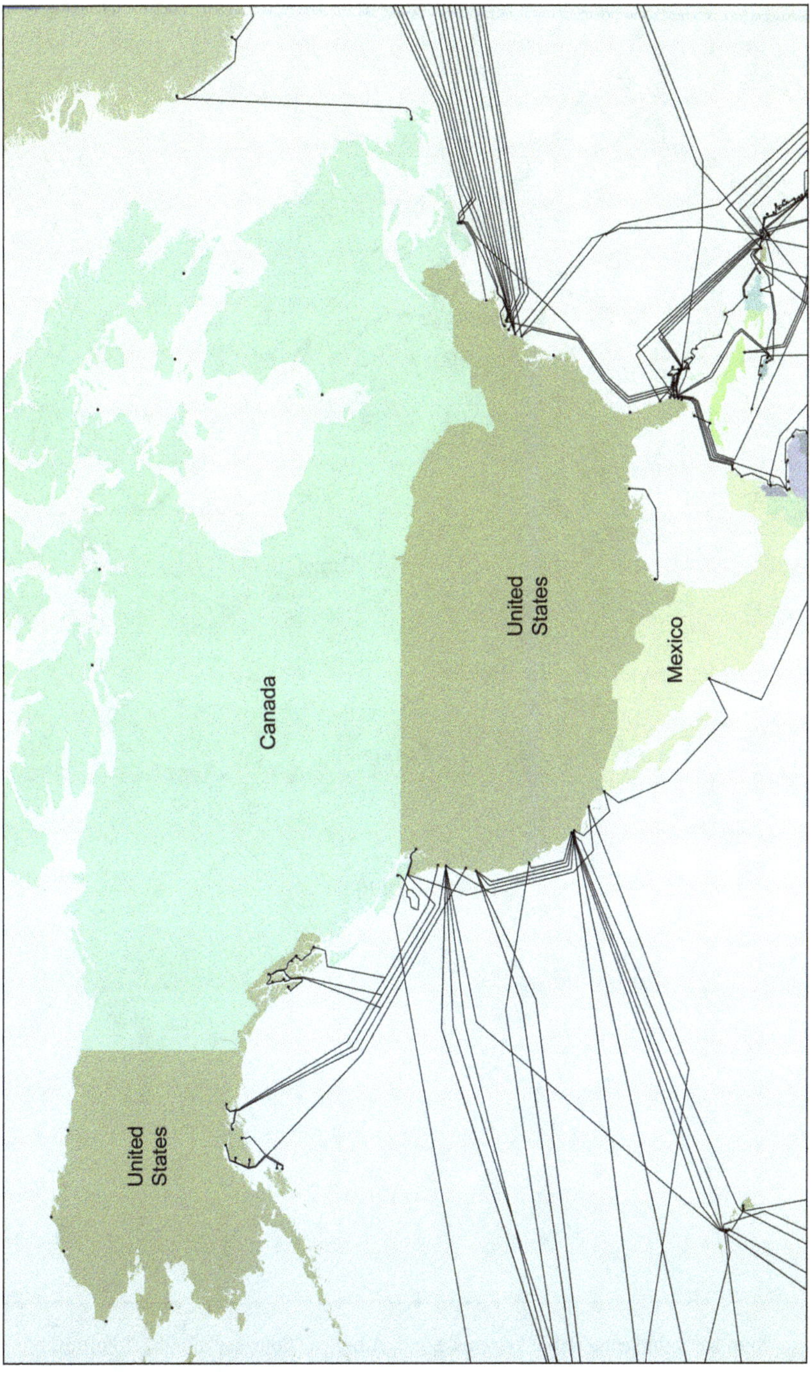

Map 1 Submarine Cable Systems, North America (Courtesy of SubTel Forum).

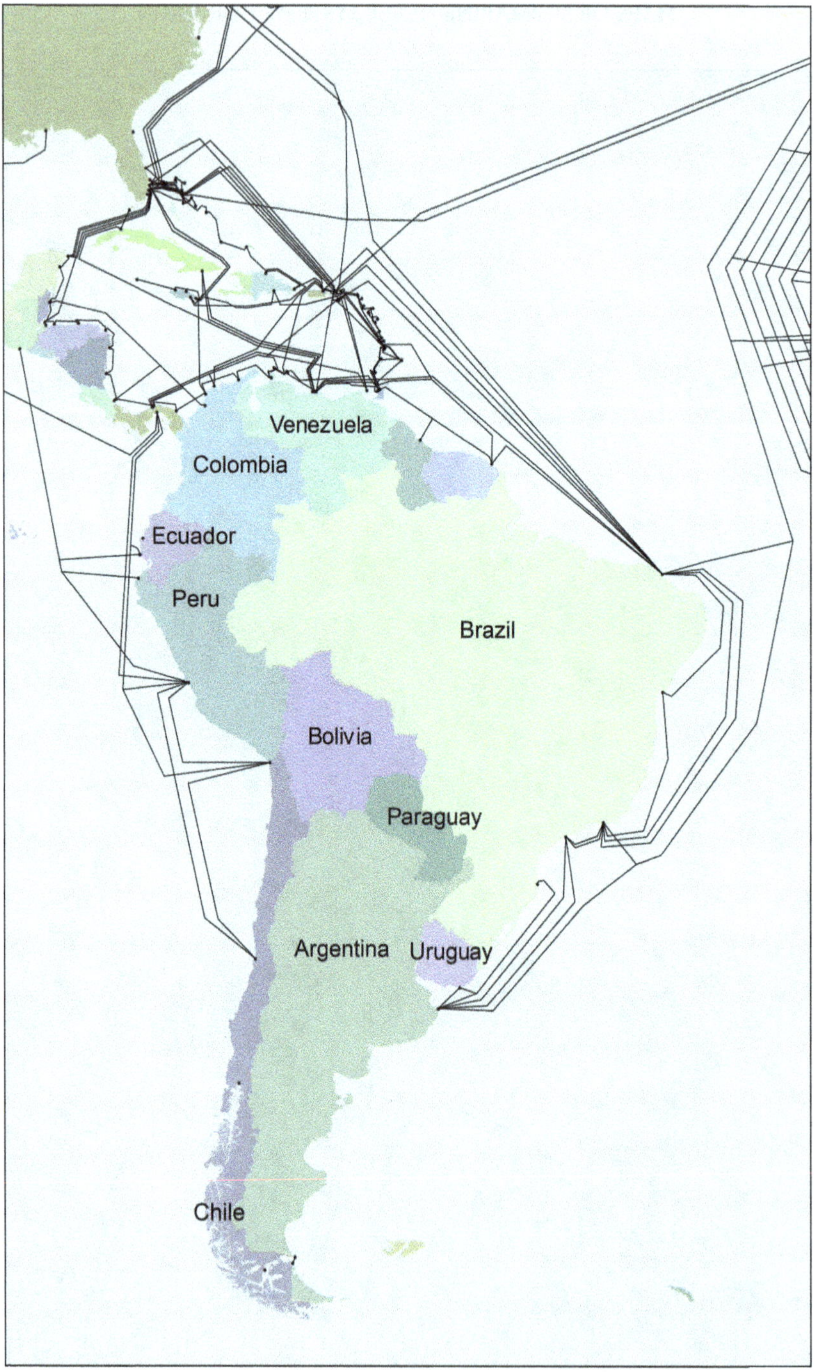

Map 2 Submarine Cable Systems, South America (Courtesy of SubTel Forum).

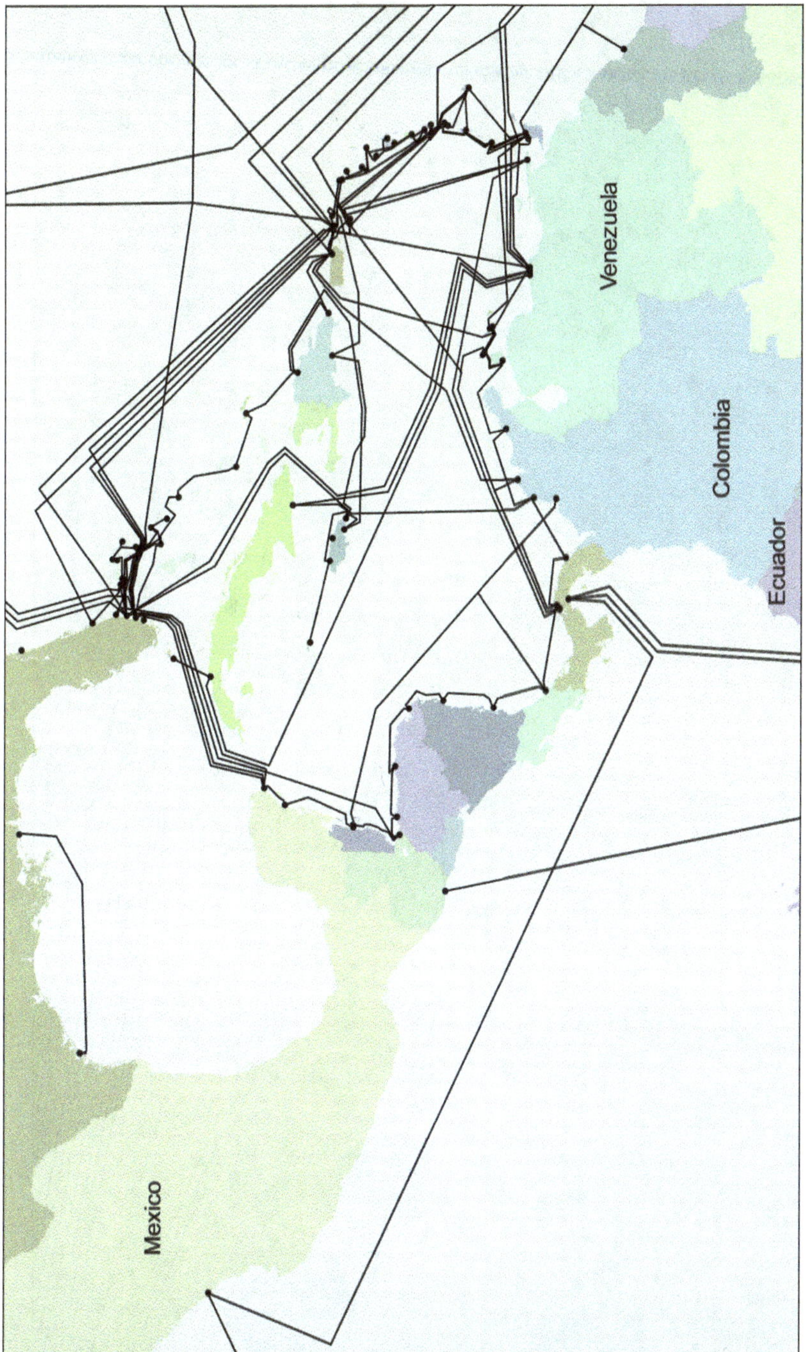

Map 3 Submarine Cable Systems, Central America (Courtesy of SubTel Forum).

Map 4 Submarine Cable Systems, Europe (Courtesy of SubTel Forum).

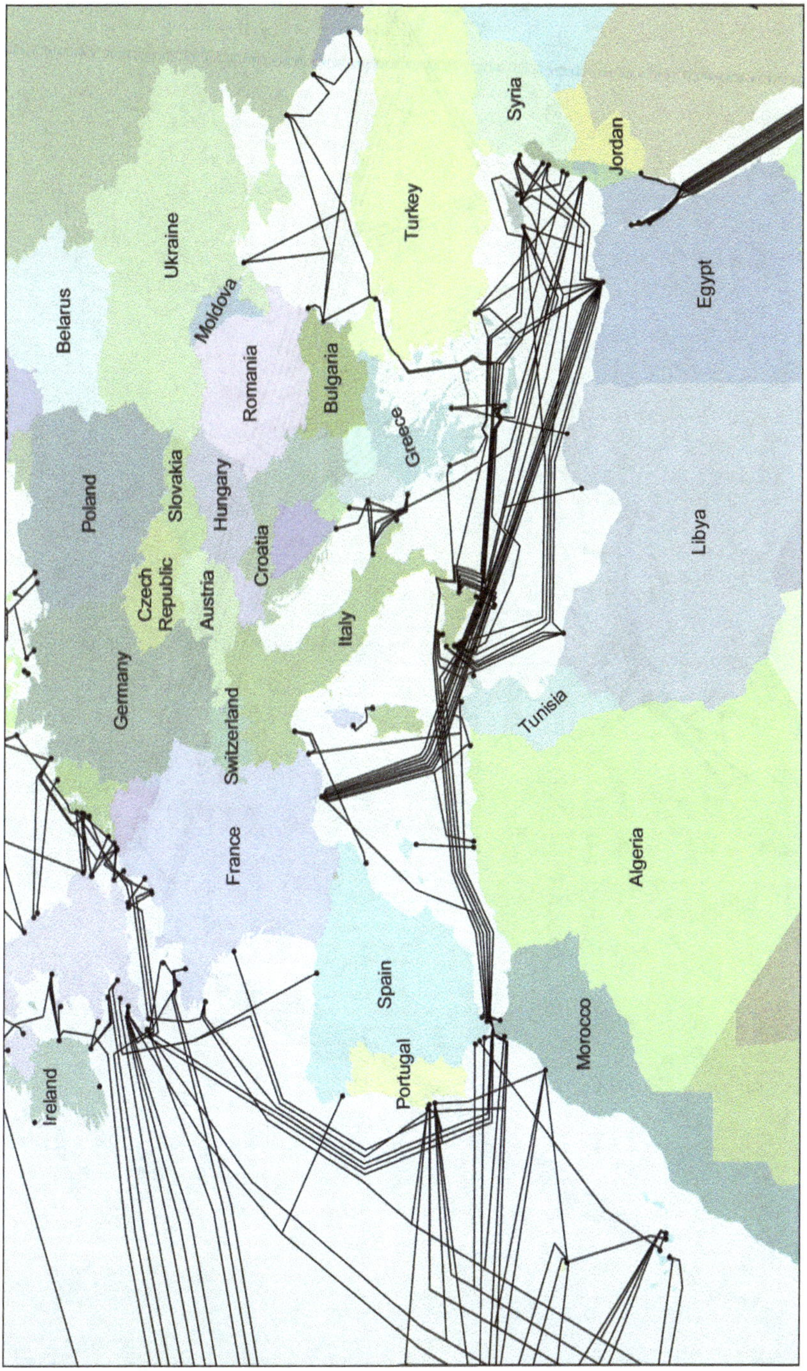

Map 5 Submarine Cable Systems, Europe/Middle East (Courtesy of SubTel Forum).

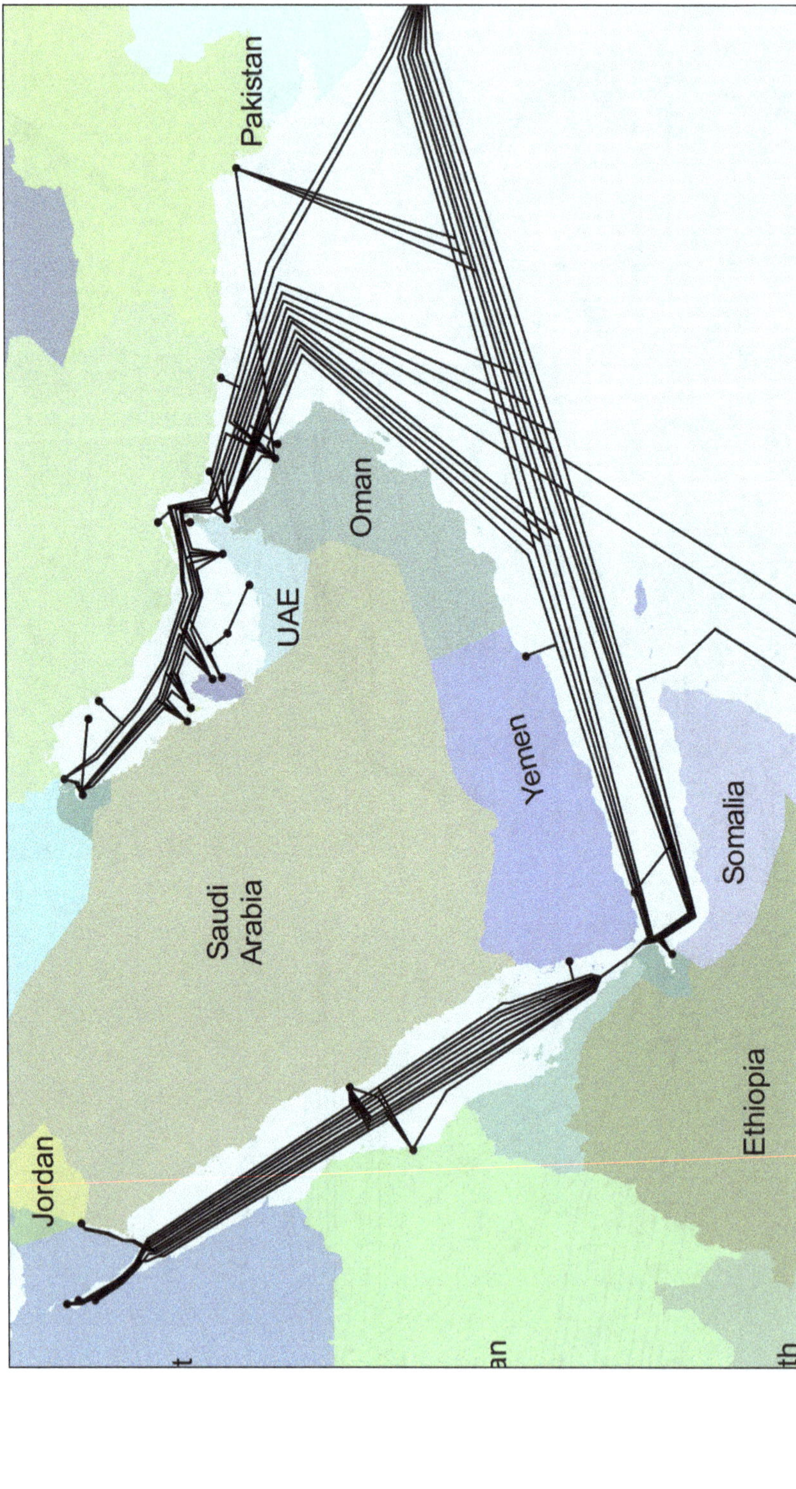

Map 6 Submarine Cable Systems, Middle East/Africa (Courtesy of SubTel Forum).

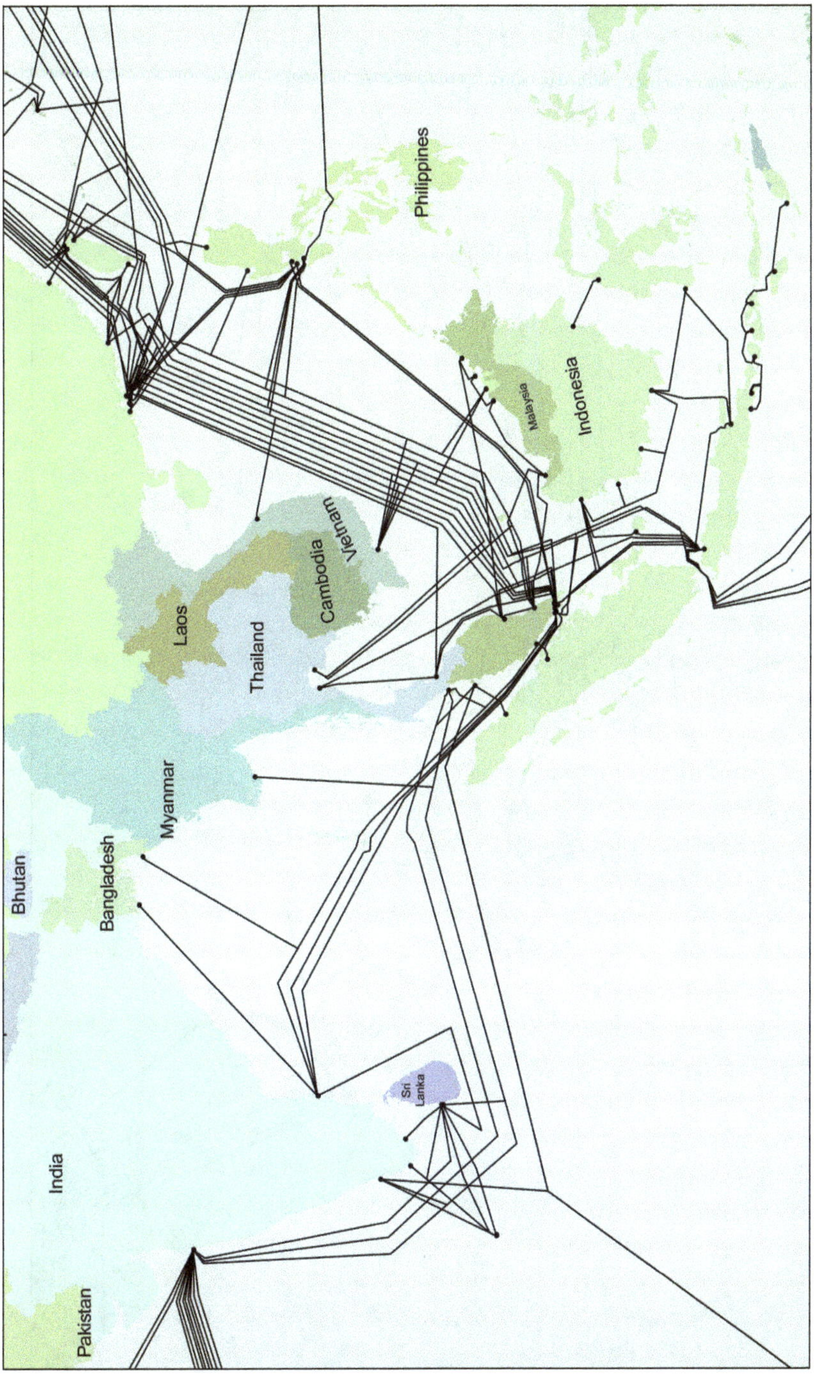

Map 7 Submarine Cable Systems, South Asia/Southeast Asia (Courtesy of SubTel Forum).

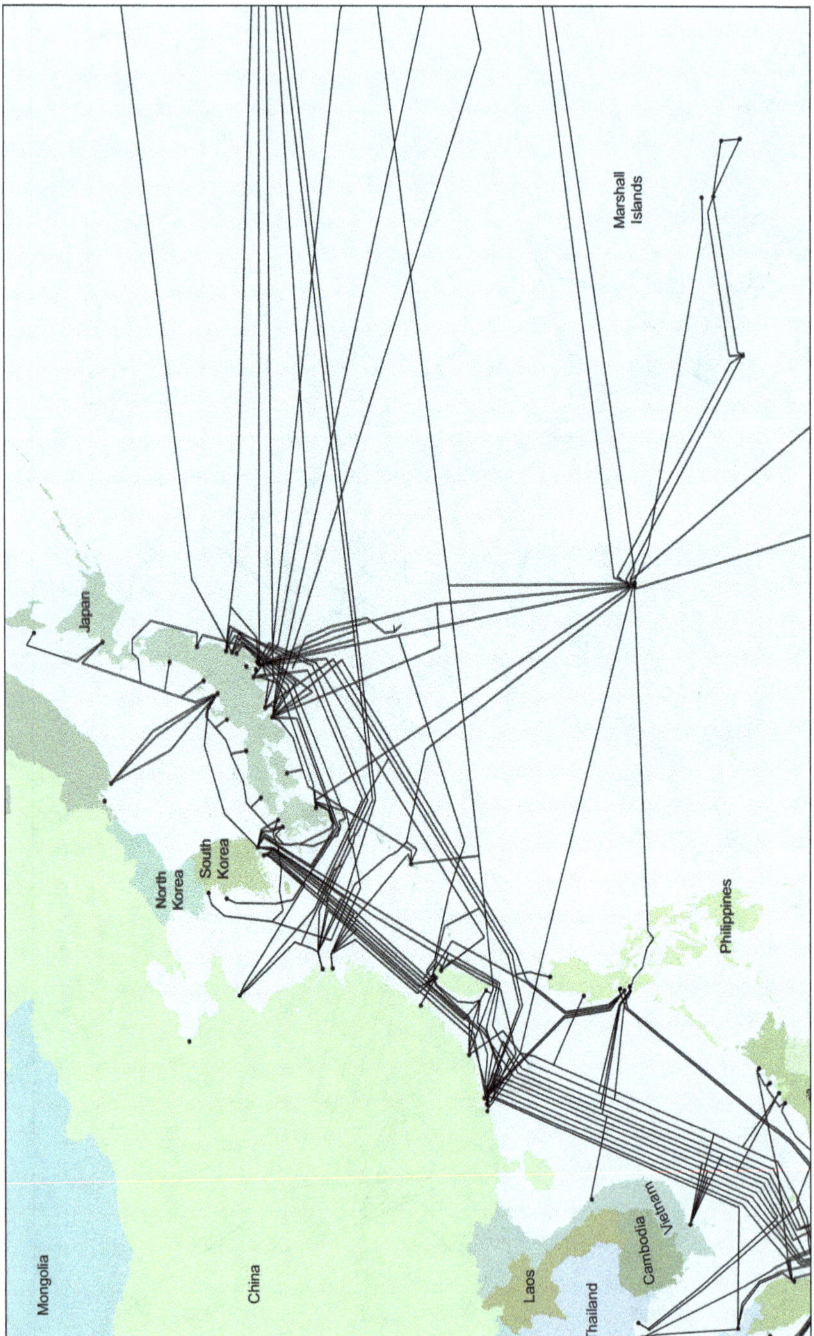

Map 8 Submarine Cable Systems, Southeast Asia/East Asia (Courtesy of SubTel Forum).

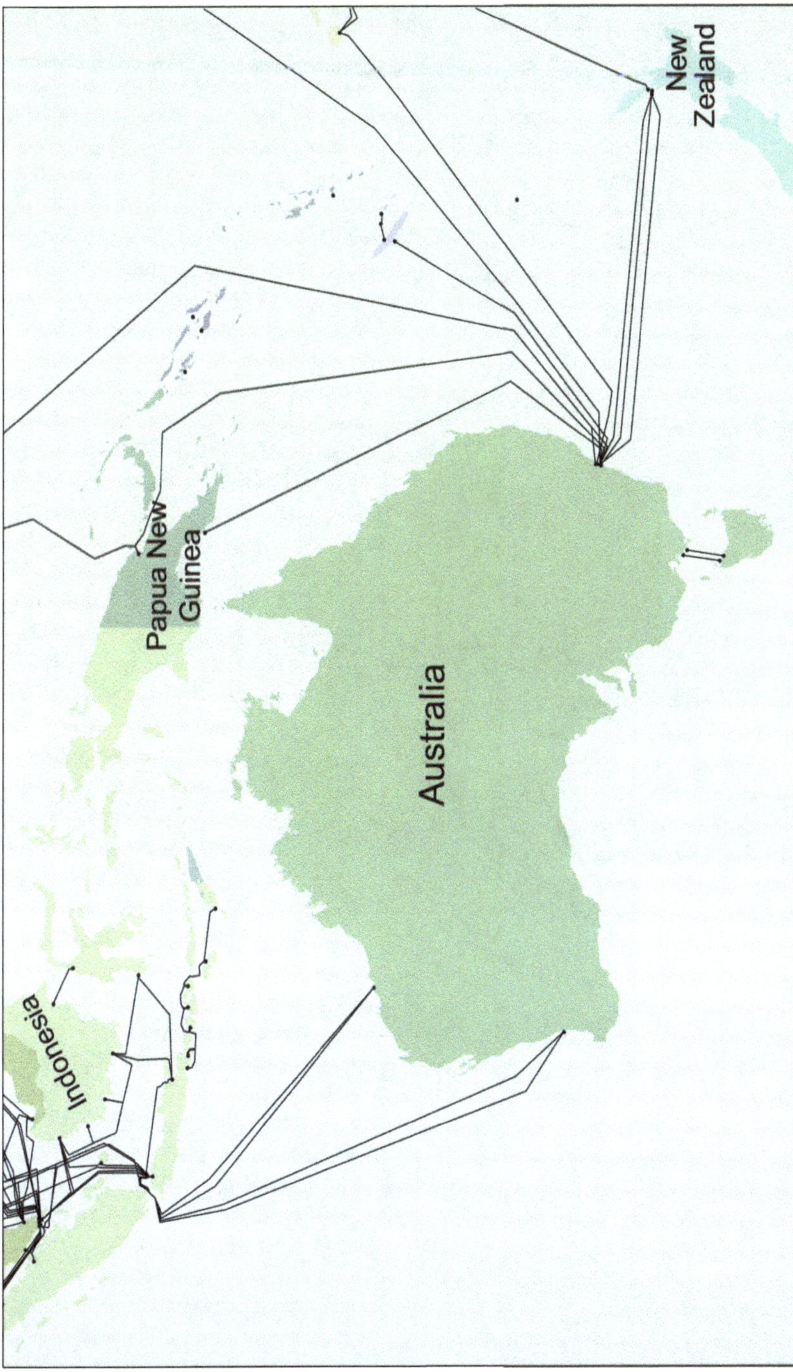

Map 9 Submarine Cable Systems, Australia/New Zealand (Courtesy of SubTel Forum).

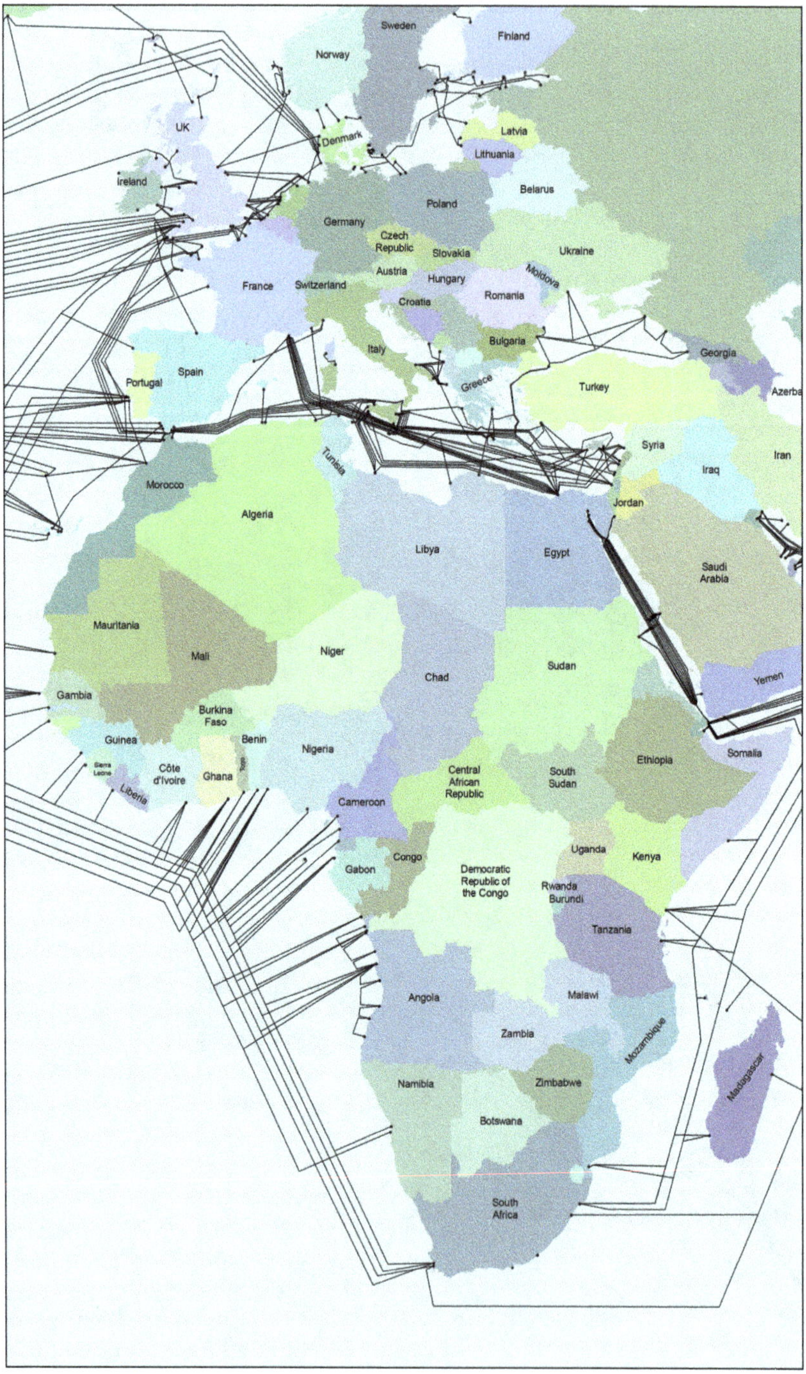

Map 10 Submarine Cable Systems, Africa/Europe/Middle East (Courtesty of SubTel Forum).

INTRODUCTION

WHY SUBMARINE CABLES?

Douglas Burnett, Tara Davenport and Robert Beckman

"Cyberspace, in the physical form of undersea fiber-optic cables, carries an even greater value for trade [than shipping goods] through financial transactions and information".
Greenleaf, J. and Amos, J., "A New Naval Era" U.S. Naval Institute Proceedings, June 2013, at 17.

Submarine Cables: The Handbook of Law and Policy has been a project long under discussion between the Editors, and after a year of hard work it has finally come to fruition. Before delving into individual chapters, the Editors believe it is important to explain why they felt that there was a need for a book on submarine cables, and what they hope the Handbook will achieve.

The Importance of Submarine Cables as Critical Infrastructure

Submarine fiber optic cables are the foundation of the world's telecommunications systems. They are laid on the seabed, are often no bigger than a garden hose, and transmit huge amounts of data across oceans. The world's reliance on submarine cables cannot be underestimated. Facebook, Twitter and other social media all utilize submarine cables. Each day, the Society for Worldwide Interbank Financial Telecommunications (SWIFT) transmits 15 million messages via submarine cables to more than 8300 banking organizations, securities institutions and corporate customers in 208 countries and/or entities. The Continuous Linked Settlement Bank located in the United Kingdom is just one of the critical market infrastructures that rely on SWIFT as it provides global settlement of 17 currencies with an average daily US dollar equivalent of approximately USD3.9 trillion. The United States Clearing House Interbank Payment System (CHIPS) is another system that processes over USD1 trillion per day to more than 22 countries for investment companies, securities and commodities exchange organizations, banks and other financial institutions.[1] It is not surprising, therefore,

[1] S. Malphrus, "Undersea Cables and International Telecommunications Resiliency" 34th Annual Law of the Sea Conference, Center for Ocean Law and Policy, University of Vir-

that the Staff Director for Management of the Federal Reserve observed in relation to submarine cable networks that "when the communication networks go down, the financial sector does not grind to a halt, it snaps to a halt".[2] The same can be said for most industries enmeshed in the global economy through the Internet including shipping companies, airlines, banks, supply chain, and manufacturing industries.

The global cable network is composed of approximately 213 or so separate, diverse, and independent cable systems totaling about 877,122 km of fiber optic cables.[3] Indeed, one only has to refer to the maps of different regions in the world in the beginning pages of this Handbook to see how extensive the submarine cable network has become. The majority of countries now rely on submarine cables for their telecommunication needs. Australia and Singapore for example, each rely on several cables landing on their shores for over 99 per cent of their international communications. It has been reported that the indirect economic costs of a fault in all the landing points in Australia would amount to USD3,169 million, mostly due to the loss of international internet traffic.[4] Similarly, the indirect economic costs of a fault in all the landing stations in the Republic of Korea would be approximately USD1,230 million.[5] The same would be true of Japan, which has approximately 20 international cable systems. The list goes on.[6] With the laying of submarine cables along the east coast of Africa in 2009 to 2010, this last major group of States now has access to the world's submarine cable network. As of mid-2012, only 21 nations and territories remain isolated from fiber connectivity and many of these have connecting cable projects underway.[7]

---

ginia, 20 May 2010, available at http://www.virginia.edu/colp/pdf/Malphrus-Presentation.pdf (last accessed 14 June 2013).

2 S. Malphrus, Board of Governors of the Federal Reserve System, First Worldwide Cyber Security Summit, EastWest Institute, Dallas, Texas, 3–5 May 2010.

3 See, International Cable Protection Committee Ltd, ICPC International Telecommunications Cables database. An interactive world submarine cable map showing these systems (last updated October 2012) can be viewed at www.iscpc.org by accessing the Cable Data Base button on the website.

4 See APEC Policy Support Unit, "Economic Impact of Submarine Cable Disruptions" December 2012 at 42 available online at http://www.suboptic.org/uploads/Economic%20Impact%20of%20Submarine%20Cable%20Disruptions.pdf (last accessed 9 June 2013).

5 *Ibid.*

6 For a detailed list of major international submarine cable systems, please see *Submarine Cable Almanac* Issue 5 (Submarine Telecoms Forum, February 2013) available at http://www.subtelforum.com/Almanac-Issue5.pdf (last accessed 9 June 2013).

7 Submarine Telecoms Forum Inc, *Telecoms Industry Report 2012* at 14–15. Inhabited sovereign States and territories without fiber optic connectivity include: Somalia, Saint Helena, Ascension, and Tristan da Cunha (British Overseas Territory); Christmas Island (Australian External Territory), Montserrat (British Overseas Territory); Saint Pierre and Miquelon (French Collecivité d' Outre-mer); Easter Island (Chilean Special Territory), Falkland (Malvinas) Islands (British Overseas Territory), Cook Islands (Self-Governing State in Free Association with New Zealand), Kiribati, Nauru, Niue (Self-Governing State

Despite the widespread reliance on submarine cables for our every day needs, it is remarkable to note that when most people think about international communications they mistakenly assume that satellites are the primary medium of modern international communications. While it is true that satellites were predominantly used up until the first trans-Atlantic fiber optic cable was laid in 1988, submarine cables have now overtaken satellites. Presently, 97 per cent of international communications are carried on a relatively small number of fiber optic submarine cables. Satellites are still responsible for some data traffic but the tremendous volume of data carried on lower cost modern fiber optic submarine cables dwarfs the limited capacity of higher cost satellites. Additionally, the technical transmission delays and other quality limitations inherent in satellites make them comparatively marginal for continuous transmission of high-speed voice, video, and data traffic. For example, if the cables (which are approximately 40 mm, i.e. the diameter of a beer bottle cap) connecting the United States to the world are cut, it is estimated that only 7 per cent of the total United States traffic volume could be carried to its destination using every single satellite in the sky.[8] There is no doubt that "these unseen and unsung cables are the true skeleton and nerve of our world, linking our countries together in a fiber-optic web".[9]

Telecommunications represent only part of the value of modern submarine cables, and submarine cables are increasingly being used for other purposes. International submarine *power* cables are growing in importance.[10] With improved technology which reduces power loss, high voltage direct current (HVDC) submarine cables, such as the 370 km Basslink interconnector linking mainland Australia with the state of Tasmania, and the 580 km NorNed cable between Norway and the Netherlands, have been successfully operating for a number of years. The United Kingdom and Iceland governments are presently in talks to lay the foundation for a 1500 km submarine HVDC power cable between the two countries. A 900 km HVDC cable between the United Kingdom and Norway is also under discussion.[11] Many coastal States also use submarine cables to operate offshore

---

in Free Association with New Zealand), Norfolk Island (Australian External Territory), Palau, Pitcairn Islands (British Overseas Territory), Solomon Islands, Tokelau (New Zealand Dependent Territory), Tonga, Vanuatu, Wallis and Futuna (French Collecivité d' Outre-mer).

8 The testimony of D. Burnett before the Senate Foreign Relations Committee on the United Nations Convention on the Law of the Sea (Treaty Doc. 103-39), 4 October 2007, S. Hrg. 110-592, pp. 143–144, available through link at http://www.access.gpo.gov/congress/senate/senate11sh110.html (accessed 14 June 1013).

9 Statement of Ambassador Vanu Gopala Menon in "General Assembly Concludes Annual Debate on Law of the Sea Adopting Two Texts Bolstering United Nations Regime Governing Ocean Space, its Resources, Uses" Press Release, 7 December 2010, available online at http://www.un.org/News/Press/docs/2010/ga11031.doc.htm (last accessed 10 June 2013).

10 Chapter 13 of the Handbook deals with power cables.

11 See "UK in Talks with Iceland over "volcanic power link" BBC News, 12 April 2012, available online at www.bbc.co.uk/news/uk-politics-17694215 (last accessed 9 June 2013).

wind farms, utilizing both array cables to interconnect offshore wind turbines and export cables to channel the collected electrical power from the wind farm to shore.[12] Denmark, Germany and the United Kingdom have well established offshore wind farms as a result of the utility of submarine cables. Tidal, wave and subsea current generators tied by cables to shore are also being trialed in various locations in the Pacific northwest of the United States and Canada.[13] In addition, coastal States have also seen offshore energy exploitation of oil and gas improved by the efficiencies introduced when offshore exploration platforms are linked to each other by undersea fiber optic cables.[14] Norway and the United States are examples where this cable use is operational. Norway's Statoil uses an array of fiber optic cables to connect floating oil platforms to shore for data transfer.[15] BP's 1216 km Gulf Fiber system, largely impervious to hurricanes and operational since 2008, connects seven fixed platforms to a central shore control center with nodes available for adding additional platforms in the future.[16]

Finally, submarine cables are being used in growing numbers for scientific purposes.[17] In a 2009 survey, the International Cable Protection Committee (ICPC) identified 193 ocean observation sites and areas worldwide, including at least 34 that plan to or are currently using submarine cables for data transmission and power in the world's oceans.[18] The 500 mile Neptune system with multiple scientific nodes off of British Columbia is a standout operational example, and a planned US cabled observatory system is intended to link to this system.[19] Japan has pioneered the use of submarine cable systems to monitor and detect tsunamis.[20]

---

[12] Chapter 16 of the Handbook examines submarine cables used for offshore energy including wind farms.

[13] See Renewable Northwest Project, available online at http://www.rnp.org/node/wave-tidal-energy-technology and Natural Resources Canada available online at http://www.retscreen.net/ang/power_projects_ocean_current_power.php (last accessed 9 June 2013).

[14] Chapter 16 of the Handbook discusses submarine cables used for offshore oil and gas platforms.

[15] See "European Drilling Outlook," *Drilling Contractor*, July/August 2007, at 25, available online at http://www.drillingcontractor.org/dcpi/dc-julyaug07/DC_July07_Statoil_revised.pdf (last accessed 8 June 2013).

[16] See BP Gulf of Mexico Fiber Optic Network, available online at http://www.gomfiber.com/ (last accessed 8 June 2013).

[17] This is discussed in Chapter 14 of the Handbook.

[18] ICPC Ocean Observation Sites and Areas, 2009, see www.iscpc.org. The survey results were compiled by Professor Lionel Carter, Victoria University, Wellington, New Zealand, the ICPC International Marine Environmental Advisor (IMEA).

[19] See Neptune Canada, available online at http://www.neptunecanada.ca/about-neptune-canada/neptune-canada-101/ and Interactive Oceans, available online at http://www.neptune.washington.edu/index.jsp (last accessed 8 June 2013).

[20] C. Manoj *et al.*, "Can undersea voltage measurements detect Tsunamis?" (2006) *Earth Planets Space* 58, 1–11; R. Monastersky, "The Next Wave" 2012 *Nature* 483, 144–146.

It is evident from the above discussion that from the time that the first submarine telegraph cable was laid in 1850 between Dover and Calais to the present day, the many astonishing uses of submarine cables has far exceeded anyone's expectations. It is fair to say that they have now emerged as one of the most important uses of the oceans. However, as with every ocean activity, the critical issue is how submarine cables can co-exist with other competing uses of the ocean, of which there are many. In this regard, international law, and in particular, the law of the sea, plays a crucial role.

Submarine Cables and International Law[21]

From time immemorial, the oceans have been claimed for the exclusive use of a small number of States. However, such notions of exclusivity were inexorably weakened by the idea that the ocean was *res communis* and that freedom of the seas was in the general community interest.[22] Over the years, the interaction between particular claims and the rejection or acceptance of such claims by the international community have refashioned and refined a body of rules and principles, known as the law of the sea. It has been said that the historic function of the law of the sea has been that of "protecting and balancing the common interests, inclusive and exclusive of all peoples in the use and enjoyment of the oceans, while rejecting all egocentric assertions of special interests in contravention of general community interest".[23]

The need for this balance between competing uses is no better illustrated than by submarine cables. Submarine cables have always faced challenges that are typical of the issues that the law of the sea aims to minimize, namely, the conflict between coastal States and non-coastal States over 'inclusive uses' of the ocean (such as navigation and submarine cables) which benefit the international community and 'exclusive uses' of the ocean by coastal States. Indeed, as early as 1884, States recognized the need to protect this infrastructure from other uses of the seas and adopted the Convention for the Protection of Submarine Telegraph Cables (1884 Cable Convention).[24] The provisions in the 1884 Cable Convention have

21 The international legal regime governing submarine cables is dealt with at various points in the Handbook, but Chapter 3 gives a comprehensive overview.

22 For example, the seminal work of Dutch jurist and philosopher Hugo Grotius *Mare Liberum*, which advocated freedom of the seas particularly for maritime trade, was a response to the monopoly on trade in the Far East by the Kingdom of Portugal. See R.R. Churchill and A.V. Lowe, *The Law of the Sea* (3rd ed., Manchester University Press, 1999) at 203.

23 M.S. McDougal and W.T. Burke, *The Public Order of the Oceans: A Contemporary International Law of the Sea* (Yale University Press, 1963), 1.

24 Convention for the Protection of Submarine Telegraph Cables, adopted 14 March 1884, TS 380 (entered into force 1 May 1888) (1884 Cable Convention). The provisions of the

significantly shaped the rights and obligations of States *vis-à-vis* submarine cables set out in subsequent law of the sea conventions such as the 1958 Geneva Convention on the High Seas,[25] the 1958 Convention on the Continental Shelf,[26] and the 1982 United Nations Convention on the Law of Sea (UNCLOS).[27] The common thread running through these conventions was the desire to establish a

> legal order for the seas and oceans which will facilitate international communication, and will promote the peaceful uses of the seas and oceans, the equitable and efficient utilization of their resources, the conservation of their living resources, and the study, protection and preservation of the marine environment.[28]

To achieve this utopian idea of a legal order that accommodated the varied uses of the oceans, the Geneva Conventions and UNCLOS recognized that coastal States had certain rights and jurisdiction in specific areas, but these had to co-exist with traditional freedoms that all States were entitled to exercise, and vice versa. With regard to submarine cables, the Geneva Conventions and UNCLOS sought to strengthen the international communications regime by, *inter alia*, preserving the freedom to lay and repair submarine cables but at the same time requiring that these freedoms be exercised with due regard to the rights and jurisdiction of coastal States. Further, the Geneva Conventions and UNCLOS also oblige States to adopt legislation to protect submarine cables from other competing uses.

While the framework established by the above-mentioned conventions, for the most part,[29] adequately balances competing uses and interests in relation to submarine cables, it is just that—a framework. Its success depends on the effective interpretation and implementation by the relevant stakeholders, including international organizations, national governments and industry. Therein lies the

---

Cable Convention are generally accepted as customary international law, see *Restatement of the Law (Third): The Foreign Relations Law of the United States Vol 2* (American Law Institute Publishers, 1987) § 521, comment f (1986). As at 2 April 2013 there are 41 State parties to the 1884 Cable Convention. A complete copy of the 1884 Cable Convention is contained in Appendix 3.

25 1958 Convention on the High Seas, adopted 29 April 1958, 450 UNTS 11 (entered into force 30 September 1962). As at 2 April 2013 there are 63 State parties. The United Nations Convention on the Law of the Sea (see below note 27) supersedes this treaty for States that are parties to both.

26 1958 Convention on the Continental Shelf, adopted 29 April 1958, 499 UNTS 311 (entered into force 10 June 1964). As at 2 April 2013 there are 58 States parties. The United Nations Convention on the Law of the Sea (see below note 27) supersedes this treaty for States that are parties to both.

27 United Nations Convention on the Law of the Sea, adopted 10 December 1982, 1833 UNTS 397 (entered into force 16 November 1994) (UNCLOS). Select UNCLOS provisions are contained in Appendix 3.

28 Preamble to UNCLOS.

29 There are some gaps in the international legal regime governing submarine cables, which will be dealt with in Chapter 12 on Protecting Submarine Cables from Intentional Damage: The Security Gap.

problem—the interpretation and implementation of this framework has fallen short of what was envisaged by the drafters.

Global and National Policies on Submarine Cables

Not infrequently, States adopt policies and regulations that undercut the viability of submarine communication cables as critical international infrastructure upon which the internet and the global economy is based. For example, despite the fact that submarine cables are vulnerable to numerous threats, such as those presented by fishing, shipping, resource exploration and exploitation activities, as well as deliberate damage, many States have not adopted measures to ensure their protection.[30] For many States, negligent or deliberate damage to submarine cables in any maritime zone is not an offence under their national legislation. This is despite it being an obligation under UNCLOS to adopt such legislation if the damage occurs in the exclusive economic zone or high seas.[31]

Further, because cable operations, such as the laying, repair and maintenance of cables, are usually carried out by foreign vessels in maritime zones under national jurisdiction, many States have adopted laws and regulations which impede the effective laying, repair and maintenance of cables.[32] For example, repairs to damaged cables, essential to the integrity of a telecommunication system serving various States, are often subject to onerous permit requirements, delaying the repair of cables and costing millions of dollars to cable operators. The cost of chartering a cable repair ship can vary between USD45,000 and USD70,000 per day. The average cost of a repair is between USD1M and USD3M, depending upon the location of the fault and the cableship, the cableship costs, and other factors.[33] Prompt repair of cables is essential not only for business reasons but also because every cable is in effect a backup cable for a damaged cable waiting to be repaired. Such cables can be used to immediately restore communication traffic by rerouting it from the damaged cable to an undamaged cable in seconds. It is this feature that allows for the resiliency of modern cable systems that generally provides for continuous global communication by cables, notwithstanding the 200 or so cable faults that occur worldwide annually from contact by fishing gear, anchors, or natural hazards such as earthquakes.[34]

---

30 This is more fully examined in Chapter 11 on the Protection of Submarine Cables from Competing Uses.

31 UNCLOS Art 113.

32 Chapters 4, 5 and 6 discuss the various challenges in law and policy in cable operations.

33 D. Burnett, "Recovery of Cable Ship Repair Cost Damages from Third Parties That Injure Submarine Cables" (2010) 35 *Tulane Maritime Law Journal* at 108.

34 *Ibid.*, at 108.

International and regional organizations have also on occasion adopted policies that undermine the integrity of the international telecommunications systems. For example, the OSPAR Commission, established to protect the marine environment in the Northeast Atlantic Sea, has devised 'best practices' on cable operations that reflect little understanding of the processes involved in the laying and repairing of cables.[35] While such efforts are no doubt motivated by admirable intentions to protect the marine environment, they appear to overlook the fact that submarine cables have a negligible footprint on the seabed. As noted above, the diameter of a modern submarine fiber optic cable is about the diameter of garden hose[36] and the impact on the marine environment is benign.[37] Similarly, recent proposals of the International Telecommunications Union (ITU) to adapt telecommunication cables to a dual use climate monitoring application are another example of regulators acting with inadequate knowledge of the cable industry and the international law applicable to cables.[38] The dual use of submarine cables for both telecommunications and marine scientific research raises complex issues as to whether the laying and repair of such cables are subject to coastal State consent (all marine scientific research in zones under national jurisdiction is subject to coastal State consent) or is a freedom of the sea.[39]

The above discussion is a snapshot of some of the issues facing submarine cables. These issues, and other challenges (all of which are discussed in greater detail in the chapters of this Handbook) underscore the fact that, many a time, regulations or policies governing submarine cables are a consequence of mistaken beliefs and knowledge gaps regarding submarine cables. They are often promulgated with little or no understanding of submarine cables and cable operations, marine engineering, seamanship and international law. In circumstances such as these, the potential benefits that submarine cables can provide to the international community have been unnecessarily compromised.

---

[35] Guideline on Best Environmental Practice (BEP) in Cable Laying and Operation (Agreement 2012–2) (OSPAR 12/22/1, Annex 14). This is more fully explored in Chapter 7 on the Relationship between Submarine Cables and the Marine Environment.

[36] A description of the physical characteristics of modern submarine cables can be viewed in the power point presentation 'About Submarine Cables' and a video which can be viewed at www.iscpc.org by accessing the Publications button on the website.

[37] L. Carter *et al.*, "Submarine Cables and the Oceans: Connecting the World" Report of the United Nations Environment Program and the International Cable Protection Committee, 2009 at 26. Available online at http://www.unep-wcmc.org/medialibrary/2010/09/10/352bd1d8/ICPC_UNEP_Cables.pdf. This report compiles and analyzes the environmental experience with cables in the marine environment since submarine cables were introduced into the ocean in 1850 and underscores the benign impact of a modern fiber optic cable on the marine environment.

[38] R. Butler, "Using Submarine Cables for Climate Monitoring and Disaster Warning" ITU Report 2012 at 23.

[39] These issues are discussed in Chapter 14 on Marine Scientific Cables.

There are several possible reasons for this lack of awareness and understanding. First, as mentioned above, there is a general misconception that satellites are the primary providers of telecommunications. After all, the idea that a telephone call made to an overseas recipient can be broken into bits, pulsed by lasers and lightwaves through unseen cables laid on the ocean floor, and reassembled into voice form thousands of miles away, all at the speed of light, is very hard to comprehend.

Second, the submarine cable network and industry has been driven by private businesses with minimum government subsidies or intervention.[40] Between 2008 and mid-year 2012 there has been approximately USD10 billion worth of investments in new systems. Of the billions of dollars spent to finance cable systems, currently less than five per cent is provided by governments or international agencies. The 95 per cent balance is provided by private consortiums (49 per cent), carriers (32 per cent) and non-government investors (14 per cent).[41] Accordingly, governments and their officials are often unaware of what it takes to build a cable system. Similarly, cable repairs (which can costs millions of dollars) are paid for privately by the cable owners and are carried out, not by government mandate, but by contract.

Third, the way the industry has evolved means there are difficulties for cable companies to assert or advocate their rights *vis-à-vis* States who have encroached upon the freedom to lay cables or who have not adopted the necessary legislation to protect cables. While the freedom to lay submarine cables is afforded to States under UNCLOS, it is actually privately owned cableships that are exercising these rights. Further complicating the situation is the fact that submarine cable systems are typically built or owned by many different private companies from different nations. A consortium of cable co-owners typically consists of about 4 to 30 or more telecom or content companies from multiple nations that co-own an international cable system's capacity and operate the cable system together pursuant to a cable construction and maintenance agreement (C&MA). Cables, unlike ships, are not registered under any flag. There is no mechanism whereby cable companies can challenge laws and policies adopted in contravention of UNCLOS.

Fourth, because States do not appear to have anticipated or appreciated the critical nature of submarine cables to their international communications, there is often no lead agency to coordinate effective policies on submarine cables. This could be a consequence of the fact that deployment of cables affects both land and sea. National telecommunications regulators frequently only address telecommunications standardization, licensing (for landing stations), and competition issues and may not be familiar with maritime issues. Similarly, maritime

---

40 More information on the way the industry works can be found in Chapter 2.

41 Submarine Telecoms Forum Inc, *Telecoms Industry Report 2012* at 16 and 23.

agencies are usually responsible for maritime operations and may not have any inkling on the nature and importance of cables. The lack of a lead agency can lead to fragmented and short-sighted policy decisions which are not good for the industry, the State or the international community at large.

Fifth, there is also no inter-governmental organization responsible for submarine cables. This is in contrast to other public infrastructure such as shipping and aviation, the governance of which has been entrusted to specialized United Nations Agencies such as the International Maritime Organization and the International Civil Aviation Organization. The ITU is the leading United Nations agency for information and communication technology but is primarily concerned with standardization in the industry and has minimal awareness of law of the sea issues. International issues with respect to submarine cables inevitably fall through the cracks without an inter-governmental body to champion it.

Notwithstanding the above, submarine cables are not without advocates. The International Cable Protection Committee has been the principal professional body of the cable industry. ICPC membership, presently 136 members from over 63 nations, includes about 97 per cent of the owners of the various cable systems worldwide and almost all of the operators of the cable vessels that lay and maintain these systems. Since 2010, membership has been open to national governments and several governments are now represented.[42] The ICPC issues *Recommendations* available to the public regarding methods of protecting submarine cables.[43] The ICPC works with governments, organizations and other seabed users on a partnership basis to promote submarine cable security and compliance with UNCLOS. These include the International Seabed Authority, the United Nations Environment Programme, ITU, APEC, the EastWest Institute, and the Rhodes Academy on Ocean Law and Policy.

Apart from the ICPC, there are also Regional Cable Protection Committees (RCPCs) where cable companies which have commercial interests in the region come together to interface with national governments. Subsea Cables UK, the North American Submarine Cables Association (NASCA), Oceania Submarine Cable Association (OSCA) and the Danish Cable Protection Committee (DKCPC) are all examples of such RCPCs that have provided an effective forum for the cable industry to communicate their concerns to governments, and vice versa.

Indeed, the efforts of the ICPC, RCPCs and other like-minded organizations and governments have had some traction in the protection of submarine cables. Recent developments are positive with States such as Australia, New Zealand,

---

42 Australia, Malta, Singapore, the United Kingdom, New Zealand, and the United States all have government representatives as ICPC members.

43 ICPC Recommendations cover areas such as cable protection, cable and pipeline crossings, cable proximity to offshore wind farms, civil engineering projects, and seismic activities, charting of cables on navigational charts, cable protection actions, and out-of-service cables. They are free upon request from the ICPC at www.iscpc.org.

Uruguay, and Colombia adopting extensive cable protection legislation and with the United Nations calling on all States to cooperate to protect submarine cables for the first time in 2010. However, there remains a lack of awareness and understanding on the nature of submarine cables, the industry that supports their development and the international legal regime that governs them. There is clearly still some way to go.

Filling the Knowledge Gap—the Objective of the Handbook

To this end, the Handbook, the first of its kind, aims to provide a collaborative practical description of the history, development, current structure and practices of the submarine cable industry and the rich, if obscure, development of cables under international law that has allowed cables to flourish as one of the most successful uses ever of the world's oceans. It addresses the various issues that have arisen (described in brief above) in national and international policies on submarine cables and provides concrete recommendations on how some of these issues may be addressed. Ultimately, the overarching objective of this Handbook is to inform, educate and generate discussion on the governance of submarine cables.

The Editors hope that the Handbook fulfills two related goals. First, we hope that one of the consequences of the Handbook will be more productive ocean laws and policies to govern submarine cables. The fundamental assumption underpinning this is that effective ocean law and policy is only attainable when governments and policy-makers understand how the submarine cable industry has evolved, is generally organized, and how cable operations take place.

Second, we hope that the one message that readers take away from the Handbook is that of balance. Balancing the various competing interests and rights of coastal States and other States requires the relevant parties to reject 'absolute' interpretations of their respective rights and obligations. The assertion of a 'doctrinaire, absolutistic conception of freedom of the seas'[44] including the freedom associated with submarine cables, without giving due regard to the rights of coastal States, may lead to even more extreme claims and actions by coastal States.[45] Likewise, coastal States that make expansive claims to rights and jurisdiction beyond what is allowed under international law will inevitably cause strain on the regime designed to protect their interests.[46] The common interest lies in minimizing conflicts between submarine cables and competing uses, with the ultimate goal of protecting the integrity of international communications.

---

44 McDougal and Burke, *supra* note 23, at 11.
45 *Ibid.*
46 *Ibid.*, at 12.

A Reader's Guide

One of the most important features of the Handbook is that it is a unique collaboration between industry experts, legal scholars, and scientists. The authors emanate from a wide range of backgrounds, with chapters being written by marine engineers, sea captains, marine geologists, commercial business leaders, diplomats and international legal scholars. The Handbook represents a rich mosaic of multiple life experiences and represents a lifetime of work in the world's oceans and the international law of the sea.

An important consequence of bringing together lawyers, cable industry experts and scientists is that this Handbook caters to a wide audience. It is not purely a legal tome or technical discourse, but a unique combination of both which aims to give readers the practical insight necessary to enhance understanding and shape policy. The Handbook will appeal to the following categories of readers (in no order of importance):

Students, Academics and Lawyers

Students and academics, particularly those involved in legal scholarship on competing uses of ocean spaces will find the Handbook a useful resource for their own research and understanding. This is also true of lawyers involved in the practice of international law, maritime law and telecommunications.

Government Officials and Policy-Makers

Presently, government officials/policy-makers have very few sources of information available to them with respect to issues surrounding submarine cables and the interests of different stakeholders in protecting cables and regulating activities associated with them. This Handbook is intended to provide a readily accessible and comprehensive overview of all of the issues from a domestic law perspective, an international law perspective, and an industry perspective. It is therefore relevant to the work of numerous Ministries and Departments within governments including telecommunications regulators, navies, maritime agencies and foreign affairs departments.

The Cable Industry

The Editors also hope that the Handbook will also be of use to industry as it continues to work in the ocean environment. Remarkably, given the dependence of the modern world on submarine cables for its critical international infrastructure, there are no degree programs or majors at the undergraduate or graduate level in submarine cable systems. Historically, this has always been the case. The question that necessarily follows is how did the industry, both on a national and international basis, develop and train its highly skilled international work force of engineers, ship officers and crews, commercial leaders, and skilled workers?

The answer is that companies historically developed extremely well structured apprenticeship and training programs that provided career training from a young age through to senior management. The training was always hands-on and merit was an essential qualification for advancement. Lessons learned in the challenging ocean environment were passed on in the form of continual refinements in ships, remotely operated vehicles, equipment, tools and procedures. Because of the international nature of the business, the training and 'lessons learned' were shared, even among competitors, through formal organizations such as the ICPC, and other professional bodies like SubOptic, the International Council on Large Electric Systems (CIGRE) and through innumerable joint ventures installing and operating submarine cable systems. These international working relationships fostered a level of cooperation among cable companies that is rare among land based industries. Every cableship and cable system operator lives with the knowledge that helping a competitor today is wise, because tomorrow they may be the one needing a favor. Formal legal disputes are uncommon in the industry; the practical need to work together to keep international communications and power uninterrupted is recognized as being more important.

But in the last 20 years or so, companies have faced difficult economic choices and the reality of escalating and rigorous market competition. The result is that many of the in-house apprenticeships and training programs have become victim to cost-cutting. Instead, companies are living off of their earlier investment in human capital. This has worked well, except that the industry now finds itself relying on a skilled but definitely aging workforce without a dependable pipeline of trained replacements. One of the motivations behind this Handbook is recognition of the need to train new workers in international law of the sea to keep the industry strong and skilled as it deals with ever changing commercial and ocean environments. Towards this end, it is hoped that this Handbook will be a valuable training and educational tool for the industry.

The Structure of the Handbook

Submarine Cables: The Handbook of Law and Policy contains 16 chapters divided into five Parts.

Part I provides readers with essential background information on submarine cables and the cable industry. Chapter 1 gives a general overview of the development of submarine cables, beginning with submarine telegraph cables and ending with submarine fiber optic cables. The technical and historical insights provided in this chapter are fundamental for any effective understanding of submarine cables. Chapter 2 gives much needed information on how the submarine cable industry works, including an overview of the different players in the industry (cable owners, suppliers and special interest groups), how a submarine cable system comes to life and the ownership structure of submarine cable systems.

Part II on 'The International Law on Submarine Cables' consists of one chapter, Chapter 3, which traces the development of the international legal regime governing submarine cables starting with the 1884 Cable Convention, and culminating with UNCLOS. Chapter 3 will discuss the relevant provisions of these conventions and will also provide the reader with an understanding of the competing uses of ocean spaces and how international law seeks to balance the interests of various stakeholders.

Part III on 'Cable Operations—Law and Practice' will provide information regarding the law and practice of individual aspects of the 'life-cycle' of submarine cables in five separate chapters. It will address: (1) the planning and surveying of cable routes (Chapter 4); (2) the manufacture and laying of cables (Chapter 5); (3) submarine cable repair and maintenance (Chapter 6); (4) the relationship between submarine cables and the marine environment (Chapter 7); and (5) dealing with out-of-service cables (Chapter 8). For each step of the 'life-cycle' there will be discussion and analysis of the nature of the activities that are undertaken, the international law that governs the activity, the law and policy challenges implicit in conducting the activity, and the proposed way forward for the future.

Part IV on 'Protecting Cableships and Submarine Cables' will address issues regarding the protection of submarine cables and vessels engaged in cable operations. Chapter 9 will give an overview of the international law on the protection of cableships engaged in cable operations and will then highlight the various issues that arise, including the disregard for safe working distances and the threat to cableships from piracy and armed robbery attacks. Chapter 10 will examine how natural occurrences such as earthquakes, typhoons and climate change impact submarine cables and the steps that the industry can take to mitigate such threats. Chapter 11 discusses the various threats to cables from competing uses such as shipping, fishing and resource exploration and exploitation and the steps States and the cable industry can take to protect cables from these threats. Finally, Chapter 12 will address the urgent security gap that currently exists with respect to the measures available in international and domestic law to protect submarine cables from deliberate damage from terrorists and propose a way forward for law and policy makers.

The last Part of the Handbook, Part V will look at other types of submarine cables, such as power cables, marine scientific research cables, military cables and cables used for offshore energy. While some of the issues raised in these chapters are similar to the issues raised in respect of submarine communication cables, these special purpose cables also raise different challenges for law and policy makers, which will be highlighted in each chapter.

Many of the chapters contain images or diagrams intended to aid readers in their understanding of the processes described in each chapter. In addition, the Appendices of the Handbook also contain invaluable information. Appendix 1 contains a comprehensive timeline on the submarine cable industry with significant milestones in the development of submarine cables. Appendix 2 contains charts which depict the major submarine system suppliers from 1850 to 2012, and how they have amalgamated or divided to form today's most important submarine cable supply companies. Both Appendix 1 and 2 have been provided by Stewart Ash. Appendix 3 contains extracts of the relevant international conventions, including the 1884 Cable Convention (which is reproduced in its entirety) and pertinent provisions of UNCLOS. We encourage readers to refer to the actual provisions when reading each chapter as this will enhance their understanding.

Ultimately, the Editors and the various contributors hope that the Handbook will provide the foundation for meaningful engagement between the industry, academics, government officials and ocean policy decision makers. It is our collective aspiration that such engagement will engender further discussion, collaboration and cooperation on issues in ocean governance that are of increasing importance to the use of submarine cables in the world's oceans.

PART I

BACKGROUND

CHAPTER ONE

THE DEVELOPMENT OF SUBMARINE CABLES

Stewart Ash

Introduction

Ever since the human race began to form itself into communities, the ability to communicate over long distances, or *telecommunications* has been important. Warning beacons, smoke signals and the use of different colors / designs of flag were some of the early ways in which limited information could be transmitted over distance. However, the most reliable method for detailed communication was a written message carried either by a runner, pigeon, or man on a horse. It was not until the end of the eighteenth century that mechanisms were devised to transmit and receive information. The first notable system was devised by brothers Claude and Rene Chappe in 1794 and operated between Paris and Lille. This system was called the *semaphore* or *optical telegraph*. It comprised a series of line-of-sight towers, on top of which were placed adjustable arms that could be positioned to transmit 196 codes. Lamps on the arms allowed the system to operate at night. Since then, technology has developed significantly. This Chapter traces the historical development of submarine cables from its inception to the widely-used technology that it is today.*

* This Chapter draws on knowledge gained by the author in a career in the submarine cable industry spanning more than forty years and from extensive research of private archive material held in the Cable & Wireless Archive, the Porthcurno Museum Archive, the Telcon Archive, STC Submarine Systems Archive, the IEE Archive (the IEE became the IET in March 2006), the British Museum, and the private papers of Lord Pender.

I. Genesis and Evolution of the Telegraph Era

Electricity and the Telegraph

The principles of electricity were formulated by Stephen Gary in 1720, but it took more than 100 years to discover the necessary elements for an electric system capable of transmitting and receiving information over significant distances. Experiments into electricity and magnetism took place in the United Kingdom and Europe over the first quarter of the nineteenth century.[1] These experiments culminated in Cooke and Wheatstone being awarded a patent in 1839 in the United Kingdom and then opening the first commercial electrical telegraph system in 1843. Samuel Morse, thanks in large part to the work of Alfred Vail,[2] achieved the same things in the United States in 1840 and 1845.

The electric telegraph was the miracle of the age and its use spread quickly. Charles Dickens is said to have commented that "the world changed forever once information could travel faster than a man on a horse".

Genesis and Evolution

Various experiments to submerge electrical conductors in sea water were conducted in the United Kingdom and the United States, but it was not until 1845, when Michael Faraday suggested gutta percha as a water proof insulator, that a practical cable became possible. The first attempt to lay a commercial submarine cable took place on 28 August 1850, when the paddle steamer *Goliath* laid an insulated copper wire between Dover and Calais. This event is generally accepted as the genesis of the submarine cable industry.[3]

Over the subsequent 163 years the industry has evolved and developed to a point where 99 per cent of all international telecommunications traffic is carried on submarine cables.[4] This industry history divides neatly into three eras, these being:

- The Telegraph Era (1850–1950)
- The Telephone Era (1950–1986)
- The Optical Era (1986–to date)

---

1 S. Ash *et al.*, *From Elektron to 'e' Commerce: 150 Years of Laying Submarine Cables* (Global Marine Systems Ltd, 2000) Section 1.

2 S. Ash, "Back Reflection" (2012) 63 *Subtel Forum* at 41.

3 *Submarine Electric Telegraph Between Dover and Calais* (The London Illustrated News, 7 September 1850).

4 An overview of the major submarine system suppliers from 1850–2012 is provided in Appendix 2 of this Handbook.

Figure 1.1 Image of the *Goliath*, taken from a booklet produced by The Telegraph and Construction and Maintenance Co. Ltd on the occasion of the visit of the Delegates of the International Telegraph Conference to the Gutta Percha Works, Wharf Road, City Road London, on 16 June 1903. (Image courtesy of Atlantic-Cable.com website)

II. The Telegraph Era (1850–1950)

The Telegraph Era is largely a British story. By the centenary of this industry some 469,500 nm of submarine cable had been laid. Over 90 per cent had been manufactured in factories in London and 82 per cent by one company, the Telegraph Construction & Maintenance Company (Telcon), from its factory at Greenwich (now Alcatel-Lucent Submarine Networks).[5]

Crossing the English Channel

The 1850 Dover to Calais cable was a naïve concept; the insulated copper cable, manufactured by the Gutta Percha Company in London, was given no external protection and was only weighed down by lead weights every sixteenth of a mile. Although some signals were transmitted between Dover and Calais, the cable soon failed due to the effects of abrasion off the coast of Calais[6] and was abandoned. Just over 12 months later a new Dover to Calais cable was installed, this

[5] *The Telcon Story 1850–1950* (The Telegraph Construction & Maintenance Company, 1950) at 173.

[6] *Submarine Electric Telegraph Between Dover and Calais, supra* note 3.

time the cable was protected by external steel wire. After some patent disputes the cable was finally manufactured on the premise of Wilkins and Weatherly in London, to a design based on a soft core steel wire patent owned by Robert S. Newall.[7] The cable went into commercial service on 13 November 1851 and despite several repairs gave satisfactory service for over a decade.

The Atlantic Telegraph (1854–1864)

Having successfully traversed the English Channel, the next step was to lay a cable across the deeper waters of the Irish Sea. This was achieved in 1853 with a cable from Donaghadee to Port Patrick. The major prize however, was a cable across the Atlantic. Since the advent of the electric telegraph many schemes had been suggested on both sides of the Atlantic. In 1854, after a meeting between Cyrus W. Field and Frederick Gisborne[8] the famous *Atlantic Telegraph* project became a reality.[9] Cyrus Field formed the New York, Newfoundland and London Telegraph Company and travelled to Britain to promote his project. In 1856 the Atlantic Telegraph Company was formed and within a few days raised the necessary capital. The cable core was insulated by the Gutta Percha Company and the external armoring was shared equally between R.S. Newall & Company in Birkenhead and Glass, Elliot & Company in Greenwich. The cable lay from Ireland began in August 1857, but after 334 nm had been deployed the cable broke and was lost. Another 900 nm was ordered and in June 1858 the laying operation recommenced. This time two ships, the *Agamemnon* and *Nigeria*, met in mid-ocean, their cable ends were joined together and the ships sailed away from each other. After many trials and tribulations the cable ends were brought ashore in Valentia, Ireland and Trinity Bay, Newfoundland. There was much rejoicing on both sides of the Atlantic and over 400 messages were sent before 20 October, at which time the cable failed, never to work again.

There was a public outcry on both sides of the Atlantic, with the Boston Courier publishing an anonymous letter that suggested the entire project had been part of an elaborate stock fraud, "Was the Atlantic cable a humbug?"[10] This failure was quickly followed by the collapse of the Red Sea cable project, in which investors also lost large sums of money. In response to public concern the British Government and the Atlantic Telegraph Company established a joint committee to investigate the reasons for the failures. The committee heard evidence from

---

7 R.S. Newall, *Facts and Observations Relating to the Invention of the Submarine Cable and to the Manufacture and Laying of the First Cable Between Dover and Calais in 1851* (E. & F.N. Spon, 1882) Items 12–16.

8 D.R. Tarrant, *Atlantic Sentinel: Newfoundland's Role in Transatlantic Cable Communications* (Flanker Press Ltd, 1999) at 17.

9 Ash *et al.*, *supra* note 1, Section 2.

10 See http://atlantic-cable.com/Article/1859Humbug/index.htm (last accessed 6 June 2013).

many experts, including Professor William Thomson (later Lord Kelvin). The committee's report was finally published in April 1861,[11] and was described as "the most valuable collection of facts, warnings, and evidence ever compiled concerning submarine cables".[12] It concluded that ocean telegraphy was not as simple as previously thought and there was much still to learn. The report set out a series of recommendations for the construction of submarine cables, the methods for laying them and the methods for testing them, during both production and installation. One of the most valuable recommendations made in the report was a call for the standardization of measurement of electric current and resistance. The report subsequently formed the basis of the first set of standards for submarine cable systems, many of which have survived to this day.

The Telegraph Construction and Maintenance Company (Telcon)

In 1864 Telcon was formed by the merger of the Gutta Percha Company and Glass, Elliott & Company. Telcon based its production at the Glass, Elliott site in Greenwich. Its chairman was Scotsman John Pender, who approached the Atlantic Telegraph Company with a proposal to lay a new Atlantic cable. The £500,000 capital was raised by John Pender and Daniel Gooch; Isambard Kingdom Brunel's[13] great iron ship the *Great Eastern*[14] was chartered and the project got underway. In July 1965 the *Great Eastern* set out, but after laying 1186 nm out from Valentia, Ireland the cable parted and could not be recovered.[15] A further £600,000 was raised by Pender and Gooch by setting up a new company, the Anglo-American Telegraph Company (Anglo), and in July 1866 the *Great Eastern* set sail again. This time the cable was successfully laid between Valentia and Hearts Content, Newfoundland. The *Great Eastern* also recovered the end of the 1865 cable and by 8 September had completed a second Atlantic cable.

Global Expansion

With the Atlantic tamed, submarine cables systems quickly expanded around the globe, largely due to the efforts of John Pender. In 1868 John Pender stood down as chairman of Telcon in favor of Daniel Gooch. Pender saw great opportunity

---

11 C. Wheatstone *et al.*, "Report of the Joint Committee Appointed by the Lords of the Committee of Privy Council for Trade and the Atlantic Telegraph Company to Inquire into the Construction of Submarine Telegraph Cables together with the Minutes of Evidence and Appendix" (Board of Trade and the Atlantic Telegraph Company, 1861) at 1.

12 S. Ash, "Back Reflection" (2009) 43 *Subtel Forum* at 47.

13 L.T.C. Rolt, *Isambard Kingdom Brunel: A Biography* (Longmans Green, 1957) at 304.

14 J. Dugan, *The Great Iron Ship* (Harper & Brothers, printed by Kingsport Press Inc, 1953) at 162.

15 W.H. Russell, *The Atlantic Telegraph* (Day & Son Ltd, 1865) at 94.

Figure 1.2 The *Great Eastern*, an illustration by Robert Dudley taken from The Atlantic Telegraph by W.H. Russell. (Image courtesy of Atlantic-Cable.com website)

in laying cables to interconnect the British Empire and he resigned from Telcon to pursue his grand plan. Commercial services between London, Bombay and Madras were in operation by 1870.[16] Services were extended to Penang, Singapore and Hong Kong (via Saigon) by 1871. By the end of 1872, services had also been extended to Darwin, Australia and over land to Adelaide.

Pender's business strategy was to spread the significant short term risk of laying these cables over a number of independent companies. Once the companies were established he consolidated them. During 1872–1873 Pender merged four companies to form the Eastern Telegraph Company Ltd. Then, in 1873, three companies were merged to form the Eastern Extension, Australasia and China Telegraph Company Ltd. In the same year Pender also formed the Brazilian Submarine Telegraph Company Ltd and the Western & Brazilian Telegraph Company Ltd. In 1873 the Globe Telegraph and Trust Company Ltd was incorporated. The Globe Trust was formed in order to further spread the risk of cable laying. Shares in the Globe Trust were offered in exchange for shares in the Eastern and Western companies. In 1887, the companies adopted the collective name of the Eastern and Associated Companies. By the time of Pender's death in 1896 the Eastern Group possessed a virtual monopoly over worldwide communication by

---

16 J.C. Parkinson, *The Ocean Telegraph to India: A Narrative and a Diary* (William Blackwood and Sons, 1870) at 271.

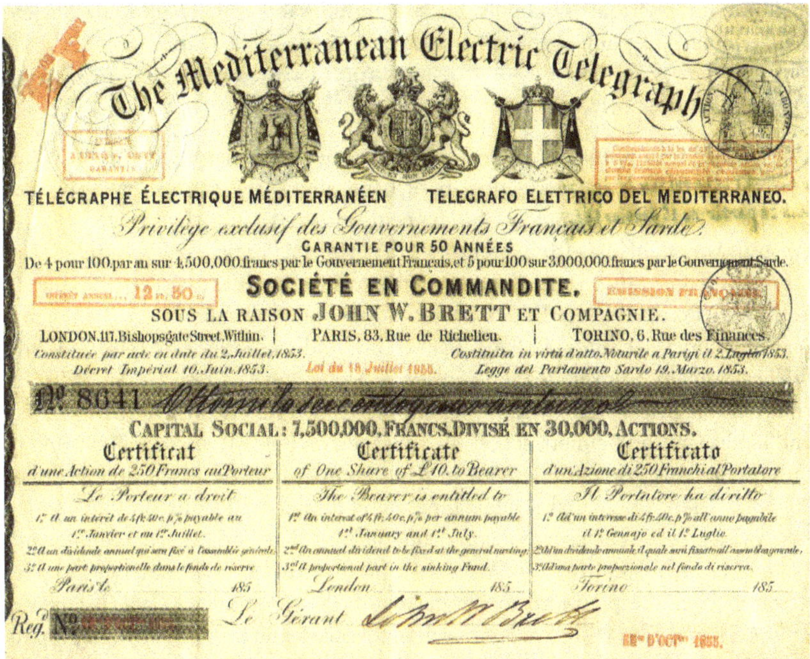

Figure 1.3 A share certificate for the Mediterranean Telegraph Company, 18 July 1853. (Image courtesy of Atlantic-Cable.com website)

telegraph cable, and the companies that he formed went on to become the basis of Cable & Wireless Ltd.[17]

Monopoly Supply

Whenever Pender's companies built a new cable, whether for a new route or duplicating cables on existing routes, the construction contract for the work was always awarded to Telcon. R.S. Newall & Co left the submarine cable business in 1870 and for around 20 years the entire world supply of submarine cable was manufactured in factories on the River Thames in London. On the North bank there was Hooper's Telegraph Works Ltd, the India Rubber Company, Gutta Percha Company and W.T. Henley Telegraph Works & Company, and on the south side there was Siemens Brothers in Charlton, and Telcon further up river in Greenwich. Due to Pender's contracts Telcon quickly outstripped its competitors,

17 K.C. Bagelhole, *A Century of Service: A Brief History of Cable and Wireless Ltd* 1868–1968 (Anchor Brendon Ltd, 1969) at 16.

maintaining a virtual monopoly on submarine cable supply until La Société Générale des Téléphones opened a cable factory in Calais, France in 1891.[18]

Competition and Cartels

During the 1870s a number of new cables had been laid across the Atlantic, but most of these were either owned by or came under the control of the Anglo-American Telegraph Company (100 per cent British owned) which, together with Western Union, held a virtual monopoly over trans-Atlantic telegraph traffic. The other companies that owned trans-Atlantic cables, such as the French Atlantic Telegraph Company, the Direct United States Cable Company, La Compagnie Française du Télégraphe de Paris à New York and the American Telegraph and Cable Company were under the control of Anglo and Western Union. With no competition, prices were high. However in 1883, two Americans, John W. Mackay and James Gordon Bennett, set up the Commercial Cable Company. The primary purpose of this company was to provide competition to Anglo and Western Union.[19] The Commercial Cable Company contracted Siemens Brothers to lay two trans-Atlantic cables connecting London with New York, via Nova Scotia and Waterville in Ireland. In 1886, Western Union started a price war by dropping its tariff from 40c per word to 12c per word, the Commercial Cable Company dropped its rates to 25c per word but then followed suit, forcing Anglo to drop its rates. The result was that no company was making money and in 1888 agreement was reached between the parties to fix tariffs at 25c per word. This agreement was invalidated in the United States by the Sherman Anti-Trust Law in July 1890. This pricing war was also the catalyst that led to the winding up of the French Atlantic Telegraph Company in 1894.

The Pacific Cable

After the boom years of the 1870s and 1880s, the 1890s was a quiet period for the manufacture of cable. Siemens Brothers laid a new trans-Atlantic cable for the Commercial Cable Company, but there were very few other projects.

The British Government had been debating the need for a trans-Pacific cable to connect its colonies and dominions since the Colonial Conference in Ottowa in 1887. Eventually, on 5 June 1896, the Imperial Pacific Cable Committee met for the first time in London to move the project forward. It progressed slowly amid much debate and concern about the threat of the new Marconi wireless telegraph.[20] It was not until December 1900 that Telcon was awarded the supply

---

[18] S. and E. Curveiller, *Cent Ans de Câble 1891–1991* (Alcatel Câble, 1991) at 13.

[19] S. Ash, "Back Reflection" (2013) 70 *Subtel Forum* at 40.

[20] H. Barty-King, *Girdle Round the Earth: The Story of Cable and Wireless and its Predecessors to Mark the Group's Jubilee 1929–1979* (William Heinmann Ltd, 1979) at 114.

contract, with Royal Assent for the project finally being received on 15 August 1901. In December the same year Marconi made his first radio telegraph transmission across the Atlantic.

The laying of the cable from Australia and New Zealand, via Norfolk Island, Fiji, and Fanning Island to Vancouver in Canada was completed in October 1902. At 3458 nm the span between Bamfield and Fanning Island was the longest single submarine link ever laid. In parallel with the British Government project, a consortium of private submarine telegraph companies formed the Commercial Pacific Cable Company in order to build a private cable from San Francisco, via Honolulu, Midway and Guam to Manila in the Philippines. With these two cables in place a telegram (or cable) could be sent from London to Sydney or Auckland in just one hour.

Emergence of Radio and Industry Decline

The taming of the Pacific marked the zenith of submarine telegraphy and encouraged the British Government to consider the potential benefits of nationalizing its submarine telegraph companies. Cable manufacture remained high during the first decade of the twentieth century, as early experiments with radio telegraph proved to be unreliable. However, with the start of World War I submarine cable manufacture was largely switched to making 'trench cable' for connecting military field telegraphs and telephones. Marconi went back to the drawing board to improve the reliability and security of radio telegraph, primarily for transmission between London and Royal Navy ships in the Mediterranean. As a result of his success, in 1919, Marconi was granted an operating license and commercial radio (wireless) telegraph became a viable commercial competitor to cables. Short wave radio could send telegraph signals at three times the speed of telegraph cables, using a fifth of the power and at a twentieth of the cost. The only benefit that cables could offer was security, which remained essential for government and military traffic. The glory days for submarine telegraph cables were over and in order to survive, in 1929, the Eastern Associated Companies were forced to merge with the Marconi Wireless Telephone Company Ltd to form Imperial and International Communications Limited. Renamed Cable & Wireless Ltd in 1934,[21] it remained a private company until it was nationalized by the British Government in 1948.

The twin forces of the great depression and competition from radio telegraph hit cable manufacture hard and by 1930 only two companies remained in the United Kingdom, these being Telcon and Siemens Brothers. In 1935 they merged to form Submarine Cables Ltd (SCL). Submarine telegraphy continued into the 1960s and, although some improvements continued to be made in cable design

---

[21] *Ibid.*, at 240.

and transmission techniques, the rise of radio technology meant that the industry ground to a halt and then went into decline awaiting a paradigm shift in transmission technology.

III. The Telephone Era (1950–1986)

Invention of the Telephone

In 1854, Charles Bourseul wrote a paper on the use of electricity for transmitting and receiving speech, while in the same year Antonio Meucci produced the first device to demonstrate this. Six years later Johann Philipp Reis built a device that could transmit musical notes and indistinct speech. He called his device the *telephone.* In 1871 Meucci set up an electrical communications system between rooms within his Staten Island home and submitted a *patent caveat* to the United States Patents Office. He renewed this twice, but by 1874 lacked the funds for a further renewal and the caveat lapsed. Meucci's lack of funds allowed Alexander Graham Bell to be awarded his infamous Patent 174 465 on 7 March 1876.[22] Many people still believe that the main idea for Bell's transmitter was stolen from the patent caveat of Elisha Gray that was filed the same day as Bell's patent application. Whatever the truth, it was Bell who made the telephone a commercial success. His first telephone exchange was opened in New Haven, Connecticut in 1878 and over the next decade the telephone spread across America and Europe. By 1890, the first telephone service had been opened in Japan.

Early Submarine Telephone Cables

It was the British Post Office, in 1891, that laid the first submarine telephone cable of any note across the English Channel. This system used a telegraph cable design that was limited to relatively short distances due to the distorting effects of the cables capacitance. To overcome this problem, two technological breakthroughs were required. The first was the research of Oliver Heaviside into the 'skin effect' of telegraph signals, leading to his patent of the coaxial cable in 1880. Telcon had been granted a patent for a submarine telegraph cable with a copper helically wrapped outer conductor in 1895; however, this idea was not exploited until 1921, when Telcon manufactured three coaxial cables and laid them between Havana and Key West. The success of this project encouraged American Telephone & Telegraph (AT&T) to approach the British Post Office with the idea of a trans-Atlantic telephone cable in 1928. Bell Laboratories had developed their own coaxial cable design and, as there was no United States manufacturer, offered the manufacturing rights to Telcon; the offer was refused. Bell turned to the

---

[22] S. Ash, "Back Reflection: Intellectual Property Disputes Nothing Changes" (2011) 56 *Subtel Forum.*

German company Norddeutsche Seekabelwerke; which eventually produced 111 nm of coaxial cable that was laid between Havana and Key West in 1930, with the telephone service going into operation on 6 June 1930.

The second technological breakthrough came in 1933, when Imperial Chemical Industries (ICI) laboratories discovered polyethylene, solving the problem of high capacitance in gutta percha insulated cables. This material had a lower dielectric constant than gutta percha, it was tougher, more easily processed, non-hygroscopic and most importantly, cheaper. Polyethylene became available for experimental cable manufacture in 1938, but its use was restricted to military cables during World War II (WW II). The first cable of this type was laid across the English Channel in 1945.

After WW II, polyethylene was made available for civilian use. The French manufacturing capability in Calais had been destroyed during the war. The manufacture of submarine cables in Japan, which had started in 1915, grew steadily based on the needs of its domestic market, and in 1941 a major factory was opened by the Nippon Submarine Cable Company but as a result of WW II this facility was also out of action. The Americas had yet to enter the industry and so, once again, submarine cable manufacture was a British monopoly.

Submerged Amplifiers (Repeaters)

In 1947, the first submarine cable using polyethylene was manufactured and laid by SCL between the Netherlands and the United Kingdom. The cable was 1.7 inch air and polyethylene spaced coaxial cable and was capable of carrying 84 voice channels. The low loss polyethylene cables allowed submarine telephony over medium distances, but to cross oceans amplification of the signal was essential. The idea of including housings in the submarine cable had first been patented in 1865, but the technology had not been forthcoming. To insert an amplifier into a sub-sea housing raised a number of major technical problems, such as how to enclose the amplifier in a water-tight casing but still get access to the transmission path, how to integrate the housing into the cable, how to provide power to the amplifier and, because it would be based on thermionic valves, how to dissipate the heat. Most importantly, all the amplifiers had to be reliable so they would not have to be recovered and replaced. These problems took time to resolve, but the first submerged amplifier housing was introduced into the already laid Anglesey to Port Erin coaxial cable in 1943.[23] By 1947 a new coaxial system containing a submerged amplifier had been laid between the United Kingdom and Germany, and shortly afterwards other systems followed to the Netherlands and Denmark. Because the North Sea was relatively shallow, the effects of hydrostatic pressure

---

[23] H.H. Schenck and L. Waldick, *1990 World's Submarine Telephone Cable Systems* (US Department of Commerce, National Telecommunications and Information Administration, 1991) System Reference 1 at 87.

were not overly significant. However, to lay a system across the Atlantic they presented far greater problems.

Bell Laboratories developed a flexible housing that could be laid by passing it through the standard cable-laying machinery of the time and which contained a uni-directional valve amplifier. The amplifier was powered by direct current sent down the cable. Design work was completed by 1941, but due to WW II a full scale trial of this repeater design was not conducted until 1950 when twin coaxial cables were laid between Havana and Key West.[24] The cable was manufactured by Simplex in the United States and marked the start of US involvement in submarine cable manufacture. Each cable contained three flexible housings and between them the two cables provided 24 × 4 kHz two-way voice channels. The installation of the system marked the beginning of the submarine *Telephone Era.*

In parallel with submerged amplifier development, the ailing submarine telegraph industry was looking for ways to improve the performance of its cables. Western Union developed a submerged housing that contained circuitry to detect incoming signals and regenerate or 'repeat' them. The first of these regenerators was successfully inserted into the 1881 American Telegraph cable in 1950. Over the next decade several more of these devices were inserted into existing submarine telegraph systems. They were named 'submerged repeaters', and although no new systems were ever installed that included them, for some reason the name for the watertight housing survived the demise of the technology.

Atlantic Telephone One

British development of repeaters resulted in an in-line, rigid housing which had room inside it for filters that allowed bi-directionally transmission over a single cable. The British design could provide up to 60 × 4 kHz voice circuits, over twice that of the United States design and, being bi-directional, it only required a single cable. However, because of the rigid housing, it would not pass through the cable machinery and so the cableship would have to be stopped in order to deploy each repeater. In deep water this procedure was viewed as having a high risk of throwing cable loops on the seabed, thus increasing the risk of cable faults. For the first Trans-Atlantic Telephone cable, TAT-1,[25] this risk was considered too high and therefore the Bell design was used with bandwidth increased to 36 × 4 kHz voice channels. TAT-1 went into service in 1956.

---

24 *Ibid.*, System Reference 5 at 91.

25 *Ibid.*, System Reference 30 at 112.

Transoceanic Systems

The British continued to work with the rigid housing and by the time the next trans-Atlantic cable, CANTAT-1,[26] was installed in 1961, the problem had been resolved. The solution was a by-pass rope attached to the cable in front of and behind the repeater. The by-pass rope passed through the cable machinery and the repeater was carried passed on a trolley. This remained the method for deploying repeaters until the introduction of the Linear Cable Engine (LCE) in 1971. CANTAT-1 also presented another problem for the rigid housing design. CANTAT-1 was the first system to use, in deep water (>1000 m), lightweight (LW) cable which had the strength member on the inside of the cable structure. Up until then, the strength of the cable had been provided by external armor wires. The coaxial cable design at that time was one inch (0.990") in diameter and it was believed that supporting the weight of the repeater in the catenary to the seabed would cause excess strain on the cable or cause cable run-away. A method was needed to relieve this additional strain. The answer was to attach parachutes to the repeaters when they were deployed, the theory being that the parachute would open in the water column and bear some of the weight of the repeater during its descent to the seabed. These parachutes were successfully trialed in Loch Fyne in 1960 and were used on all British manufactured repeaters deployed in deep water until the introduction of a new, stronger, one and a half inch (1.47") cable design in 1968; after which time the practice was abandoned.

The first telephone cable across the Pacific went into service between Australia, New Zealand, Fiji, Hawaii and Canada in 1963. It was called COMPAC[27] and provided 80 × 3 kHz voice channels. It contained 322 repeaters, all in rigid housings, manufactured by SCL and Standard Telephones & Cables Ltd (STC). The cable was manufactured by the same two companies at Greenwich and at STC's factory in Southampton, UK, that had been opened in 1956.

System Capacity Increases

CANTAT-1 had marked the end of the flexible housing design and from that time on manufacturers in France, Japan, the UK and the US all adopted the rigid housing. For the remainder of the Telephone Era, repeater mechanical design changed very little. Semiconductor transistors replaced thermionic valves in the early 1970s and system transmission capacity was improved through several iterations. The 80 circuits of CANTAT-1 soon became 160 circuits (1.2 MHz); this was followed by system designs of 5 MHz, 12 MHz, 14 MHz, 30 MHz, 36 MHz and finally 45 MHz. The highest capacity submarine telephone system ever built was PENCAN 3,[28] which

---

26 *Ibid.*, System Reference 48 at 130.

27 *Ibid.*, System Reference 59 at 144–145.

28 *Ibid.*, System Reference 192 at 306.

was installed in 1977. It was designed to support 71 supergroups or 5680 × 3 kHz voice channels. This represented a 23 fold increase in capacity since 1950, but the increase was only achieved by reducing the spacing between repeaters. Pencan 3 had a repeater spacing of 2.7 nm (5.01 km). The last telephone cable to be laid across the Atlantic was TAT-7[29] in 1983, which provided 4246 × 3 kHz voice channels and contained 67 repeaters spaced at 5.1 nm (9.46 km). This relatively low capacity combined with increased cost due to the number of repeaters, meant that submarine telephone systems came under threat from satellite communications, even though submarine cables had lower latency than geostationary satellites, were not prone to echo and offered much higher security. During the 1970s and early 1980s satellites gradually became the predominant mode of transmission for international telephony. In 1986 the Telephone Era came to an end with the laying of the last STC 14 MHz system between India and the United Arab Emirates[30] and, in the same year, the deployment of the first fiber optic systems.

The Market Model

The Telephone Era was characterized by telephone companies, often government owned, that had a monopoly over the international telephone traffic to and from their respective countries. Between them these companies purchased and operated the telephone cables, unlike the telegraph model in which a single company owned the entire asset. For example, TAT-1 was procured by a consortium comprising AT&T, the British Post Office (now BT), the Canadian Overseas Telecommunications Corporation (now TATA Communications (Canada)) and the Eastern Telephone and Telegraph Company (then a Canadian subsidiary of AT&T). These companies needed a way to define their rights and obligations under their collaboration, and to do this the first version of what is known as a Construction and Maintenance Agreement (C&MA) was developed. This is now the standard contract model for all consortium based cable systems.

Cable Protection and Maintenance

Throughout the Telephone Era, almost all submarine systems were laid on the seabed (not buried). This made them vulnerable to ship anchors and trawling. System owners soon recognized that they needed to be able to repair their cables quickly and required cableships on standby ready to sail at short notice. Some major investors in submarine cables, such as AT&T, British Telecom, Cable & Wireless and France Telecom, had their own fleet of cableships which were used for both laying and repair of cables. However, as most cables were owned by a

---

29 *Ibid.*, System Reference 255 at 367.

30 *Ibid.*, System Reference 282 at 396.

consortium and not all consortiums had members with cableships, it was necessary to find a way to share the cost of this service. The owners' solution was to form clubs covering geographical areas. The first of these was set up in 1965 to cover the North Atlantic region. This was, and still is, the Atlantic Cable Maintenance Agreement (ACMA).[31] Under this agreement cable owners share the cost of having a cableship(s) on standby 24/7 ready to repair a cable. The cost of the ship(s) on standby is divided amongst the cable owners in proportion to the amount of cable being protected for them. When a fault occurs, the cable owner concerned takes control of the ship and pays for the cost of the repairs. There are a number of other 'maintenance zones' around the world that operate on the principles developed by the ACMA. What was special about the Telephone Era was that the ships that provided the repair service in these zones were all owned by subsidiary companies of major submarine cable owners.

The Optical Era (1986–)

The origins of fiber optic transmission can be traced back to 1966 when Dr Charles Kao and Dr George Hockham made a revolutionary discovery:

> A fibre of glassy material constructed in a cladded structure with a core diameter of about λ° and an overall diameter of about 100 λ° represents a practical optical waveguide with important potential as a new form of communication medium… compared with existing co-axial cable and radio systems, this form of waveguide has a large information capacity and possible advantages in basic material cost.[32]

As we will see 'large information capacity' proved to be a massive understatement. What Hockham and Kao had established was that the attenuation of glass fiber was not a fundamental property of the material but was caused by impurities. If a sufficient number of these impurities could be removed then attenuations could be reduced to a few decibels per kilometer or even less. This was easier said than done and it was not until the late 1970s that experimental and then commercial terrestrial fiber optic systems went into operation. Development work continued and by 1980 the first sea trial of a fiber optic submarine system containing a repeater was conducted by STC in Loch Fyne, Scotland.[33]

---

31 See generally http://www.acmarepair.com/ (last accessed 6 June 2013).

32 G. Hockham and C. Kao, "Dielectric-fibre Surface Waveguides for Optical Frequencies" (1966) Volume 113(7) *Proceedings of the Institute of Electrical Engineers* at 115–1158.

33 Schenck and Waldick, *supra* note 23 System Reference 225 at 334.

A Level Playing Field

At the end of the Telephone Era the British remained the world leaders in the supply of submarine systems through STC, having absorbed SCL in 1970. The other major suppliers were: in the United States, AT&T Submarine Systems Inc (SSI); in France, Alcatel Submarcom, the marketing division selling Alcatel Câble & CIT products; and in Japan, Fujitsu or NEC Corporation repeaters / electronics, were sold with Ocean Cable Company (OCC) cable. Although STC was the market leader, the switch to fiber optic technology gave the other suppliers the opportunity to catch up, and as a result substantial investment in research and development was undertaken globally. The first commercial repeatered submarine system installed was a 300 km domestic system, FS-400M,[34] laid in 1986 between the islands of Honshu and Hokkaido in Japan. In the same year Alcatel installed its first repeatered system from France to Corsica,[35] AT&T SSI installed an experimental system Optican-1 in the Canary Islands and STC installed the first commercial international fiber optic system, UK–Belgium No 5.[36] The Optical Era had begun.

New Technologies

Unlike previous eras, the manufacturers had set out from the start to develop systems that could cross the deepest oceans, so the submarine cable and repeater designs were already in place for the next step, namely a system across the Atlantic Ocean. However, TAT-8[37] involved a number of new technological issues. Firstly, two fibers are required for two-way transmission and, because a cable can contain more than two fibers, it is possible to split transmission paths into different cables. This required a new submerged housing called a Branching Unit (BU). The BU separated the fiber paths and contained switching circuitry to manage the configuration of the system power feed. Secondly, TAT-8 was supplied jointly by three suppliers, and although each company had achieved the same technical performance their design approaches had differed slightly and significant integration engineering was required, including development of a method for joining together the different supplier's cable designs. This issue of jointing different cable designs led to the establishment of the Universal Jointing Consortium in 1989.[38] Finally, the first generation repeaters were designed to detect and regenerate the incoming light pulse. This required high power feed currents and this, combined with the new design of LW cable, presented new electrical

[34] *Ibid.*, System Reference 275 at 389.
[35] *Ibid.*, System Reference 279 at 393.
[36] *Ibid.*, System Reference 278 at 392.
[37] *Ibid.*, System Reference 286 at 400.
[38] See http://www.ujconsortium.com/ (last accessed 6 June 2013).

issues concerning personnel safety and surge protection. These issues and other lessons were learned slowly and, on many occasions, painfully.

Cable Burial

Fiber optic technology was beginning to deliver system owners with an unimagined increase in cable system capacity but, at the same time, commercial fishing was becoming more intensive, trawlers were getting larger and were operating in greater water depths. This combination made system security increasingly significant. In the UK, British Telecom International (now BT) conducted a thorough investigation into the risks posed to submarine cables from external aggression in the English Channel, North Sea and on the Atlantic Continental Shelf.[39] From this study BT, in collaboration with Soil Machine Dynamics (SMD), developed a new design of ship towed cable plow. This plow underwent successful sea trials in May/June 1985 and was first used to install the UK-Belgium No 5 system in 1986. It is now standard practice to bury cable in water depths up to 1000 m, and sometimes beyond, to protect them against fishing activity.

Capacity Expansion

First generation optical repeaters initially operated at a transmission wavelength of 1310 nm and a digital line rate of 280 Mbit/s. By 1990 technology had advanced and the transmission wavelength for TAT-9[40] had moved to 1550 nm with a digital line rate of 565 Mbit/s providing 80,000 × 64 kbit/s voice channels. This was now in excess of capacity available via satellite and, once again, submarine cables became the dominant international telecommunications medium. This must have been beyond Hockham and Kao's wildest dreams; but more was to come.

Optical Amplifiers

The development of the second generation of optical repeaters can be traced back to 1986, when the Erbium Doped Fiber Amplifier (EDFA) was first demonstrated by Professor David Payne and his team at Southampton University (UK). In simple terms, the EDFA consists of a length of optical fiber, doped with erbium, which when excited by a pump laser amplifies the incoming transmission signal. The EDFA is much simpler and more reliable than regenerative circuitry and offers direct amplification independent of the signal line rate. It also allows for greater spacing between repeaters.

From its initial invention it took several years for Payne's group, and a parallel development team at Bell Laboratories, to produce a reliable EDFA that could

---

39 C. Cole and R. Struzyna "Protection and Installation Techniques for Buried Cables" (1986) Volume 5, Part 2 *British Telecommunications Engineering* at 130–132.
40 Schenck and Waldick, *supra* note 23 System Reference 331 at 455.

be manufactured in volume. The first trans-Atlantic optically amplified systems were TAT-12 and TAT-13, creating a ring network; these systems utilized a transmission wavelength of 1550 nm with a line rate of 5 Gbits/s on two fiber pairs. They went into operation in 1996. Around this time, the amount of data transmitted on submarine systems began to exceed voice traffic and the convention of expressing system capacity in 64 kbit/s voice channels was abandoned.

The start of the Optical Era coincided with the start of the "dot com" boom and the deregulation/liberalization of the telecommunications industry worldwide. Competition between telecoms companies, the increasing demand for capacity from the internet and the willingness of banks to fund submarine cable projects, created an environment that sent the industry into a boom period. With so many traditional carriers and start-up companies wishing to build new cable systems, system suppliers looked to find ways to distinguish themselves.

Wave Division Multiplex

During early experiments it was found that the EDFA could simultaneously amplify signals at two or more wave lengths, Wave Division Multiplexing (WDM), something that was not possible with regenerative systems. WDM was quickly developed to offer 16 wavelengths per fiber pair. The ability to reduce the spacing between wavelengths was then developed for terrestrial systems, giving birth to Dense Wave Division Multiplexing (DWDM), which was quickly taken up by the submarine cable industry. This gave suppliers the opportunity to develop and offer systems with more and more capacity on a single fiber pair. Because of the EDFA, the concept of the 'transparent pipe' became popular; the idea that the capacity of a fiber system is only limited by the equipment connected to each end. This is, of course, an over simplification, as system design is always contingent on current knowledge and the available technology. All submarine systems were, and still are, designed to have a specific 'design capacity' which is based on the technology available at the time. Generally they are equipped at a lower capacity, allowing for growth over their theoretical 25 year design life. However, in a relatively short timescale, the available capacity on a fiber pair for an optically amplified system had moved from one wavelength (λ) @ 5 Gbit/s in the mid-1990s to an industry standard offering of 64λ × 10 Gbit/s = 640 Gbit/s, by the year 2000.

Greater System Design Capacity

The total capacity of a submarine cable is a function of line rate, wavelength, and the number of fiber pairs in the cable. For repeatered systems, the number of fiber pairs is constrained by the number of amplifiers that can be accommodated in the repeater and that can be powered through the cable. From its inception, the repeatered system model had been built around a maximum of four fiber pairs per system but, during the boom, design and development was undertaken for six and eight fiber pair repeaters. For repeaterless systems there was

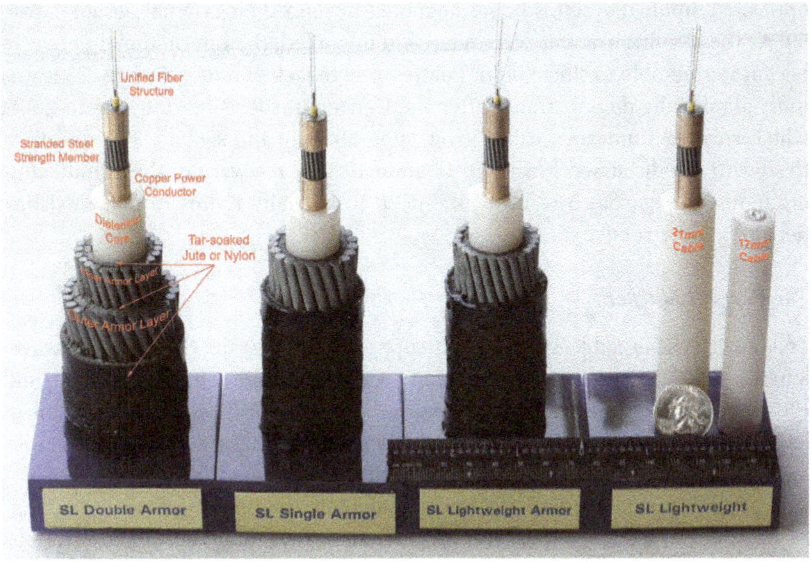

Figure 1.4 Modern telecommunications fiber optic cables. (Photograph courtesy of L. Hagadorn)

no such constraint and one system was installed between the United Kingdom and Belgium that contained 96 fiber pairs. The EDFA also played a major role in the development of repeaterless systems, extending the distance that could be spanned through the use of transmit and receive amplifiers plus Remote Pumped Optical Amplifiers (ROPA). This made fiber connectivity possible to islands and remote locations where the cost of repeatered systems could not be commercially justified.

From Boom to Bust

In the year 2000 the submarine cable industry was on the crest of a wave, buoyed by what was, in retrospect, insane optimism that the exponential growth in demand would continue forever. The boom had created a market in which the provision of new capacity rather than the demand for capacity was spiraling. It had been expected that bandwidth hungry applications would create demand for huge capacities on the major routes, and although demand did continue to rise at a steady rate, history has shown that the forecasts for growth were excessively optimistic. The bubble burst and for the next five years very few new systems were procured. The growth in demand was taken up by commissioning of previously unlit[41] capacity or through advances in technology which resulted in

[41] The term 'unlit' refers to fibers within a cable system not equipped with electro-optic transmission equipment.

optically amplified systems being upgraded beyond their original design capacity. As the existing systems were filled and upgraded, the international networks became vulnerable to single point failures, due to lack of route diversity. This was most graphically demonstrated after the Hengchun subsea earthquake in 2006, which severed numerous cables over large areas of the seabed. The effects of this event are discussed further in Chapter 10. The recovery of the manufacturing industry began as a result of the need to diversify routes to make existing networks more robust.

The Current Market

Today submarine cable systems can support > 100λ @ 10 Gbit/s, multi-wavelengths @ 40 Gbit/s or a combination of the two on a single fiber pair and 100 Gbit/s line rate technology is soon to be deployed. Repeaterless systems of up to 400 km in length can be accommodated with ease and more 'heroic' systems of 400–500 km in length, are possible.

Beyond direct telecommunications applications, submarine fiber optic technology has been deployed as part of scientific arrays designed to allow various studies of the oceans floor. In addition, submarine fiber optic cable systems are a proven solution for providing communications to off-shore assets in the oil and gas industry. The provision of high capacity, low latency communications allows increased automation and reduced manning to be built into the infra-structure

Figure 1.5 A modern cableship, *Ile de Bréhat*, laying shore end of cable off the Canary Islands. (Photograph courtesy of Alcatel-Lucent)

design, resulting in savings on both capital and operational expenses, so submarine cables are now being factored into the earliest design studies for new off-shore developments. For both of these applications a new technology, the Optical Add Drop Multiplexing (OADM) BU, has been developed. The first systems to utilize this technology were the BP Gulf of Mexico Fiber Optic Network System,[42] interconnecting offshore platforms in 2008, and the Neptune Scientific Observatory[43] in 2009.

Conclusion

The history of the submarine cables industry demonstrates that we owe much to the ingenuity, tenacity and vision of private individuals and companies. There is no reason to believe that the future will be any different.

[42] Information on the Gulf of Mexico Fibre Optic network is available at http://bpgulfofmexicofibre.com/ (last accessed 6 June 2013).

[43] Information on the Neptune Canada system is available at http://www.neptunecanada.ca/ (last accessed 6 June 2013).

CHAPTER TWO

THE SUBMARINE CABLE INDUSTRY: HOW DOES IT WORK?

Mick Green

Introduction

The first submarine cable was laid between Dover, in the United Kingdom, and Calais, in France in 1850. The cable was promoted by two brothers, Jacob and John Brett, who formed a company called the English Channel Submarine Telegraph Company, together with four shareholders to fund the project. The private commercial model employed for the cable continues to this day.

The English Channel Submarine Telegraph Company continued to lay submarine cables until 1868 when nationalization of the United Kingdom's telegraph system commenced and its assets were passed to the British Post Office (BPO). Although the BPO continued to install regional telegraph cables, the development of intercontinental telegraph cables was the domain of the entrepreneur and continued to be so for the next 100 years. It was the telephone era, which commenced in the 1950s, that saw the formation of the first consortium to construct a telephone cable across the Atlantic. This consortium comprised the BPO, the American Telephone and Telegraph Company (AT&T), the Canadian Overseas Telecommunications Corporation and the Eastern Telephone and Telegraph Company (a subsidiary of AT&T). Together these companies agreed to build the first trans-Atlantic telephone cable, TAT-1. The consortium model continues to be the most frequently employed model for major cable systems.

I. Who is Involved and What are their Roles?

Cable Owners

The first submarine cables were either privately owned, developed as a result of the vision of entrepreneurs such as the Brett brothers and Cyrus Field, or State owned, following nationalization of telegraph companies. Private telecommunication companies in many States emerged after the introduction of privatization in the 1980s.[1] Essentially three types of companies invest in, and become owners of, a submarine cable; namely, the telecommunication operator (State or private), the non-telecommunication company and an investment bank, either directly or through a *special purpose vehicle*. The telecommunication operator utilizes its share of capacity within its own network and potentially wholesales any surplus capacity.[2] *Capacity* is the light transmission characteristics of a cable that directly translates to the ability to transmit bits of information per second and results in the voice, video and data traffic communicated by the cable system. Non-telecommunication companies[3] may also seek to invest in the cable if they have high capacity demands for their private network and can justify direct investment in a cable rather than purchasing capacity on the wholesale market. Finally an investment bank can also own capacity in a cable as collateral for debt finance provided to complete the cable. In addition to these three types of companies, a submarine cable supplier may also own capacity in a cable as collateral for vendor finance to complete the cable. Recently, internet content providers such as Google have joined traditional telecommunications companies to become cable co-owners in consortium models.

Suppliers

A broad group of suppliers service the industry for the construction, operation and maintenance of submarine cables.

System Suppliers

The principal group of suppliers are the *full system suppliers* who design, plan, manufacture, install and commission a complete submarine cable. This group manufactures all elements of the system, including submarine cables, submerged repeaters and terminal equipment.

A secondary group of suppliers include those who specialize in upgrading (increasing the capacity) of existing submarine cables by replacing the existing transmission equipment with newer technology to increase the capacity and also to provide different interfaces.[4]

---

1 The emergence of private telecommunications companies during this time occurred largely in the United Kingdom and in many western European countries.

2 Examples of these types of telecommunications operators include France Télécom and Cable & Wireless.

3 For example, Google.

4 An interface refers to the physical equipment (either electrical or optical) between the submarine cable system and the equipment that extends the capacity inland. A range of different transmission rates are available.

Figure 2.1 Cableship *Tyco Reliance* proceeding at top speed to a cable repair. (Image reproduced with permission from Tyco Electronics Subsea Communications LLC (TE SubCom), © TE SubCom 2013 all rights reserved)

Finally there are suppliers who solely manufacture submarine cables and supply to either the full system suppliers or directly to the telecommunication operators.

Marine Service Suppliers

Underpinning the submarine cable manufacturers are the marine service providers. These providers supply the specialist vessels used to survey the seabed along the planned route for a new cable and also the specialist cableships used to install and maintain new cables. The specialist vessels have the capability to precisely control their position along their given route and to bury the cable on the seabed of the continental shelf to protect it from third party damage. The specialist cableships are also used to repair damaged or faulty submarine cables. The specialized repair vessels are equipped to locate and recover damaged cables at a fault location, test and insert new sections of cable, relay the cable to the seabed and rebury it if necessary. Their relatively large crews are highly trained and experienced.[5]

Cable Joint Suppliers

Key to these repair operations are the suppliers of the joints and associated equipment required to replace damaged sections of cable with new cable. The most

[5] For more information on repair and maintenance processes, please refer to Chapter 6.

common type of joint is supplied by the Universal Jointing Consortium (UJC), which provides the technology, equipment and procedures to enable a joint to be constructed between different types of cable by almost any cableship worldwide. The UJC is comprised of a cableship operator and full system providers.

The UJC only supplies joints for fiber optic systems. No similar device exists for power or high voltage direct current (HVDC) cables.[6] This is one reason why repairs to power cables take longer and involve significantly higher costs than repairs to telecommunications cables. Only the HVDC manufacturer or its representatives are authorized to carry out power cable repairs. Additionally, there are perhaps only three power cable vessels in the world equipped for HVDC cable repairs and installation.

Special Interest Groups

The submarine cable industry includes special interest bodies, such as the International Cable Protection Committee (ICPC), who play an essential role in providing leadership and guidance on issues related to submarine cable security and reliability. Since its formation in 1958 by BT and Cable & Wireless (C&W), the ICPC membership has grown to over 136 members from more than 63 countries. Members include owners and operators of submarine cables, submarine cable system suppliers, submarine cable suppliers, survey companies, cableship operators and governments, as well as several HVDC cable system owners. Ninety-eight per cent of installed fiber optic cables are owned and operated by ICPC members. Further, virtually all the marine service suppliers that own and operate the cableships used to install and repair these systems are members of the ICPC. The ICPC issues Recommendations (available upon request to the public) which are based on the consensus view of measures that have been successful for improving cable protection and reducing the risk of damage to cables. The ICPC also encourages scientific research into the relationships between submarine cables and the environment. This work has involved collaboration with the United Nations Environment Program (UNEP) and other scientists studying the ocean environment. The ICPC also works to strongly encourage adoption, implementation and compliance with the United Nations Convention on the Law of the Sea (UNCLOS) by its members and national governments. To this end, it has worked with the International Seabed Authority, the International Telecommunication Union, the Rhodes Academy on Oceans Law and Policy, national governments and selected non-governmental organizations, such as the EastWest Institute, to address UNCLOS compliance issues and exchange views regarding practical measures to improve cable security and the sharing of the seabed with other users. Mechanisms of cooperation relied upon range from formal memoranda

---

[6] For more information on power cables, please refer to Chapter 13.

of understanding to 'track two' diplomatic workshops and other seminars and information-exchange presentations.

II. How Does an International Submarine Cable System Come to Life?

A new submarine cable is conceived to satisfy one or more of the following drivers:

- capacity demand to meet forecast growth on an existing route;
- connectivity demand to new places; or
- political demand for new routes to support economic development.

Capacity demand is the driver for a new cable when all opportunities to upgrade the Design Capacity[7] of an existing cable have been exhausted and the demand for capacity is forecast to exceed available capacity within the timescales required to develop, manufacture and install a replacement cable. Capacity demand can also be the driver when a cable is no longer able to provide capacity with the required type of interface.

Connectivity demand is the demand for new connection points (landings servicing national network nodes or hubs, sometimes referred to as POPS (points of presence) within the same country. The demand is usually driven by the need to provide terrestrial diversity from an existing cable landing station and/or diversity for a landing point party. Connectivity demand is also the demand for new routes between countries already connected by an existing cable, for the purpose of providing undersea diversity. Diversity is an important practical means of improving submarine cable network reliability, as it allows communication to continue even when one cable system suffers a fault by rerouting traffic from the damaged cable to an undamaged system. This diversity can be obtained by anything from a few tens of kilometers separation over essentially the same route, to a new route traversing different oceans, for example a Europe–Asia cable that travels via the Mediterranean Sea, Red Sea, Indian Ocean versus a Europe–Asia cable that travels via a polar route.

Political demand is the connection of a country to the 'submarine cable network' and is driven by the strategic planning of individual or multiple governments, usually to support economic development. An example is the Eastern Africa Submarine Cable System; a fiber optic cable system deployed along the east coast of Africa, linking South Africa and Sudan via a series of landing points. This cable system commenced commercial services in 2010.[8]

---

[7] Design Capacity is the term utilized to indicate the maximum capacity that the initial submarine cable system is intended to support.

[8] *See* http://www.eassy.org/index-2.html (last accessed 6 June 2013).

III. Different Types of Commercial Model

Submarine Cable Consortiums

A submarine cable consortium is a collection of companies that 'club' together to fund the design, construction and maintenance of a new cable, hence the commercial model is sometimes referred to as a 'club cable'. The volume of capacity allocated to each of the consortium members is not always proportionate to their individual levels of investment. The levels of investment can depend upon whether the consortium member is a landing party at the end of the cable, a landing party at the end of a branch, or a non-landing party. The levels of investment are typically split into tiers, with the tiered structure generally establishing the minimum levels of investment required to participate in the project, to land a branch of the cable or to land either end of the cable. Each tier can also attract a different capacity weighting, meaning the higher the level of investment, the higher the tier and the higher the volume benefit in capacity. The consortium produces a Construction and Maintenance Agreement (C&MA). This is an agreement between the consortium members specifying how they will work together to construct and maintain the cable throughout its operational life and beyond.

Development Phase

The company who initially identified the driver for a new cable (see Part II above) will determine the key requirements for developing it. The key requirements will include factors such as the proposed configuration, the capacity/technology, the completion date and the commercial principles for the cable. This initiating company then identifies other companies who may potentially have an interest in becoming a landing party or investing in the cable at a significant level. These companies are referred to as the 'initial parties' or 'anchor parties'. The initial parties formalise their intention to participate in ongoing development of the new cable with a Memorandum of Understanding (MOU). The MOU sets out the main activities to be undertaken during the development phase, including determination of capacity allocation principles, use of the cable, system configuration, Operations and Maintenance (O&M) costs, principles for capital recovery for landing stations and will define how each of the initial parties will participate in funding the development of the cable.

During the development phase a number of important decisions will be made regarding the cable system such as, the configuration of the cable, the route it will follow, the technology to be employed, implementation timescales, initial capacity of the cable and the project budget. However, the key factors in the development phase are the forecast day-one capacity and the potential levels of investment for each of the consortium members. The aggregated day-one capacity determines the transmission equipment required for the initial period of operation, which is a key input for the construction of the project budget and for determining whether there are sufficient members in the consortium to fund the project. The development phase of the cable also includes the production of an Invitation to Tender (ITT), which comprises the technical specifications and the proposed terms and conditions of the supply contract. The technical specifications include the consortium's requirements for design capacity, day-one equipage, upgradability, connectivity, performance, reliability, interconnection to backhaul, equipment features, quality assurance, training, documentation, manufacturing, testing, installation, commissioning, and operation and maintenance of the submarine cable system. The ITT is issued to potential suppliers and a preferred supplier will be selected by tender process. The culmination of this phase of the cable system will be the signing of the C&MA by all members of the consortium, followed by signing of the Supply Contract.

Construction and Maintenance Agreements

The C&MA is the binding document that the consortium parties sign which governs all of their dealings with the cable system. The C&MA is the primary governance contract for the cable system. It describes the configuration of the cable, the obligations of the parties and how the consortium will be managed. It details how the key components such as the wet segment, cable stations and O&M costs will be paid, how the system can be used and how capacity can be sold. Each C&MA is unique and includes matters relevant to the particular cable. Whilst the C&MA binds all of the signatories, it will not necessarily be a comprehensive or exhaustive document. Interpretation of the C&MA can at times be difficult and subject to review and analysis by the consortium members of different nationalities, companies, and experience. In order to facilitate common international understanding, it is not uncommon to have the controlling language of the C&MA to be English.

The C&MA establishes a Management Committee that oversees the cable system and ensures that the intent of the C&MA is maintained in practice. Decisions made by the Management Committee are generally carried by a simple voting majority, although changes to the C&MA will require unanimous agreement.

The C&MA is not a static document and updated versions may be produced at any time, subject to the agreement of all of the signatories. Revised versions of the C&MA are Amendatory Agreements and may introduce changes and/or additions to the original document. The amendments may reflect minor changes, such as a change to a party's name, but may also include more fundamental changes.

The C&MA also includes a number of schedules and annexes that provide additional information. The schedules typically include a list of the parties to the agreement, voting rights and investment shares, O&M costs and procedures, capital shares and allocated capacity. The annexes usually include the terms of reference for the sub-committees and other information pertaining to the cable system, such as configuration diagrams, capacity allocation methodology and investment levels. Frequently an annex will specify, for tax or local considerations, that the portion of an international cable in territorial seas be owned by the corresponding landing party with the ownership of the cable in between and outside the territorial seas of the landings to be co-owned by all of the parties to the C&MA.

Capacity Distribution and Allocation

Companies invest in a submarine cable in exchange for the acquisition and use of capacity. The early submarine cables tended to be point to point, of a fixed bandwidth size and the allocation of capacity was fairly straightforward. The available capacity was finite and the costs easily apportioned. However, as systems became more complex in their configuration and more flexible in their usage, for example allowing portability, it became more difficult to allocate capacity. Portability and flexibility could easily result in congestion of some segments and therefore the Network Administrator would model traffic usage and manage the release of capacity for use to prevent blockages. The Network Administrator's duties and responsibilities are normally specified in an annex to the C&MA. In practical terms, it may be useful to think of the Network Administrator's function in this regard as that of the traffic police, enforcing C&MA rules and managing traffic to avoid interruptions and to maintain maximum efficiency for the cable system.

Capacity on a system exists in various guises:

- *Allocated capacity*—this is the capacity that a consortium party receives in return for its investment. The capacity can be further subdivided:
  - (i) *Assigned capacity*—this reflects capacity that is assigned by a party between specific paths on the cable system;
  - (ii) *Reserve capacity*—refers to capacity that has not yet been assigned but is available for assignment; and
  - (iii) *Pool capacity*—this is capacity that remains the property of the party but the consortium has the right to assign it for other uses, such as Indefeasible Right of Use (explained below) sales or for other third party uses.

Capacity can generally shift between these categories at any time, as specified in the C&MA and coordinated by the Network Administrator, however, cables vary and not all of the categories of capacity will be available on each cable system.

- *Common reserve capacity*—common reserve capacity is not part of allocated capacity but rather reflects a consortium pool from which capacity can be used for restoration purposes. For example, it can be set aside for mutual restoration between two different submarine cables and will be used by the other cable in the event of failure or *vice versa*. It can also be used for commercial restoration of another cable in the event of failure. The consortium parties own this capacity but do not have access to it unless it is released and distributed.

Management of the Submarine Cable during Installation versus Operation

A consortium cable establishes a working group structure that is defined within the C&MA and is led by a Management Committee. There are two distinct elements to the working group structure. The first element is supported by representatives of the consortium members who will land the cable and be responsible for managing the installation of the cable and establishing the O&M capability for the cable on behalf of all the owners. This element is led by what is usually termed as the Procurement Group. The Procurement Group only remains in existence during the construction of the cable and during any subsequent upgrades. The manpower and travel costs for meetings incurred for this element are borne by all of the owners.

The second element of the working group structure comprises a number of sub-committees. These sub-committees are responsible for the operational and financial performance and for other matters relating to assignment and restoration of the cable. The committees of the working groups are open to attendance by all members of the consortium. The committees continue to operate throughout the life of the cable and are comprised of representatives from all members of the consortium. Meetings are generally held once per year for the purpose of reviewing the operation of the cable. The representatives bear their own costs for travel and participation in the committees.

Landing Party Obligations

A landing party has two obligations. The first is termed the Maintenance Authority (MA), which refers to the responsibility for the O&M of the portion of the cable system that is laid between their own cable landing station and an agreed point within the cable. Different landing point MAs will be responsible for other sections of the cable. In a branched cable system the landing parties may be the MA for the cable up to the point of the branching unit, with the two trunk (end) stations being responsible for the trunk from their landing point up to an agreed point towards the middle of the segment.

This responsibility includes proactive activities required to prevent faults, such as liaison with fishermen, monitoring shipping activities close to the cable, reviewing activities planned by other industries, beach inspections, spare cable and joint stock audits, power plant routines and so forth. It also includes maintaining capability for repair of equipment in the cable station, the land cable, or the sea cable in the event of failure. The MA is responsible for managing a repair in the event of a cable failure. Frequently the C&MA also authorizes the MA to pursue damages from the owner of third party vessels or others who cause damage to cables. This action is normally undertaken through the relevant Protection and Indemnity (P&I) Club.[9] This authorization is essential, because it can result in legal action to arrest vessels and this must be rapid in order to obtain security for the damage claim prior to the ship leaving a jurisdiction.

Private Submarine Cables

A private submarine cable will be initiated by the same drivers as a consortium cable, but with two notable differences. The primary goal of the private cable is to supply the wholesale market and debt finance will normally be required to fund the construction. Private cables will manifest in several guises; for example a company may be formed by one or more entrepreneurs for the purpose of constructing a cable using their own equity, or several operators may together form a company and provide equity to become shareholders. The first case scenario solely supplies the wholesale market, whereas the second can also supply the network and retail needs of the shareholders.

Special Purpose Vehicles (SPV)

In some circumstances it may be preferable to form an SPV, which is an amalgamation of two or more parties within a consortium. The SPV will be used to facilitate finance from external sources to fund the cable and provide benefits to its members. Companies can elect to join an SPV in order to avoid having to provide upfront investment. The SPV purchases capacity on behalf of its members and is likely to enjoy the most attractive volume discounts, as it is effectively buying in bulk. SPV members can then draw down on the capacity as they require it, subject to the rules of the SPV. This provides substantial benefits to the members in terms of financing, although they are precluded from participating in management decisions regarding the cable as decision-making is devolved to the SPV. It may be possible for parties to form a 'mini-SPV' and enjoy some of the bulk-buying benefits without having to source external finance. The SPV arrangement is, however, reasonably uncommon and is an appropriate model only in limited circumstances, such as when a relatively large number of parties may experience difficulty in funding their investment.

Specialist Companies

From time to time private companies have sought to enter the cable-building industry, particularly when market conditions have been favourable. Their aim has been to establish a cable, probably financed by bank loans, in order to sell capacity to companies and use the sales revenue to repay the loans and make a return on the investment. Although this type of capacity is relatively more expen-

---

9 P&I Clubs are mutual insurance associations of vessel owners which provide various types of marine insurance. See also 87, 259, 262 n. 34, 269–270.

sive for the industry, the advantage is that it requires no upfront investment and capacity is available on demand.

Single Company Networks

A somewhat rarer occurrence is that of companies deciding to build their own infrastructure or acquiring existing systems in a bankruptcy process without financial assistance sourced from other entities. This clearly requires substantial finance and effort, but once established can create a dominant position. Some companies have created their own global network in this manner, using the network both for their own services and also selling surplus capacity to help fund the build.

IV. Financial Arrangements

Finance for a subsea cable is basically either sourced internally or externally. This can be a considerable investment, since a modern transpacific system may cost around USD1 billion. Shorter systems can run into the hundreds of millions of dollars. The traditional consortium system requires the parties to provide all of the finance needed to build and support the cable. An exception to this is where an SPV exists within the consortium. This form of internal financing ensures that capacity is provided at cost, although it can require large sums of upfront investment, depending on the investment criteria, and long lead times before the parties will see a return.

Any cable that requires external financing will have relatively more expensive capacity, as the debt needs to be financed and profits made in order to justify the build.

Some cables have been built on equity stakes, where telecommunication companies form a separate company with equity that is used to build the cable. This is a variation of the consortium cable where the equity remains in the hands of telecommunication companies, although there are often other third party costs incurred in running the system and company.

V. Funding for Cable Stations

Bringing the submarine cable ashore into landing stations and having suitable access to those stations is a key part of every cable system. Usually the cable station is built and owned by the designated terminal party for that country and financing is provided outside of the consortium. However, the costs of using the landing station are passed on to the parties through agreed mechanisms specified in the C&MA. Usually this involves setting a depreciation term for the cable station and allocating the capital costs through that period. These costs are then borne by the parties who use the landing station in direct proportion to their utilization of the facility. Once the cable stations have fully depreciated, i.e. when the repayment term is over, no additional costs for their use will be levied.

Some cable systems attempt to fund the cable stations through the initial investment, though only few succeed. Although it is preferable to fund the stations in this manner, it requires the parties to raise additional funds at a time when they are already burdened with financing the cable build and lay. Using an existing cable landing station may also facilitate permitting and landing licenses, depending upon the country involved. For example, landings in California in the United States are notoriously difficult to permit and can take years and cost millions of dollars. Using an existing landing station, if available, reduces permitting requirements, cost and delay risks.

For some private and other systems, the owners select landing parties in the relevant countries and these parties build and operate the station in return for obtaining capacity in the cable. In this way the cable owners are able to subcontract the shore side of the system.

VI. Commercial Management During Construction and Operation

Building a cable system requires the same degree of management and technical expertise regardless of whether the system is a private or consortium system.

Consortium Submarine Cables

Sub-Committee Structure

The Management Committee is the forum in which decisions are made regarding the most important aspects of the cable system. Reporting into the Management Committee are a number of sub-committees that are responsible for providing expert guidance and offering advice to the Management Committee as to how the system should be managed. The Management Committee generally provides for a voting mechanism with the sub-committees and seeks to reach decisions through consensus. Cable systems vary with regard to their reporting structure but the most frequent model is provided in Figure 2.2.

Roles and Generic Responsibilities

The responsibilities of these sub-committees are set out below:

- *Procurement Group*—this is the key initial group to be formed. Its activities are central to the build of the system and for negotiating with suppliers and developing the *supply contract.* Its technical input is essential for building a sound and reliable cable system. Once the cable is built, the role of the procurement

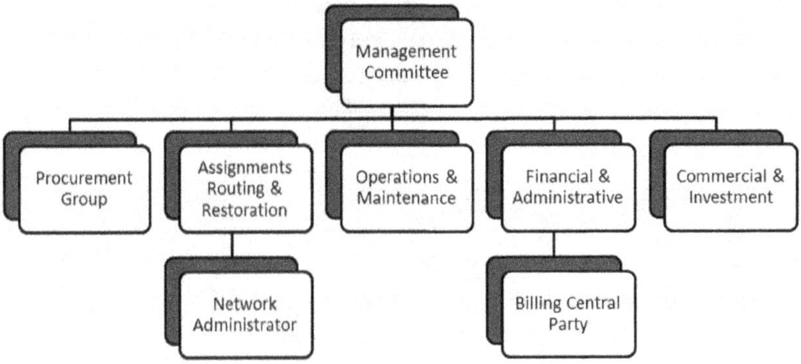

Figure 2.2 Typical reporting structure for consortium submarine cables.

group will diminish, although in practice the group moves on to other activities, such as upgrade work.

- *Assignments, Routing and Restoration*—this sub-committee oversees the technical operation of the cable, optimizing usage of capacity, traffic modelling, restoration and other 'in use' matters relating to capacity.
- *Operations and Maintenance*—this sub-committee is responsible for ensuring the operational integrity of the system, maximizing the availability of capacity and producing the O&M budget.
- *Financial and Administrative*—this sub-committee oversees all financial aspects of the system, including managing and balancing budgets.
- *Commercial and Investment*—this sub-committee attends to all the commercial activities for the cable, including the interpretation of the C&MA, the initial investment model and any ongoing commercial matters.
- *Network Administrator*—the Network Administrator often reports to the Assignments, Routing and Restoration sub-committee, although this may vary between systems. It is responsible for interfacing with the cable users, issuing capacity assignments and keeping records of usage.
- *Central Billing Party*—the central billing party usually reports to the Financial and Administrative sub-committee and is responsible for issuing bills and collecting payments from parties with respect to capital and O&M payments. It is also responsible for paying suppliers in accordance with the billing schedule.

Private Submarine Cables

Company Structure

There is a considerable amount of variation in the structure of privately owned and managed cable systems. Some act only as a supplier, whereas others establish a quasi-consortium arrangement, with purchasers of capacity having customer forums in which they are ostensibly afforded some degree of input into the

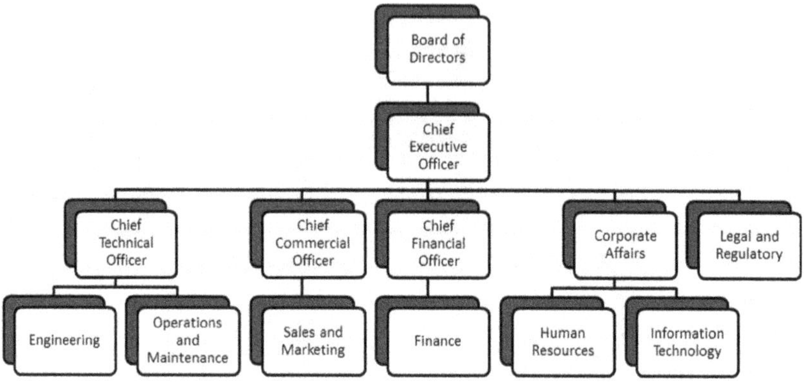

Figure 2.3 Typical company structure for a private submarine cable.

running of the system. A private system has different drivers and incentives to a consortium based system, with one essential difference being the need to sell capacity and raise revenue, which requires a sales force and marketing group. This is a major difference from the consortia model, in which there is little or no incentive to make any capacity sales. A typical structure for a private submarine cable is provided in Figure 2.3.

Customer Interfacing

A key difference between private and consortium systems is that private systems require a customer interface in order to maximize sales, generate revenue and ultimately profit. Whilst the consortium cable bases its costs on actual costs incurred, the private system sells capacity for the maximum that the market will bear. Although this leads to more expensive capacity, the private system does not have to rely on raising external finance and is thus able to build systems that a consortium may be precluded from building.

Once capacity has been acquired, telecommunication companies do not generally discern a difference between the consortia and private systems. One area in which there may be a disadvantage is the lack of flexibility and portability of capacity which may be evident in private systems. Often the capacity is purchased on an Indefeasible Right of Use basis, which lacks portability and has fixed O&M charges. As a result, the capacity can be a financial burden once the purchaser's need for it has expired.

VII. Operations and Maintenance Arrangements

Operations and Maintenance Organization

Consortia Systems

In a consortium cable the Maintenance Authorities are jointly responsible for the Operation and Maintenance of the cable through the O&M sub-committee. Together they ensure that there is the capability to undertake repairs to the undersea parts of the cable. This is usually achieved by entering the system into a cable maintenance agreement (CMA). In addition, the Maintenance Authorities also ensure that the equipment in the cable stations is maintained and repaired pursuant to agreements with the system supplier. A trend over recent years has been to set up a single Network Operations Center (NOC) to coordinate maintenance of the system on a day-to-day basis.

Private Systems

A private system will have a part of its organization that manages the operation and maintenance of the system, using similar arrangements to those used by the consortia systems except that a private system is more likely to enter the cable into a private maintenance agreement than a zone maintenance agreement for the repair of the undersea parts of the cable.

Cable Maintenance Agreements[10]

Zone Cable Maintenance Agreements

A zone cable maintenance agreement (zone CMA) consists of a large number of cable system owners, both consortium and private, who enter into a common commercial and technical agreement with a single or multiple maintenance provider/s. The cable system owners are also bound by an agreement between themselves. These agreements are supported by dedicated repair ships, usually based in fixed strategically placed depots in base ports where system spares are stored for maintenance purposes.

Private Maintenance Agreements

A private maintenance agreement consists of a single cable system, either consortia or private, or multiple cables belonging to a single owner, who enter into a sole commercial and technical agreement with a maintenance provider. A single vessel or a small number of vessels that are dedicated or otherwise permitted to undertake outside work usually support these agreements. During periods in

10 See also Chapter 6 for information on Cable Maintenance Agreements.

which these vessels are unavailable, additional vessels from the maintenance provider's fleet may replace them. Private maintenance agreements may have a dedicated depot for maintenance spares but may also operate by having strategic spares located on one or another of the maintenance vessels.

Comparison of Services Provided under Zone and Private Maintenance Agreements

The members of the zone CMA benefit from having full visibility and control over the ships within the agreement. In a private maintenance agreement the members may not have the same visibility and control over the ships that service their cables, as it is likely that the maintenance provider will have similar agreements with other cable owners who share the common maintenance assets. Although private maintenance agreements may not have the same level of control over the ships within their agreements, in times when their dedicated vessel is not available there may be the option to utilize another ship from the maintenance provider's fleet to support their contract, whereas under a zone CMA a member will have to wait for an appropriate vessel to become available in circumstances where all the assets are utilized.

Key points of a zone CMA are as follows:

- Run by members for the members
- Outside work by permission of members
- Outside work provides standing charge rebate to members
- Vessels dedicated to maintenance operations
- Priority usually given to longer cable systems
- New cables join by members' agreement
- Price per km variable depending on cable km in agreement

Key points of private CMA are as follows:

- Operated by the ship operator for the members
- Outside work may be managed by the ship operator
- Ship operator retains all financial benefit of outside work
- Ship operator decides whether new cables can join and share repair vessels
- Price per km may be substantially reduced thus offsetting the outside work rebate

Zone versus Private CMA

The Atlantic Cable Maintenance Agreement (ACMA)[11] is an example of a successful non-profit cooperative subsea maintenance agreement consisting of more than

---

[11] ACMA is a zone cable maintenance agreement. See Figure 6.1.

60 members. ACMA members are companies responsible for the operations and maintenance of undersea communications in the Atlantic Ocean, North Sea and south eastern Pacific Ocean. The entire ACMA maintenance operation, including ships, remotely operated vehicles and technical specialists, is under independent contract to dedicated third party service providers, with formal key performance indicators, such as cableship mobilization, spares loading and operational timings. Accordingly, ACMA members also benefit from a quality and efficient service provided by an independent fleet with its facilities dedicated to the sole use of the ACMA members.

VIII. Types of Capacity

As the global submarine cable network developed, capacity was always bought and used on a bilateral basis and thus for a cable linking country A with country B, country A would own the cable laid from its landing point to a notional midpoint and country B would own the remainder. The reason for this practice was that at that time telecommunication companies generally only had licences to operate and own infrastructure capability in their home country. So all capacity was *jointly assigned capacity* and traffic and revenues were shared proportionately. At the time, most cables had not even considered the possibility of having *wholly owned capacity* (WAC).

As the telecommunications market began to liberalize, companies started to obtain licences and develop business in countries other than their home country. They sought to offer an end-to-end service and, of course, retain most of the revenue rather than share it. Submarine cable companies began to offer the opportunity to buy WAC, although in many cases there were conditions attached to its use which did not really assist in its deployment. As some companies saw the availability of WAC as a threat to their domestic markets, its introduction was slow and laboured. Gradually however, it began to become more freely available within existing systems and eventually became a pre-requisite for new cables being built. Some of the older cable systems still impose restrictions on how WAC can be used, which is at odds with the rules of some of the newer systems.

Types of Capacity Acquisition

Ownership

The traditional consortium cable sells capacity on an 'ownership' basis, which means that a party is a co-owner within the consortium and has rights and liabilities in proportion to its ownership share. It has a say in how the system is managed and a vote with a weighting that accords with its ownership share. During the C&MA construction phase the co-owner is able to influence, to a degree, the rules of the system and therefore the co-owning parties have an opportunity to create the type of system that they want. The drawback to this structure is that decisions are made by committee, which can be slow and require unsatisfactory compromises.

Indefeasible Right of Use

Private systems usually sell capacity on an Indefeasible Right of Use (IRU) basis. Whilst this method still gives the purchaser the right to use the capacity for a fixed term, generally for the life of the cable, there are some subtle differences. The capacity tends to be fixed between specific points and lacks portability, although some improvements have been introduced more recently. When capacity is sold on an IRU basis, purchasers do not have any input into the cable system during the construction phase and generally have little say in how the cable is operated and managed, which is a disadvantage, although the rules related to the use of the IRU tend to be clearer. IRUs can also be bought from consortium cables, if ownership status is no longer available, and also from other telecommunication companies. In cases where capacity is bought from other carriers, the terms and conditions of the sale are a private arrangement although the purchaser will still be bound by the rules of the consortium.

Lease

Capacity can also be made available by carriers and private cable systems on a lease basis, which may operate for any period agreed between the lessor and the lessee. The benefit of a lease, as opposed to an IRU, is that the capacity can be acquired for a shorter duration and there is no indefinite liability. The disadvantage is the higher cost of a lease. Frequently the cost of a three year lease is equivalent to an IRU purchase and therefore the carrier has to assess which option for acquiring capacity would be more viable.

Restoration Capacity of Cables

Any type of cable failure is a serious issue, both in terms of the traffic and services that can be lost and also the time that may be required to execute a proper repair. Any repair that requires the mobilization and voyage of a cable repair ship to the scene of a cable fault invariably takes time, therefore cable systems adopt a variety of protective measures to reduce the impact of cable failures.

Cable systems have always had restoration arrangements with other cables in order to provide mutual assistance if one cable suffers from a fault. The advantage of this arrangement is that services can be restored very quickly, however, in order to provide coverage there must be other cables landing nearby and owners willing to share restoration capacity. This makes the exercise of organizing restoration facilities very complicated and potentially quite costly for parties requiring restoration. Recognizing the critical nature of prompt and seamless restoration, each cable system will have a skilled specialist Restoration Liaison Officer (RLO) to plan restoration, conduct exercises with other systems, and to handle restorations when the need arises from a cable fault or repair on a 24/7 basis. Cable system owners may also differentiate the costs of services provided to their customers based in part on the degree and speed of restoration specified in the relevant contract.

In order to overcome the problems of availability and cost involved in restoring cables, a ring configuration was developed so that any problem detected in one half of the system would be alleviated by routing the traffic the other way around the system. This was an ideal solution as restoration was readily available at no extra cost and there was virtually no delay in its implementation. The main problem with ring systems, however, was their build cost, as the system effectively employs two cables. Companies needing to reduce costs following the dot com boom often found that a single strand configuration was a much more viable prospect. Telecommunication companies also began to develop their own mesh networks which did not require in-system restoration and so the cost of restoration could not be justified. A mesh network essentially uses capacity on a number of cables owned by the same cable owner. The common ownership of capacity on several cables allows the cable owner to restore its traffic on a damaged cable by shifting the traffic to other capacity it owns on other systems. In addition to being cheaper to build than a ring system, a mesh approach can provide greater redundancy, reliability, and reduce the risk of service interruption.

Some cables are still able to provide limited in-system restoration, although this depends entirely on the configuration.

IX. Cable Station Access

Cable stations are usually owned by the landing party of the respective countries in which the cable lands. Under the old bilateral model, cable owners were not concerned with the operation of another country's landing station, since the capacity would usually be held bilaterally with the landing party and they would connect up to their own backhaul. With the increased use of wholly owned capacity, parties needed to be able to egress from the distant cable station and get the capacity backhauled to their Points of Presence (POP) or customer's premises. It was therefore important that the C&MAs were drafted so as to ensure that all users could get their capacity out of the cable station and that there was some obligation on the station owner to provide access to other parties.

This can be achieved either by providing co-location space within the landing station to enable other entities to pick up their traffic onsite, or by providing for a hand-off outside of the landing station where the parties can pick up their own backhaul. Of course, it is still possible for the local entity to provide backhaul from the cable station to the destination within the country, as would have been common practice in the bilateral days, although there would be a fee for this service.

More recent cables have attempted to provide access to the cable at a POP within the country rather than at the cable station itself. This is generally seen as advantageous, although it is not necessarily viewed as such by the local dominant telecommunications provider within that country. This has resulted in POP access to cables being more an aspiration than a reality, which is not unusual as progress is often slow regarding changes that are seen as a threat to the local operator.

Conclusion

This Chapter is designed to provide government ocean-policy makers with basic information on how submarine cable businesses are organized and function in the modern era. Over time, carefully balanced and varied contractual models have been developed within the industry. These models have provided the world with a robust yet nimble communications system, a system which in turn has enabled the internet and the global economy to flourish. Despite this astonishing communications system being international in nature, it has been built with little or no government funding. The cable industry itself is constantly evolving and the search for better efficiency, lower costs and greater reliability continues. Regulators should carefully weigh national regulations in light of the international nature of the services that the cable systems support, and the dictates of UNCLOS upon which the cable business depends.

PART II

INTERNATIONAL LAW ON SUBMARINE CABLES

CHAPTER THREE

OVERVIEW OF THE INTERNATIONAL LEGAL REGIME GOVERNING SUBMARINE CABLES

Douglas Burnett, Tara Davenport and Robert Beckman

Introduction

In 1880, Sir Travers Twiss, a noted English jurist, made the following statement:

> The preliminary question, which deserves consideration, is whether the maintenance of the telegraphic sea-cables, which have an international importance, is an interest of the highest order to States, analogous to the interest of the public health and of the public revenue, which each nation is allowed by courtesy to protect beyond the strict limits of its territorial waters.
>
> If we look to the public services which the telegraphic sea-cable is now called upon to perform in time of peace, that it has become the normal instrument of communication between Governments and their envoys in foreign countries that international treaties are from time to time concluded between the nations of the two hemispheres through the medium of cable telegraphs; that through the same instrumentality approaching tempests are announced in advance to Europe from America, by which great damages and destruction to life and shipping may be averted; that no great criminal can now hope to escape from Europe to the western shores of the Atlantic Ocean with the fruits of his crime without a telegram anticipating his arrival, when he finds himself the captive of the law at the moment when he expects to set his foot upon a land of liberty;
>
> ...the answer to the question above stated must, we think, be in the affirmative, and there can be no doubt that the great arterial lines of telegraphs have become indispensable for the circulation of the political life blood so necessary to maintain the vitality of our modern international State system.[1]

The above quote captures the marvel and wondrous astonishment with which the world welcomed submarine telegraph cables and the progress they brought. It is, in many respects, the way modern society looks upon the submarine fiber

[1] T. Twiss, "Submarine Telegraph Cables" (November 1880) XLIX: XI *The Nautical Magazine* at 883–884.

optic cables that have transformed the global economy, political systems, and everyday life for the world and its citizens.

This widespread appreciation of the value and dependency on submarine telegraph cables compelled governments to come together on several occasions to try and reach agreement on an international regime governing submarine cables, with the ultimate aim of protecting this critical public good. From 1863 to 1913, the protection of submarine cables appeared on the agenda of seven international conferences.[2] Between 1884 and 1982, the international community adopted four international instruments which set out substantive provisions on the rights and obligations of States *vis-à-vis* submarine cables. These are: (1) the 1884 Convention for the Protection of Submarine Telegraph Cables (1884 Cable Convention);[3] (2) the 1958 Geneva Convention on the High Seas;[4] (3) the 1958 Convention on the Continental Shelf;[5] and (4) the 1982 United Nations Convention on the Law of Sea (UNCLOS).[6]

While the 1884 Cable Convention was a stand-alone convention dealing solely with submarine cables, the 1958 Geneva Conventions on the High Seas and the Continental Shelf (collectively referred to as the "1958 Geneva Conventions") and UNCLOS were broad-ranging instruments covering a wide range of law of the sea issues including the use of submarine cables. The 1884 Cable Convention, which presently has 40 State Parties,[7] played an instrumental role in shaping the development of the law on submarine cables. Certain critical provisions in the 1884 Cable Convention have been incorporated into both the 1958 Geneva

---

2 United Nations Documents on the Development and Codification of International Law: Supplement to American Journal of International Law, Vol 41, No 4, October 1947, available at http://untreaty.un.org/ilc/documentation/english/ASIL_1947_study.pdf (last accessed 7 June 2013).

3 Convention for the Protection of Submarine Telegraph Cables, adopted 14 March 1884, TS 380 (entered into force 1 May 1888) (1884 Cable Convention). The provisions of the Cable Convention are generally accepted as customary international law, see *Restatement of the Law (Third): The Foreign Relations Law of the United States Vol 2* (American Law Institute Publishers, 1987) § 521, comment f (1986). As at 2 April 2013 there are 40 State parties to the Cable Convention. A complete copy of the Cable Convention is contained in Appendix 3.

4 1958 Convention on the High Seas, adopted 29 April 1958, 450 UNTS 11 (entered into force 30 September 1962). As at 2 April 2013 there are 63 State parties. The United Nations Convention on the Law of the Sea (see below note 6) supersedes this treaty for States that are parties to both.

5 1958 Convention on the Continental Shelf, adopted 29 April 1958, 499 UNTS 311 (entered into force 10 June 1964). As at 2 April 2013 there are 58 States parties. The United Nations Convention on the Law of the Sea (see below note 6) supersedes this treaty for States that are parties to both.

6 United Nations Convention on the Law of the Sea, adopted 10 December 1982, UNTS 1833 (entered into force 16 November 1994) (UNCLOS). Select UNCLOS provisions are contained in Appendix 3.

7 Restatement of the Law, *supra* note 3.

Conventions and UNCLOS. Similarly, the provisions on submarine cables in the 1958 Geneva Conventions were incorporated more or less *ad verbatim* into UNCLOS. It is therefore accurate to say that UNCLOS provisions on submarine cables represent the applicable legal regime on submarine cables. This conclusion is reinforced by the fact that UNCLOS presently has 166 parties[8] and prevails, as between States Parties, over the 1958 Geneva Conventions.[9] Most of the provisions of UNCLOS (including provisions on submarine cables) can be said to bind non-parties as they are best evidence of customary international law.[10]

Notwithstanding the fact that UNCLOS now largely governs the subject-matter, the inter-relationship between the four international conventions means that each of them should be examined. To this end, this Chapter will give an overview of each of the conventions. Part I will examine the 1884 Cable Convention and Part II will give a brief overview of the three Geneva Conventions on the law of the sea. Parts III, IV and V will explain the relevant UNCLOS provisions on submarine cables, addressing cable operations (surveying of cable routes and the laying, repair and maintenance of cables), protection of submarine cables from competing uses and dispute settlement procedures respectively. Part VI will highlight other international conventions which may be pertinent to the legal regime governing submarine cables.

I. The 1884 Convention for the Protection of Submarine Telegraph Cables

The 1884 Cable Convention was the first international treaty governing submarine telegraph cables. Its genesis was the recognition by States that these cables were vital means of communication which needed to be protected. For example, following the first successful trans-Atlantic submarine telegraph cable in 1866 the United States, the cable's noted proponent, Cyrus Field, urged that the "telegraph in the air and under the water should be regarded as a sacred thing, protected by unanimous consent against all attack or damage", and proposed as an international code that:

---

8 165 States Parties, and the European Union. Status of UNCLOS, United Nations Treaty Collection, available at the United Nations Treaty collection website: http://treaties.un.org/pages/ViewDetailsIII.aspx?&src=TREATY&mtdsg_no=XXI~6&chapter=21&Temp=mtdsg3&lang=en (last accessed 2 September 2013).

9 UNCLOS Art 311(1).

10 This is also because the 1958 Geneva Convention on the High Seas, which was incorporated into UNCLOS, purported to codify existing customary international law at the time. See R. Churchill and A.V. Lowe, *The Law of the Sea* (3rd ed, Juris Publishing, 1999) at 24.

> Any person whosoever, or any nations whatever, who without authority from the owner and with the intent to injure, vex, or annoy any other person whomsoever, or any nation whatever, removes, destroys, disturbs, obstruct, or injures any oceanic telegraphic cable not his own, or any part thereof, or any appurtenance or apparatus therewith connected, or severs any wire thereof, is deemed a pirate.[11]

Under this *hostis humani generis* classification associated with pirates, the cable injurer would be subject to the summary penalty of death. Although the protection of submarine cables was held to be "an interest of the highest order to the States, analogous to the interests of the public health and of the public revenue" the United States proposal was received as overly excessive and out of step with the "milder manners of the age".[12] The proposal and the debate it engendered were nonetheless sharp indications of the gravity with which injuries to submarine cables were viewed by the nations.

From 1882, a series of diplomatic conferences were held with the purpose of establishing an international treaty to protect and foster the growth of the new technology of submarine cables.[13] The catalyst which underscored the need for such an international treaty was the damage to several British-owned submarine telegraph cables in the North Sea caused by the negligence of fishermen. Because the technology was pioneering, engineers, fishermen, and naval officers from various countries, as well as diplomats, participated in the conferences. The remarkable result was the 1884 Cable Convention. As noted above, the 1884 Cable Convention is the bedrock for the provisions on submarine cables found in UNCLOS, which is now the primary legal regime governing the subject matter.

The overarching objective of the 1884 Cable Convention was to require States to adopt national legislation to protect submarine cables. Its provisions apply "outside territorial waters to all legally established submarine cables landed on the territories, colonies or possessions of one or more of the High Contracting Parties".[14] The 1884 Cable Convention also makes clear that its provisions do not apply in wartime, rejecting the approach of an earlier 1864 treaty between France, Brazil, Haiti, Italy, and Portugal that decreed these parties would not cut or destroy cables in time of war.[15] State practice since the 1884 Cable Convention is consistent with the principle that international cables are legitimate wartime targets.[16]

---

11 *Supra* note 1, at 879–880.

12 *Ibid.*, at 890–881.

13 The fascinating account of these conferences is captured in the notes of Professor Renault, the French scholar, who was considered the *de facto* rapporteur of the conferences: See L. Renault, "The Protection of Submarine Telegraphs and the Paris Conference (October–November 1882) in Brussels and Leipzig" *International Law Review* (Merzbach and Falk, 1884).

14 1884 Cable Convention Art I.

15 1884 Cable Convention Art XV and Renault, *supra* note 13, at 2.

16 "Right to Cut Cables in War; Admiral Dewey Created a New Precedent Under the Law of Nations in Manila Bay" *The New York Times*, 23 May 1898; *Eastern Extension, Australia*

Protection of Cables from Breakage or Injury

The 1884 Cable Convention contains three provisions relating to breakage or injury of cables. First, Article II provides that it is a punishable offence to break or injure a cable, wilfully or by culpable negligence, in such a manner as might interrupt or obstruct telegraphic communications, either wholly or partially.

The term 'culpable negligence' as used in the 1884 Cable Convention is based on the holding of an English court in the first reported case involving cable damage caused by a vessel.[17] Culpable negligence refers to the precautions demanded by the ordinary experience of the mariner and the particular circumstances in which the actions of the vessel are taken.[18] The standard of prudent seamanship in avoiding fixed objects like cables applies with regard to culpable negligence.

However, if the damage caused by the vessel arose through an attempt to save the vessel or its crew, there is no liability.[19] The classic example is a vessel in foul weather that loses propulsion, drifts towards shoals, and drops its anchor to avoid being dashed upon the rocks and in this process catches a cable. In this case there is no liability, assuming the master exercised prudent seamanship to avoid the situation in the first place.

It is clear that Article II balances the need to protect submarine cables (with criminal and civil sanctions) from wilful injury or injury caused by culpable negligence against the actions of the master of a ship taken to save his vessel and crew.

The second provision dealing with breakage or injury to cables is Article IV which addresses damage to cables in the situation of cable crossings. If two cables cross and both owners enjoy the freedom to lay their cable, who bears the loss if a cable is damaged in crossing? The compromise in Article IV is that the priority lies with the first laid cable. While every cable can cross another cable, if in the course of the crossing the first laid cable is damaged, the crossing cable must indemnify the first laid cable for the cost of repairs.[20]

---

and China Telegraph Company Limited (Great Britain) v United States, 9 November 1923, 6 Rep J International Arbitration Awards 112 (Arb 1923); J. Pelzer, "False Invasion Repelled (Raid on Cienfuegos, 1898)" (1993) 10(2) *Military History*, 66–73; R.K. Massie, *Castles of Steel: Britain, German and the Winning of the Great War at Sea* (Vintage, 2007) at 77; E. Brose, *Death at Sea: Graf Spee and the Flight of the German East Asiatic Naval Squadron in 1914* (Eric Brose, 2010) Chapter XIX at 189.

17 *Submarine Cable Company v Dixon*, The Law Times, 5 March 1864, Reports Vol X, NS at 32–33.

18 1884 Cable Convention Art II; Renault, *supra* note 13 at 9, *The Clara Killiam*, 1870, Vol iii LR-3 Adm Eccl at 161–167; *The Government of the Netherlands, Post Office v G'T Manneteje-Van Dam [Fishing Cutter GO 4]*, File No 325/78 (District Court Rotterdam, decision rendered 20 November 1978), aff'd *sub nom* G't Mannetje-Post Office, File No 69 R/81 and File No rb 325/78 (The Court at the Hague, Second Chamber, decision rendered 15 April 1983).

19 1884 Cable Convention Art II and Renault, *supra* note 13 at 9.

20 1884 Cable Convention Art IV and Renault, *supra* note 13 at 11.

In a Declaration dated 1 December 1886, the States Parties clarified the application of Article II and IV when applied to ships carrying out cable repairs. The Declaration states that the imposition of penal responsibility does not apply to cases "of breaking or injuries occasioned accidentally or necessarily in repairing a cable, when all precautions have been taken to avoid such breakings or damages."[21] Civil liability as determined by a competent tribunal would apply to the vessel or cable owner who damaged the first laid cable.

In modern practice, companies planning a new cable system go to great lengths to identify any cable or pipeline that must be crossed. Advance liaison is carried out to plan for a safe crossing. In some cases there is a voluntary formal crossing agreement. In other cases there is not, because there is no requirement for such an agreement under international law. In any event, crossings typically receive careful engineering and planning scrutiny and liaison with both systems involved in a crossing so that damages to cables and pipelines are rare.

The third provision on breaking or injury to cables is Article VII. It addresses the situation where a vessel or master which, through no fault of its own, finds it has fouled a cable with its fishing gear, nets, or anchor. In this circumstance, the 1884 Cable Convention requires the vessel to sacrifice its gear or anchor to avoid the greater harm of disrupting international communications.[22]

In return, the 1884 Cable Convention requires that the cable owner indemnify the vessel for the replacement cost of the sacrificed gear.[23] The Convention then details the procedure for a vessel to claim indemnity by filing witness statements and cost supports with the cable owner or, if not known, with the captain of the port or the coast guard within 24 hours of arrival in port. This practice is widely followed by present day cable owners who generally maintain 24 hour telephone hotlines to receive reports and provide information to masters of vessels who report that they may be fouled on a cable. Once a claim is filed, the cable owner investigates and in most cases the sacrificed gear is recovered and returned or indemnity compensation is paid. As required by international law, the indemnity is limited to the costs of the actual sacrificed gear or anchor and does not include damages for lost profits or catch.[24]

The 1884 Cable Convention requires contracting parties to take or to propose to their respective legislatures the necessary measures for ensuring the execution of the above provisions, which appears to envisage the adoption of national legislation to implement the provisions on injury to cables.[25]

---

21 25 Stat. 1424; TS 380-2

22 1884 Cable Convention Art VII and Renault, *supra* note 13 at 13.

23 *Ibid.*

24 *Ibid.*, and *Agincourt Steamship Company Ltd v Eastern Extension, Australia and China Telegraph Company Ltd* 2 KB 305 (1907).

25 1884 Cable Convention Art XII.

Protection of Cableships Engaged in Laying and Repair Activities

The 1884 Cable Convention also has provisions on the protection of cableships engaged in laying and repairing operations. Article V of the Convention requires that vessels maintain a distance of one nautical mile from a cableship laying cable which displays the appropriate day shapes or lights at night denoting its restricted maneuverability status to other vessels in the area.[26] Article VI requires vessels to maintain a safety distance of one-quarter nm from a cable repair buoy.[27]

Jurisdiction and Enforcement Provisions

In terms of jurisdiction over offences, the Convention provides that "the tribunals competent to take cognizance of infractions of the present Convention are those of the country to which the vessel on board of which the offence was committed belongs". It also recognizes that in relation to "subjects and citizens" of contracting States, the rules of general criminal jurisdiction prescribed by national laws or international treaties will apply.[28]

Article X of the Convention also allows an officer from a naval vessel from any of the States Parties to board a ship on the high seas suspected of damaging an international cable.[29] This provision was relied upon in 1959 by the United States when a USS Navy destroyer boarded a Soviet trawler suspected of cutting five trans-Atlantic cables over a two-day period.[30]

II. The Development of the Law of the Sea and its Impact on Submarine Cables

By the twentieth century, developments in technology relating to the manufacture and laying of cables had resulted in the increased use and reliance on submarine cables and the submarine cable network had been greatly expanded.[31] By the 1950s, submarine cables were not only used for telegraphic communication but were also facilitating telephonic communications, which enabled hundreds of voice calls to be made between different continents.

---

26 1884 Cable Convention Art V and Renault, *supra* note 13 at 11–12.

27 1884 Cable Convention Art VI and Renault, *ibid.*, at 12.

28 1884 Cable Convention Art XII.

29 1884 Cable Convention Art X and Renault, *supra* note 13 at 16–17.

30 The Novorossisk, Dept of State Bull, (20 April 1959) Vol XL, No 1034 at 555. Press release: The Embassy of the United States of America refers to the Ministry's Note No 17/OSA, dated 4 March 1959 concerning recent breaks in certain transatlantic submarine telecommunication cables and the consequent visit to the Soviet trawler *Novorossisk* by a boarding party from the USS *Roy 0 Hale*, which was the subject of the Embassy's aide memoire of 28 February 1959.

31 Please refer to Chapter 1 on the Development of Submarine Cables.

The vast expansion of the submarine cables network over greater areas of the seabed was coupled with an increasing use of the oceans for other purposes and ever-expanding claims by coastal States to ocean space. Because of these extensive claims to ocean space which purported to restrict the rights of other States, there was a pressing need to codify principles to govern the assertion of such claims and from 1930, starting with the Hague Codification Conference, attempts were made by the international community to adopt an international treaty on the law of the sea.[32]

While this attempt failed,[33] State practice continued to develop with various claims made by States to maritime zones of varying extents. In particular, the 1945 Truman Proclamation by the United States, whereby the US claimed jurisdiction and control over an area of seabed contiguous to its coast, precipitated a spate of similar claims by other States. This laid the foundation for the development of the continental shelf regime under international law which recognized that coastal States had certain exclusive rights over the seabed under the high seas. However, the lack of uniformity in continental shelf claims made by States revived international efforts to codify the law of the sea in order to ensure a clear basis and defined limits for maritime claims.

The Work of the International Law Commission (ILC) and the 1956 ILC Draft Articles

The task of codification was entrusted by the United Nations to the International Law Commission (ILC).[34] The ILC met eight times in the course of five years and eventually produced a series of Draft Articles on Law of the Sea in 1956, which contained seventy-three articles with commentaries covering the territorial seas, contiguous zone, the high seas and the continental shelf.[35]

Given the increasing importance of submarine cables to the international community, it was no surprise that the ILC spent some time considering submarine cables during its sessions. With regards to the protection of submarine cables, there was considerable debate during the ILC sessions on whether to include the provisions of the 1884 Cable Convention in any codification attempts on the law of the sea. This was part of a larger debate on whether the ILC should attempt to codify all aspects of maritime law, particularly when the subject was regulated

32 See generally D. Rothwell and T. Stephens, *The International Law of the Sea* (Hart Publishing, 2010) at 4–5.

33 The attempt failed due to a lack of agreement on the breadth of the territorial sea.

34 Please refer to the Summary of the Work of the International Law Commission available online at http://untreaty.un.org/ilc/summaries/8_1.htm (last accessed 7 June 2013).

35 Please refer to the Summary of the Work of the ILC, *ibid.* The Draft Articles and the commentaries can be found in the *Yearbook of the International Law Commission,* Volume II, UN Doc A/CN.4/SER.A/1956/Add.1, (1956) at 256–301.

by a convention.[36] Ultimately, only *three* articles on the protection of submarine cables from breakage or injury (dealt with above) in the 1884 Cable Convention were incorporated into the ILC Draft Articles.[37] This was on this basis that these three articles were essential principles on the law of the sea and were consequently necessary to include in any codification efforts.[38] Accordingly, the ILC only adopted Article II (protection of cables beneath the high seas from breaking or injury through wilful action or culpable negligence), Article IV (indemnification obligation for breaking or injury of cable by another owner) and Article VII (indemnification obligation for ship owners for sacrifice of equipment) from the 1884 Cable Convention.

In addition to three provisions in the 1884 Cable Convention, the ILC Draft Articles also contained an article which required States to regulate trawling so as to ensure that all fishing gear used shall be constructed and maintained as to reduce any danger or fouling of submarine cables or pipelines.[39]

The 1956 ILC Draft Articles also addressed the *freedom to lay submarine cables.* Unlike the provisions on the protection of cables, the freedom to lay cables had not previously been codified in any international agreement, possibly due to the fact that the right of States to lay cables had never been questioned.[40] However, the ILC agreed that it was important to include provisions on the freedom to lay cables in any convention on the law of the sea.[41] Accordingly, the ILC Draft

---

36 *Yearbook of the International Law Commission,* Volume I, UN Doc.A/CN.4/Ser.A/1951 (1951) at 363.

37 Articles II, IV and V of the 1884 Cable Convention were incorporated in Articles 27, 28 and 29 of the 1958 High Seas Convention. Copies of these articles are contained in Appendix 3.

38 There were initial misgivings that the provisions on the protection of submarine cables proposed for adoption were too detailed and that the ILC should only state general principles. However, the ILC ultimately adopted three provisions from the 1884 Cable Convention based on the rationale that the articles chosen contained essential principles: see *Yearbook of the International Law Commission,* Volume I, UN Doc A/CN.4/Ser.A/1955 (1955) at 20–21.

39 See ILC Draft Articles, Art 64. This was based on a resolution adopted at a 1913 Conference in London convened by the British Government to adopt further measures for the protection of submarine cables, see M. McDougal and W. Burke, *The Public Order of the Oceans: A Contemporary International Law of the Sea* (Yale University Press, 1962) at 846–847.

40 Indeed, the 1884 Cable Convention deals solely with the protection of submarine cables and does not address the freedom to lay cables because at the time, "it was evident that freedom of use was conceded by all and that the real concern was to adopt measures for protecting cables from other, sometimes physically incompatible, uses of the ocean." See McDougal and Burke, *ibid.,* at 781.

41 In 1950, the ILC first recognized the principle that all States are entitled to lay submarine cables on the high seas: see *Report of the International Law Commission on its Second Session, Official Records of the General Assembly, Fifth Session, Supplement No 12* (A/1316), UN Doc No A/CN.4/34 (1950) at 384. When it was first discussed in the ILC

Articles recognize that freedom of the high seas includes the freedom to lay submarine cables and pipelines[42] and that "subject to its right to take reasonable measures for the exploration of the continental shelf and the exploitation of its natural resources, the coastal State may not impede the laying or maintenance of such cables or pipelines".[43]

1958 Geneva Conventions on the Continental Shelf and the High Seas

In 1956, after submitting its Draft Articles to the General Assembly, the ILC recommended that the General Assembly should summon an international conference of plenipotentiaries.[44] The first Conference on the Law of the Sea was held in 1958. The ILC Draft Articles were used as the basic negotiating text and were divided up between four Committees which then undertook short debates. As a result of this process, four separate conventions emerged, namely the 1958 Geneva Conventions on (1) the Territorial Seas and Contiguous Zone, (2) the High Seas, (3) the Continental Shelf and on (4) Fishing and the Conservation of the Living Resources of the High Seas.

The 1958 Convention on the Continental Shelf and the 1958 Convention on the High Seas contained provisions on the protection of submarine cables and the freedom to lay cables. The provisions on submarine cables ultimately adopted in these Conventions were not the subject of contentious debate—significant debate was instead reserved for issues such as the breadth of the territorial sea and fisheries jurisdiction. Both Conventions adopted, more or less *ad verbatim*, the provisions on cables set out in the ILC Draft Articles,[45] with a few modifications. With regards to the protection of cables, the three provisions from the 1884 Cable Convention on the protection of cables (Articles II, IV and VII) and incorporated in the ILC Draft Articles were adopted in the High Seas Convention. This was despite the concern of the United States that the adoption of only some of the detailed provisions of the 1884 Cable Convention rather than the whole Convention would undermine its effectiveness.[46] The United States eventually

at its second session, it was even commented that, as the right to lay submarine cables had never been questioned, there was no need to explicitly mention it in any convention on the topic. However, the rest of the Commission agreed that while the principle of freedom to lay submarine cables had never been challenged, it was important to include it in any convention on the issue: see Comments of Judge Hudson and Mr. Spiropolous, *Yearbook of the International Law Commission*, Volume I, UN Doc A/CN.4/Ser.A/1950 (1950) at 199.

42 ILC Draft Articles on the Law of the Sea Art 27(1).

43 ILC Draft Articles on the Law of the Sea Arts 61(2) and 70.

44 Please refer to the Summary of the Work of the ILC, *supra* note 34.

45 For example, Arts 27, 61, 62, 63, 65 and 70 of the ILC Draft Articles were incorporated in Arts 2, 26, 27, 28 and 29 of the 1958 High Seas Convention, and Art 4 of the Continental Shelf Convention.

46 McDougal and Burke, *supra* note 39, at 846–847.

withdrew its objections on the assurance that the adoption of these three articles in the High Seas Convention would not undermine the 1884 Cable Convention.[47]

Interestingly, the US also objected to the ILC Draft Article concerning the regulation of trawling and fishing equipment[48] on the basis that it would be undesirable to impose such an obligation without providing for uniformity in the regulation to be adopted.[49] This was not included in the 1958 Conventions.

With respect to the freedom to lay cables, both the High Seas Convention and Continental Shelf Convention recognized that the freedom to lay cables is a high seas freedom[50] and that subject to its right to take reasonable measures for the exploration of the continental shelf and the exploitation of its natural resources, the coastal State may not impede the laying or maintenance of such cables or pipelines.[51]

1982 United Nations Convention on the Law of the Sea

Soon after the adoption of the 1958 Geneva Conventions, the General Assembly called for a second conference on the law of the sea to focus on two issues which remained unresolved in the Geneva Conventions, namely the breadth of the territorial sea and fishery limits.[52] The second Conference on the law of the sea, held in 1960, ultimately failed to adopt a convention.

However, the 1960s saw a slew of developments which called into question the durability and effectiveness of the 1958 Geneva Conventions for ocean governance. Soon after the Geneva Conventions were concluded, there were significant technological advances in numerous areas, such as the exploration and exploitation of seabed resources, fishing practices, and deep-sea mining. Such technological developments meant that States could exploit ocean resources farther afield than ever before and were consequently making more expansive claims in the sea. This was especially the case with respect to claims to exclusive fishing zones, which ranged in breadth from 50 nm to 400 nm, in some cases.[53] The process of self-determination for many formerly colonized States also brought about a shift in influence within key international forums such as the United Nations and a new energy for revisiting many of the customary laws which these former

---

47 *Convention on the Law of the Sea; Hearings before the Foreign Relations Committee of the Senate, 86th Cong, 1st Sess* at 92, 106 (1960). Article 30 of the High Seas Convention which stated that its provisions would not affect conventions or other international agreements already in force as between States Parties to them, contributed to the withdrawal of the objections by the United States.

48 ILC Draft Articles on the Law of the Sea Art 64.

49 McDougal and Burke, *supra* note 39 at 846.

50 1958 High Seas Convention Arts 2 and 26.

51 1958 High Seas Convention Art 26.

52 Rothwell and Stephens, *supra* note 32 at 9.

53 *Ibid.*, at 10.

colonies had been unable to participate in formulating. The call by Ambassador Arvid Pardo from Malta to the United Nations General Assembly for the seabed and ocean floor to be declared the "common heritage of mankind" captured the imagination of the international community and set in place a process which would ultimately culminate in the adoption of UNCLOS in 1982. In 1970, to address the outstanding matters unresolved from earlier codification attempts and in order to develop new law for emerging issues, it was decided that a third United Nations Law of the Sea Conference would be convened.

Negotiations for UNCLOS officially commenced in 1973 and spanned 9 years, although the Seabed Committee had undertaken much of the groundwork in the preceding years. The agenda for the Conference covered an extremely wide range of issues and it was acknowledged throughout the negotiations that the outcome would need to be a 'package deal' if it were to be widely accepted.[54] Accordingly, the Convention adopted in 1982 reflected a series of carefully constructed compromises which were intended to maintain a balance between the rights of the coastal State and the rights of other States in the utilization of the oceans. It did this by demarcating zones of juridical competence: the territorial sea, the contiguous zone, archipelagic waters, continental shelf, exclusive economic zone (EEZ) and high seas where different rights and obligations were extended to coastal States and other users of the sea.

There is no doubt that the provisions on submarine cables in UNCLOS sought to strengthen a "legal order for the seas and oceans which [would] facilitate international communication", as recognized in the Preamble to the Convention. This recognition of the importance of submarine cables may seem surprising given that during the UNCLOS negotiations in the 1970s and early 1980s, reliance on submarine cables for telecommunications had decreased. It was satellites rather than cables that carried the bulk of data information, the former being considered more cost-effective than the latter. Indeed, the creation of the Internet and development of fiber optic cables which revolutionized telecommunications only occurred in 1986 after the adoption of UNCLOS.[55] However, the cable industry, particularly in the United States, the United Kingdom and Australia, were in contact with their respective delegations during the negotiations of UNCLOS and were carefully monitoring the ways in which the developments would impact future submarine cable development.[56] Fiber optic research and technology had progressed during the 1970s and the cable industry was well-aware of the poten-

---

54 *Ibid.*, at 13.

55 Please refer to Chapter 1 on the Development of Submarine Cables. Also see T. Friedman, *The World is Flat: A Brief History of the Twenty-First Century* (FSG Books, 2005) at 71–75.

56 See Letters dated 31 January and 10 April 1980 between United States Ambassador E. Richardson and F. Tuttle (AT&T Long Lines).

tial impact that such technology could have on communications.[57] They were therefore cognizant of the fact that, at the very least, the regime on the protection of cables and the freedom to lay cables established in the 1958 Geneva Conventions had to be maintained. In this respect, the provisions on the protection and the laying and repair of submarine cables in the 1958 Geneva Conventions are reproduced more or less *ad verbatim* in UNCLOS, with few modifications.

The next two sections will give an overview of the UNCLOS provisions on cable operations (cable route surveys, laying, repair and maintenance) and UNCLOS provisions on the protection of submarine cables.

At this juncture, it should be noted that submarine cables presently have many uses. While they were initially developed in relation to telecommunications, submarine cables can be used for various purposes. These include cables used to exploit conventional natural resources (cabled oil and gas platforms), alternative energy (offshore wind farms and tidal current generators), marine scientific research (MSR) (cabled ocean observatories and ocean monitoring systems), international High Voltage Direct Current (HVDC) power cables between States and cables used for military purposes. This will be explored in more detail in Part V of the Handbook. For present purposes, it will be assumed that the provisions of the UNCLOS regime discussed in this Chapter apply to all types of cables, unless otherwise specified.

III. UNCLOS Provisions on Cable Surveying, Laying and Repair Operations

Part III of the Handbook gives an in-depth explanation of the different stages of cable operations. For present purposes, it suffices to say that cable operations which involve the use of vessels and which could potentially come into conflict with other uses of the sea and which are governed by UNCLOS consist of: (1) the surveying of cable routes; (2) the laying of cables; and (3) the repair and maintenance of cables. The rights and obligations of coastal States and the corresponding rights and obligations of other States in relation to cable operations will depend on whether the cable operations take place in (1) maritime zones under the sovereignty of coastal States (territorial seas or archipelagic waters), (2) maritime zones which are outside coastal State sovereignty but within their national jurisdiction (EEZ and continental shelf) or (3) maritime zones beyond national jurisdiction (high seas and deep seabed).

57 *Ibid.*

Territorial Seas and Archipelagic Waters

Under UNCLOS, a coastal State has sovereignty over a 12 nm[58] belt of sea known as the territorial sea, including the air space above and the seabed and subsoil below.[59] However, such sovereignty must be exercised "subject to this Convention and to other rules of international law".[60] Under UNCLOS, the main limit to a coastal State's sovereignty over its territorial sea is that it must allow ships of all States the right of innocent passage.[61] Article 21 allows coastal States to impose laws and regulations on innocent passage through territorial seas, however, such laws and relations are limited to certain specified subjects.

Similarly, an archipelagic State[62] has sovereignty over the waters enclosed by its archipelagic baselines known as archipelagic waters.[63] Such sovereignty is exercised subject to Part IV of UNCLOS which stipulates that foreign vessels have the same right of innocent passage through the archipelagic waters of archipelagic States that they have through territorial seas.[64] There is an express obligation on archipelagic States to "respect existing submarine cables laid by other States and passing through its waters without making a landfall" and to permit maintenance and replacement of such cables upon receiving due notice.[65] The term "laid by other States" refers not only to cables laid by States but to those laid by their nationals.[66] Given the passage of time, this provision has little practical utility since cables existing at the time UNCLOS entered into force are likely to have been retired. New cables that plan to transit archipelagic waters should obtain permission of the archipelagic State.

Coastal States and archipelagic States clearly have extensive authority to regulate ships engaged in cable operations i.e. the surveying of cable routes and the

---

58 UNCLOS Art 3.

59 UNCLOS Art 2.

60 UNCLOS Art 2(3).

61 UNCLOS Art 17. Under UNCLOS Art 19(1), innocent passage is passage which is not "prejudicial to the peace, good order or security of the coastal State". Article 19(2) sets out a list of activities which renders passage non-innocent.

62 As defined in UNCLOS Art 46.

63 UNCLOS Art 49.

64 UNCLOS Art 52.

65 UNCLOS Art 51(2). This provision was first introduced at the negotiations of UNCLOS III to take into consideration the concerns of States that the introduction of the concept of an archipelagic State would unduly hinder access to existing submarine cables in waters previously not under the sovereignty of States: see M. Nordquist *et al.*, (eds.), *The United Nations Convention on the Law of the Sea 1982: A Commentary*, Volume II (Martinus Nijhoff Publishers, 1993) (Nordquist *et al.*, Vol II, 1993) at 449. This Commentary is also known as 'the Virginia Commentary'. It only applies to existing cables, and the laying of new cables is dependent on the consent of the archipelagic State: see Churchill and Lowe, *supra* note 10 at 126.

66 *Ibid.*, Nordquist *et al.*, Vol II, 1993 at 474 [51.7(i)].

laying, repair and maintenance of cables, pursuant to their sovereignty over their territorial seas and archipelagic waters. Indeed, the authority of coastal States to regulate cable route surveys is expressly recognized by UNCLOS. For example, in territorial seas and archipelagic waters, ships carrying out "survey activities" would not be carrying out innocent passage.[67] Both coastal States and archipelagic States are allowed to adopt regulations on innocent passage relating to "hydrographic surveys" within their territorial seas or archipelagic waters.[68] During archipelagic sea lanes passage in archipelagic waters, foreign ships including hydrographic survey ships may not carry out any "survey activities" without the prior authorization of the archipelagic State.[69] While "survey activities" and "hydrographic surveys" are not defined and appear to be used interchangeably,[70] it seems reasonably clear that cable route surveys would fall within the definition of "survey activities".[71]

The Exclusive Economic Zone and Continental Shelf

UNCLOS allows coastal States to claim an EEZ and continental shelf beyond their territorial seas, where they enjoy certain sovereign rights over the exploration and exploitation of natural resources but where other States enjoy the rights of navigation and the freedom to lay and maintain submarine cables. These are considered areas outside the sovereignty of coastal States but areas in which coastal States have certain specified rights and jurisdiction.

The EEZ may extend up to 200 nm from the territorial sea baseline. The EEZ is a *sui generis* regime which is neither high seas nor territorial seas.[72] Under Article 56, the coastal State has:

> ... sovereign rights for the purpose of exploring and exploiting, conserving and managing the natural resources, whether living or non-living, of the waters superjacent to the seabed and of the seabed and its subsoil, and with regard to other activities for the economic exploitation and exploration of the zone, such as the production of energy from the water, currents and winds.[73]

Article 56 also sets out the extent of the jurisdiction of the coastal State in its EEZ. Since the EEZ is not subject to its sovereignty, the right of the coastal State to regulate activities in its EEZ is expressly provided for in this article. It states

---

67 UNCLOS Arts 19(2)(j) and 52(2).

68 UNCLOS Arts 21(1)(g) and 52(2).

69 UNCLOS Arts 40 and 54.

70 UNCLOS Art 19(2)(j) refers to survey activities, Art 21(1)(g) refers to hydrographic surveys, Art 40 states that hydrographic survey ships may not carry out survey activities without the prior authorization of the States bordering the straits.

71 T. Davenport, "Submarine Communications Cables and Law of the Sea: Problems in Law and Practice" (2012) 43(3) *Ocean Development and International Law* at 205.

72 UNCLOS Art 55.

73 UNCLOS Art 56(1)(a).

that the coastal State has jurisdiction as provided for in the relevant provisions of this Convention with regard to: (i) the establishment and use of artificial islands, installations and structures (Arts 60, 80, 87, 147, 208, 214, 246, 259); (ii) marine scientific research (Arts 87, 238–265, 297) and (iii) the protection and preservation of the marine environment (Arts 192–237).

With regards to the continental shelf, a coastal State has "sovereign rights for the purpose of exploring it and exploiting its natural resources,"[74] which includes "mineral and other non-living resources of the seabed and subsoil". The continental shelf is defined as "the seabed and subsoil of the submarine areas that extend beyond its territorial sea throughout the natural prolongation of its land territory to the outer edge of the continental margin".[75] A coastal State is allowed to claim a continental shelf up to a distance of 200 nm or if the outer edge of its continental margin extends beyond 200 nm,[76] it can claim what is known as an extended continental shelf.[77]

There are now two distinct legal bases for coastal State rights in relation to the seabed outside of territorial sovereignty. First, the EEZ gave the coastal State sovereign rights for the purpose of exploring and exploiting the *non-living natural resources of the seabed and its subsoil*.[78] Second, the continental shelf regime gave the coastal State sovereign rights over its continental shelf for the purpose of *exploring it and exploiting its natural resources* which includes mineral and other non-living resources of the seabed and subsoil.[79] The EEZ regime and continental shelf regime within 200 nm will usually apply concurrently to the same geographical area. In recognition of this, Article 56(3) provides that the rights set out in the EEZ with respect to the seabed and subsoil shall be exercised in accordance with Part VI on the continental shelf.

Freedom to Lay, Repair and Maintain Submarine Cables in the EEZ and Continental Shelf

As part of the compromise reached during the negotiations of UNCLOS, which granted coastal States extensive rights over economic resources and specific jurisdictional competences, other States were granted rights (as well as duties) in the EEZ:

---

[74] UNCLOS Art 77(1).

[75] UNCLOS Art 76(1).

[76] The outer limit of the continental margin is to be determined in accordance with the formula set out in UNCLOS Art 76(4).

[77] A coastal State can claim an extended continental shelf up to 350 nm from the baseline from which the territorial sea is measured or 100 nm from the 2500 meter isobaths: UNCLOS Art 76(5).

[78] UNCLOS Art 56(1)(a).

[79] UNCLOS Arts 77(1) and 77(4).

> *Article 58. Rights and duties of other States in the exclusive economic zone*
>
> 1. In the exclusive economic zone, all States, whether coastal or land-locked, enjoy, subject to the relevant provisions of this Convention, the freedoms referred to in article 87 of *navigation* and overflight and of *the laying of submarine cables and pipelines, and other internationally lawful uses of the sea related to these freedoms, such as those associated with the operation of ships, aircraft and submarine cables* and pipelines, and compatible with the other provisions of this Convention.
> [emphasis added.]

Article 87(1) provides that freedoms of the high seas include the "freedom to lay submarine cables and pipelines, subject to Part VI" [on the continental shelf]. Article 58(1) is explicit that the specific freedoms listed in Article 87 including the "laying of submarine cables ... and other internationally lawful uses of the sea related to these freedoms, such as those associated with the operation of ... submarine cables" are recognized in the EEZ.[80]

The maintenance and repair of cables by cableships is considered to fall under "other internationally lawful uses of the sea related to these freedoms, such as those associated with the operation of ... submarine cables" in the EEZ as stated in Article 58.[81] On the continental shelf, although Article 79(1) does not refer to the repair and maintenance of submarine cables, the rest of the provisions contained in Article 79 appear to assume that the right to lay submarine cables includes the right to maintain and repair them.[82]

Similarly, as cable route surveys are essential to cable laying operations,[83] they should also be considered to be "internationally lawful uses of the sea" associated with the operation of submarine cables.[84]

At this juncture, it should be mentioned that UNCLOS gives the right to conduct cable operations in the EEZ/continental shelf to "[a]ll States". It has been noted that the expression "[a]ll States" in Article 79 should not be read

---

80 Nordquist *et al.*, Vol II, 1993 *supra* note 65, at 565.

81 See R. Beckman, "*Submarine Cables—A Critically Important but Neglected Area of the Law of the Sea*" Paper presented at the Indian Society of International Law, 7th International Conference on Legal Regimes of Sea, Air, Space and Antarctica, 15–17 January 2010, New Delhi, at 5. Available online at http://cil.nus.edu.sg/wp/wp-content/uploads/2010/01/Beckman-PDF-ISIL-Submarine-Cables-rev-8-Jan-10.pdf (last accessed 7 June 2013).

82 UNCLOS Art 79(2) refers to the "laying or maintenance" of submarine cables and Art 79(5) refers to "repairing" existing cables. See Beckman, *ibid.*, at 6.

83 Please refer to Chapter 5 on the Manufacture and Laying of Submarine Cables.

84 See Beckman, *supra* note 81 at 8; L. Carter *et al.* "Submarine Cables and the Oceans: Connecting the World" (2009) Report of the United Nations Environment Program and the International Cable Protection Committee (UNEP-WCMC-ICPC) at 26. Available at http://www.unep-wcmc.org/medialibrary/2010/09/10/352bd1d8/ICPC_UNEP_Cables.pdf (last accessed 7 June 2013). See also J.A. Roach and R.W. Smith, *Excessive Maritime Claims* (3rd ed, Martinus Nijhoff Publishers, 2012) at 458.

restrictively as "in practice many submarine cables and pipelines are privately owned and are laid by corporations or other private entities. The term therefore refers to the right of States or their nationals to lay cables and pipelines".[85]

Obligations of States Conducting Cable Operations in the EEZ and the Continental Shelf

States that wish to survey cable routes or lay, repair and maintain submarine cables on the seabed of the EEZ/continental shelf have certain obligations under UNCLOS. First, such States must have due regard to cables or pipelines already in position and must not prejudice possibilities of repairing existing cables or pipelines.[86]

Second, States exercising the right to conduct cable route surveys, and lay and repair cables in the EEZ (and hence, on the continental shelf to the extent that it overlaps with the EEZ)[87] shall have due regard to the rights and duties of the coastal State.[88] The "rights and duties of the coastal State" refers to the rights and duties contained in Article 56 (as elaborated on in other provisions of UNCLOS), namely rights over the exploration and exploitation of: (1) living resources; (2) non-living resources; (3) other economic resources such as the production of energy from the water, currents and winds; (4) jurisdiction over artificial islands, installations and structures; (5) jurisdiction over marine scientific research; and (6) jurisdiction over the protection and preservation of the marine environment and the consequent duties that accompanies such jurisdiction.

Third, States conducting cable operations "shall comply with the laws and regulations adopted by the coastal State in accordance with the provisions of this Convention and other rules of international law in so far as they are not incompatible with this Part".[89] The question is what "laws and regulations" can the coastal State impose on cable operations in the EEZ/continental shelf.

85 Nordquist *et al.* (eds.), *United Nation on the Law of the Sea 1982: A Commentary*, Volume III (Martinus Nijhoff Publishers, 1995), (Nordquist *et al.*, Vol III, 1995) at 264.

86 UNCLOS Art 79(5).

87 On the extended continental shelf beyond the EEZ, the regime of the high seas would apply to the water column above. In the high seas, all States have an obligation to exercise their high seas freedoms such as the laying of submarine cables with due regard for the interests of other States exercising their high sea freedoms: see UNCLOS Art 87(2).

88 UNCLOS Art 58(3).

89 UNCLOS Art 58(3).

Rights of Coastal State to Regulate Cable Operations in the EEZ and Continental Shelf

Article 79(2) of UNCLOS states:

> Subject to its right to take reasonable measures for the exploration of the continental shelf, the exploitation of its natural resources and the prevention, reduction and control of pollution from pipelines, the coastal State may not impede the laying or maintenance of such cables or pipelines.

Article 79(2) suggests that a coastal State may only subject cable operations to reasonable measures for (1) the exploration of the continental shelf and (2) the exploitation of its natural resources. Article 79(2) draws a distinction between submarine cables and *pipelines*. For pipelines, a coastal State may not impede the laying or maintenance of pipelines subject to its right to take reasonable measures for (1) the exploration of the continental shelf, (2) the exploitation of its natural resources, and (3) the prevention, reduction and control of pollution from pipelines.[90] The omission of submarine cables from this last measure means that a coastal State cannot subject the laying, maintenance and repair of submarine cables to such measures. This is in recognition of the fact that submarine telecommunications cables do not cause pollution.[91] The same is true for modern international HVDC power cables that use polyethylene or non-oil based plastic insulation.[92]

Regulations which may **not** be adopted by coastal States are regulations on the delineation of the cable route. Article 79(3) of UNCLOS provides that "[t]he delineation of the course for the laying of such *pipelines* on the continental shelf is subject to the consent of the coastal State" (emphasis added). The delineation of the course for *submarine cables* is **not** subject to the consent of the coastal State and this interpretation is supported by the legislative history of this provision.[93]

---

90 The provision for prevention, reduction and control of pollution from pipelines was not included in the equivalent article of the 1958 Convention, and was added in during the negotiations of UNCLOS III. See Nordquist *et al.*, Vol II, 1993, *supra* note 65 at 912.

91 UNEP/ICPC Report, *supra* note 84. This report compiles and analyzes the environmental experience with cables in the marine environment since submarine cables were introduced into the ocean in 1850 and underscores the benign impact a modern fiber optic cable has on the marine environment. Please also refer to Chapter 7 on the Relationship between Submarine Cables and the Marine Environment.

92 Oil based insulation for international submarine power cables connecting States was generally phased out in the early 1990s in favor of non-polluting polyethylene, ethylene-propylene rubber or other superior forms of plastic insulation, see *About Power Cables* ICPC website, www.iscpc.org at Publications, (last accessed 7 June 2013), see also Chapter 13 of this Handbook.

93 See Nordquist *et al.*, Vol II, 1993, *supra* note 65 at 915. Interestingly, it was previously intended that the coastal State should have the right to control the route to be followed. In the commentary to the equivalent article, Art 70 of the 1956 ILC Draft Articles, it is stated:

UNCLOS also imposes certain procedural obligations on the coastal State when imposing resource-related measures on laying, repair and maintenance of cables in the EEZ/continental shelf. First, these measures must be "reasonable".[94] Second, in the EEZ, a coastal State must have due regard to the rights and duties of other States and shall act in a manner compatible with the provisions of UNCLOS.[95] Third, on the continental shelf, a coastal State must not exercise its rights in a manner which will infringe or result in "any unjustifiable interference" with navigation and other rights and freedoms of other States as provided for in UNCLOS.[96] Paragraph 2 reiterates a consistent UNCLOS principle that coastal States must recognize the rights and freedoms of other States that are provided for in the Convention. "It emphasizes that, in the exercise of its rights over the continental shelf, a coastal State must not infringe or cause unjustifiable interference with navigation and other rights and freedoms of other States as provided in Convention [and] [t]he categoric character of this obligation is emphasized by the use of the words '***must not***'."[97] The reference to "other rights and freedoms of other States" includes rights regarding submarine cables.[98]

It should be noted that UNCLOS preserves the rights of coastal States to regulate cable operations beyond resource-related measures on laying, repair and maintenance of cables in the EEZ/continental shelf in certain defined circumstances. First, Article 79(4) provides that nothing in Part VI (on the continental

> The coastal State is required to permit the laying of submarine cables on the seabed of its continental shelf but in order to avoid unjustified interference with the exploitation of the natural resources of the seabed and subsoil, it may impose conditions concerning the route to be followed.

See Articles concerning the Law of the Sea with commentaries, in *Yearbook of the International Law Commission*, Volume II, UN Doc. A/3159 (1956) at 299. However, Art 79(3) now makes it clear that the coastal State does not have jurisdiction over the route to be followed. This is also supported by discussions during UNCLOS II and UNCLOS III. During UNCLOS II, a Venezuelan amendment for Art 70 of the 1956 ILC Draft Articles would have expressly provided the coastal State with the right to regulate with respect to the routes to be followed but this was rejected on the basis that it failed to provide any standards for the regulations to be made. See M.M. Whiteman, "Conference on the Law of the Sea: Convention on the Continental Shelf" (1958) 52 *American Journal of International Law* at 643. At UNCLOS III, the proposal by China that the delineation of the course for laying submarine cables on the continental shelf by a foreign State be subject to the consent of the coastal State was also eventually rejected: see Nordquist *et al.*, Vol II, 1993, *supra* note 65 at 911.

94 It is not clear what is meant by "reasonable", "no more definite criterion than that of reasonableness could be established for the measures which coastal states may take, for the reason that it was impossible to foresee all situations that might arise in the application of this article" Statement by US Representative during the Eighth Session of the ILC: Whiteman, *ibid.*, at 642.

95 UNCLOS Art 56(2).

96 UNCLOS Art 78(2).

97 Nordquist *et al.*, Vol II, 1993, *supra* note 65 at 906 [78.8(c)].

98 *Ibid.*, at 907 [78.8(d)].

shelf) affects the right of the coastal State to establish conditions for cables or pipelines entering its territorial sea. This relates to the coastal State's sovereignty over its territory and territorial sea.[99] It has been said that the purpose of this provision is to ensure that:

> The restrictions in article 79 on the right of a coastal State to regulate cables on the continental shelf (where it has sovereign rights but not sovereignty) does not affect the more extensive rights of the coastal State to impose additional conditions on cables which enter its territory or territorial sea (where it has sovereignty).[100]

If coastal States impose additional conditions (apart from those related to the exploration and exploitation of their resources) on the laying or repair of a submarine cable which falls both on its continental shelf and on the seabed of its territorial sea, then the conditions would only apply to the part of the cable located in the territorial sea.[101]

Second, Article 79(4) also recognizes that coastal States still have jurisdiction over cables and pipelines constructed or used in connection with the exploration of its continental shelf or exploitation of its resources or the operations of artificial islands, installations and structures under its jurisdiction.[102] This does not apply to international telecommunication or State-to-State HVDC power cables. It would, however, apply to a fiber optic or power cable used as shore links to offshore wind farms, tidal current generators, or oil and gas platforms.

Third, cables used in connection with cabled laboratories and other scientific purposes in the EEZ are subject to coastal State permission under the regime regulating marine scientific research.[103]

The High Seas and Deep Seabed

The high seas and deep seabed are areas beyond the national jurisdiction of any State. The latter is termed "the Area" under UNCLOS and is defined as "the seabed and ocean floor and subsoil thereof, beyond the limits of national jurisdiction".[104] UNCLOS has created a complicated regime in Part XI to govern the exploration and exploitation of the mineral resources of the Area, which includes the

---

99 *Ibid.*, at 915.

100 See Beckman, *supra* note 81 at 7.

101 There has been some argument that Art 79(4) allows the coastal State to impose additional conditions on cables on its continental shelf if such cables enter its territorial sea. However, such an interpretation would defeat the purpose of allowing the coastal State to only subject the laying and repair of submarine cables on the continental shelf to "reasonable measures for the exploration of the continental shelf and the exploitation of its natural resources" as provided for in Art 79(4): see *ibid.*, at 7 and would allow the coastal State to delineate the cable route, which is expressly not allowed, see *supra* note 93.

102 UNCLOS Art 79(4).

103 UNCLOS Art 56(2)(b)(i)(ii).

104 UNCLOS Art 1(1).

establishment of the International Seabed Authority to regulate exploration and exploitation activities.[105]

The water column over the Area is considered to be high seas. Accordingly, Article 87 freedoms would apply, including the freedom to lay submarine cables. Article 112(1) of UNCLOS recognizes that States are entitled to lay submarine cables *on the bed of the high seas beyond the continental shelf* which refers to the Area.

However, there are obligations on States which lay submarine cables on the seabed/high seas. First, Article 112(2) requires States to have due regard to cables already in position and not to prejudice the possibility of repairing existing cables or pipelines. Second, Article 87(2) requires that the freedom to lay submarine cables be exercised with due regard for the interests of other States in their exercise of high seas freedoms and also with due regard for the rights under UNCLOS with respect to activities in the Area.

IV. UNCLOS Provisions on the Protection of Submarine Cables

As will be explained in Part III of this Handbook, submarine cables are susceptible to threats from competing uses of the oceans such as shipping activities, fishing activities and resource exploration and exploitation activities. UNCLOS gives States the right to enact legislation to protect cables from damage from competing uses, depending on the area of sea in which the cables are located.

Territorial Seas and Archipelagic Waters

Coastal States[106] and archipelagic States[107] have the right to adopt laws and regulations relating to innocent passage through their territorial sea and archipelagic waters in respect of the protection of cables and have a general competence to enact laws to protect submarine cables within territorial waters. However, under UNCLOS there is *no obligation* on coastal States to adopt laws and regulations to protect submarine cables within territorial waters. UNCLOS assumes that coastal States would have every incentive to have legislation to protect cables that either land in their territory or transit their territorial waters.[108]

---

105 UNCLOS Arts 1(3) and 134.

106 UNCLOS Art 21(1)(c).

107 UNCLOS Art 52.

108 See Beckman, *supra* note 81 at 12. Indeed, the 1884 Cable Convention only applied "outside territorial waters to all legally established submarine cables landed on the territories, colonies or possessions of one or more of the High Contracting Parties" (Art I) because of the assumption that Parties to the Convention would have sufficient measures in place for the protection of submarine cables within territorial waters as

EEZ/Continental Shelf

Articles 113 to 115 of UNCLOS address the protection of submarine cables on the high seas and are based on three articles in the 1884 Cable Convention, which have been dealt with above. They are also applicable in the EEZ under Article 58(2) as well as on the continental shelf.[109]

Article 113 requires States to adopt laws and regulations to provide that the breaking or injury by a ship flying its flag or by a person subject to its jurisdiction of a submarine cable beneath the high seas done wilfully or through culpable negligence, is a punishable offence.[110] Such laws and regulations must also apply to conduct calculated or likely to result in such breaking or injury.[111] However, it shall not apply to any break or injury caused by persons who acted to save lives or their ships, after having taken all necessary precautions to avoid such an occurrence. Article 113 essentially extends a State's criminal jurisdiction (usually limited to territory) over acts of breaking or injury to submarine cables done "wilfully or through culpable negligence" only to ships flying its flag on the high seas or EEZ or to their nationals who commit such acts, consistent with general principles of international law on the prescription of extra-territorial jurisdiction.[112]

"one cannot imagine a legislator taking measures in relation to the open sea but not for the territory and territorial waters": see Renault, *supra* note 13, at 6.

109 See Nordquist *et al.*, Vol III, 1995, *supra* note 85 at 270, 273 and 278.

110 Article 113 differs from Article II of the 1884 Cable Convention in that the latter stated that such criminal sanctions were "without prejudice to any civil action for damages". Article 113 would not be a bar to a civil action based on general rules of tort law but it is arguable whether Article 113 would provide the basis for an implied civil remedy in all jurisdictions. In the United States Federal Court, it was found that the Article II of the 1884 Cable Convention which had been implemented in domestic legislation through the Submarine Cable Act (1888) did not give an implied private civil remedy for submarine cables owners against parties who allegedly damage such cables: see *American Tel & Tel Co v. M/V Cape Fear* 763 F Supp. 97 (DNJ 1991).

111 Article 113 also differs from Article II of the 1884 Cable Convention in that the former added the sentence "this provision shall apply also to conduct calculated or likely to result in such breaking or injury." This sentence first appeared during discussions at UNCLOS III prompted by concerns relating to the potential for cable damage by fishing vessels anchoring to pipelines in the North Sea and exploration activities by researchers around cables. Accordingly, the sentence "conduct calculated or likely to result in such breaking or injury" "widens the scope of the provision and makes the intent or attempt to break or injure a submarine cable or pipeline a punishable offence": See Nordquist *et al.*, Vol III, 1995, *supra* note 85 at 268. It has been observed that this is an improvement over the 1884 Cable Convention where "the cable owner must wait until the damage is done before sanctions are triggered: D. Burnett, "The Importance of UNCLOS to the Cable Industry" (May 2006) 26 *Submarine Telecoms Forum* at 23, available online at http://www.subtelforum.com/issues/Issue%2026.pdf (last accessed 7 June 2013).

112 Article 113 also added the words "by a ship flying its flag or by a person subject to its jurisdiction". These words were not in Article II of the 1884 Cable Convention, and first

Article 114, which is based on Article IV of the 1884 Cable Convention, requires every State to adopt laws and regulations concerning the liability of owners of cables for the cost of repairs to existing cables which are damaged in the course of laying or repair operations.[113] The laws and regulations would only apply to persons subject to that State's jurisdiction i.e. owners who are nationals of the State. The indemnity in the case of a pipeline is limited to the actual repair costs and does not include compensation for any financial losses of the owner or the contents of a broken pipeline.[114] Again, this article illustrates a practical and common sense approach to the conflict that would otherwise arise with successive laying of cables and pipelines on the same seabed area.[115] Cable industry practices for cable crossings embrace this common sense approach.[116] With the exception of the Arabian Gulf where energy companies, often affiliated with coastal States, sometimes demand one sided and onerous crossing agreements for pipeline crossings that violate UNCLOS, this article has been successfully implemented.

Article 115, which is based on Article VII of the 1884 Cable Convention,[117] provides that every State should adopt laws and regulations to provide for an indem-

---

made their appearance in Art 27 of the 1958 Convention on the High Seas. This was to ensure that it was clear that the legislative measures referred to are applicable only to those subject to such legislation under general international law, i.e. a State could not take legislative measures against nationals of another State, only against its own ships or nationals: see McDougal and Burke, *supra* note 39 at 847; Nordquist *et al.*, Vol III, 1995, *supra* note 85 at 268. Takei notes that there is also "State practice and writings that supports universal jurisdiction over the breaking of submarine cables". See Yoshinobu Takei, "Law and Policy for International Submarine Cables in the Asia-Pacific Region," (paper presented at the 2nd National University of Singapore—Asian Society of International Law Young Scholars Workshop, Singapore, 30 September–1 October 2010) at 17–18, available online at http://www.asiansil.org/publications/2010-13%20-%20Yoshinobu%20Takei.pdf (last accessed 7 June 2013).

113 UNCLOS Art 114 limits the liability of the owner to the cost of the repairs. This "excludes any notion of liability for replacing a damaged cable or pipeline or of obligating the responsible person(s) for any financial losses incurred by the owner of the cable or pipeline as a result of the damage:" Nordquist *et al.*, Vol III, 1995 *supra* note 85 at 273.

114 Nordquist *et al.*, Vol III, 1995, *ibid.*, at 273 [114.7(b)].

115 The compromise reflected in Art 114 is directly derived from Art IV of the 1884 Cable Convention.

116 ICPC Recommendation No 9A Telecommunication Cable and Oil Pipeline/Power Cable Crossing Criteria, available upon request www.iscpc.org.

117 Article VII of the 1884 Cable Convention was followed in Art 65 of the 1956 ILC Draft Articles and subsequently adopted in Art 29 of the 1958 Convention on the High Seas. However, Art VII provided for a procedure on how an indemnity may be claimed. In order to be entitled to establish a claim to such compensation, a statement, supported by the evidence of the crew, should, whenever possible, be drawn up immediately after the occurrence; and the master must, within twenty-four hours after his return to or next putting into port, make a declaration to the proper authorities. The latter shall communicate the information to the consular authorities of the country to which the owner of the cable belongs. The reason for the omission of this procedure

nity to be paid by cable owners to ship owners whose master sacrifices an anchor, a net or any other fishing gear in order to avoid injuring a submarine cable, provided that the ship owner has taken all reasonable precautionary measures beforehand.[118] Such laws and regulations will only apply to nationals and ships flying their flag.[119] While the measures to be taken are not specified, they would have to be balanced against the obligation of fishing vessels to avoid submarine cables in the first place.[120] The indemnity, however, is limited to the sacrificed gear or anchor and does not include lost profits or catch.[121]

Articles 114 and 115 reflect the very successful balancing and practical compromise of the competing uses of submarine cables on the one hand and fishing and navigation on the other.[122] However, it should be borne in mind that apart from the treaty remedies, damages for injury to submarine cables are typically dealt with by civil suits in traditional admiralty courts where the offending vessel is subject to arrest.[123] Many in-house telecommunications attorneys have learned the hard way that effective and successful civil legal action usually depends upon prompt retention of experienced admiralty counsel when a cable is damaged by a vessel.

---

in the 1956 Draft Articles, the 1958 Convention on the High Seas and UNCLOS is that "[i]t is anticipated that more detailed guidelines will be included in the laws and regulations adopted by each State under Art 115". See Nordquist *et al.*, Vol III, 1995, *supra* note 85 at 278.

118 This is to make clear that compensation cannot be claimed if there has been any negligence on the part of the ship: see Commentary on Art 65 of the ILC Draft Articles Concerning the Law of the Sea with commentaries in *Yearbook of the International Law Commission*, Volume II, UN Doc. A 3159 (1956) at 294.

119 See Nordquist *et al.*, Vol III, 1995, *supra* note 85 at 278.

120 *Ibid.*, at 277.

121 *Ibid.*, at 272; *Agincourt Steamship Company Ltd. v. Eastern Extension, Australasia and China Telegraph Company Ltd.* 2 K.B. 305, 310 (1907).

122 The compromise reflected in Art 113 and 115 is directly derived from Art II and VII of the Cable Convention and are widely followed as the custom and practice of the cable industry.

123 *The Government of the Netherlands, Post Office v G'T Manneteje-Van Dam* [*Fishing Cutter GO 4*], File No 325/78 (District Court Rotterdam, decision rendered 20 November 1978), aff'd *sub nom* G't Mannetje-Post Office, File No 69 R/81 and File No rb 325/78 (The Court at the Hague, Second Chamber, decision rendered 15 April 1983); 9 Whiteman, *Digest of International Law* 948 (Dep't State 1968) (Alex Pleven); *AT&T Corp v Tyco Telecommunications Inc* 255 F Supp 2d 174 (SDNY 2003); *American Telephone and Telegraph v MV Cape Fear*, 763 F Supp 97 (DNJ 1991), rev'd on other grounds, 967 F 2d 864 (3rd Cir 1992); Arbitration between Concert Global Network Services Ltd, in its own capacity, and as co-maintenance authority of submarine cable system TAT-10, as Claimant and Tyco Telecommunications (US) Inc as Respondent, (Arb. New York, SMA 3779, 2002).

High Seas and Deep Seabed

The obligations on States Parties in Articles 113, 114 and 115 of UNCLOS discussed above would apply to cables located underneath the high seas and on the deep seabed.

V. Dispute Settlement and UNCLOS

Finally, in the context of disputes of competing uses in the EEZ or upon the continental shelf, it is important to recognize that the laying and maintaining of submarine cables enjoys the highest level of protection under the UNCLOS dispute resolution provisions.[124]

Under UNCLOS, disputes concerning the interpretation or application of UNCLOS with regard to the exercise by a coastal State of its sovereign rights and jurisdiction are subject to compulsory procedures entailing binding decisions set out in Section 2 of UNCLOS (Section 2 Procedures). Disputes on whether a coastal State has acted in contravention of the provisions of UNCLOS with regard to the laying of submarine cables or other internationally lawful uses of the sea would be subject to Section 2 Procedures. Similarly, disputes on whether a State has acted in contravention of the laws and regulations adopted by the coastal State in conformity with UNCLOS are also subject to Section 2 Procedures. A State Party could bring a claim against another State Party which, for example, imposed excessive regulations on the laying or repair of cables in its EEZ or breached coastal State laws and regulations on the laying or repair of cables.

The question is who is going to refer disputes relating to submarine cables to dispute settlement under Section 2. Cable companies, who have every interest in bringing such a claim and who are in effect exercising the rights of States,[125] are precluded from using the UNCLOS dispute settlement mechanisms, as these mechanisms are only open to States. Cable system owners are usually telecommunications carriers or a consortium of telecommunications carriers which are either fully or partially privatized. Such owners would need to persuade the States in which they are incorporated to bring a claim under UNCLOS. Another obstacle is the fact that cables are usually owned by a consortium of companies incorporated in different States and determining the appropriate State may be challenging.

---

124 UNCLOS Art 297(1) [Limitation of applicability of Section 2].

125 Although note that the Virginia Commentary found no issue with the fact that States were given the right to lay cables but such rights were in actual fact exercised by private companies (see Nordquist *et al.*, Vol III, 1995, *supra* note 85 at 264). However, it is unlikely that such private companies would be able to avail themselves of the dispute settlement mechanisms in UNCLOS.

Cable installers which lay, repair and maintain cables either own the cableships conducting the cable operations or charter such ships. The flag State of the cable laying or repair vessel could challenge, for example, regulations of a coastal State requiring permits for laying or repairs outside of territorial sovereignty.[126] However, this would require cable installers either to register ships or to charter ships with flag States that have the political will and interest to challenge excessive regulations on their behalf, which may prove difficult.

VI. Other Relevant International Conventions

Apart from the UNCLOS provisions on submarine cables, there are also other international conventions that may apply to submarine cables and cable operations. The 1972 Convention on the International Regulations for Preventing Collisions at Sea (COLREGS) provide that a vessel engaged in laying, servicing or picking up a submarine cable ("a cableship") is considered a "vessel restricted in their ability to manoeuvre".[127] The COLREGS contain provisions on the signals and sounds to be exhibited by a cableship so that other vessels are aware of what it is doing.[128] They also require both power-driven vessels, and vessels engaged in fishing to keep out of the way of such vessels.[129] This will be further examined in Chapter 9 on Protecting Cableships Engaged in Cable Operations.

Similarly, the 1972 Convention on the Prevention of Marine Pollution By Dumping from of Wastes and Other Matter[130] and its 1996 Protocol[131] may also be relevant to the abandonment of cables on the seabed when they have reached the end of their operating life. This will be further examined in Chapter 8 with respect to out-of-service cables.

Conclusions

Each of the international conventions referred to above is a reflection of the period in which it was drafted and the interests, at that time, of the States involved in negotiating the balance between the right to lay submarine cables

---

126 See Beckman, *supra* note 81 at 12.

127 1972 Convention on the International Regulations for Preventing Collisions at Sea, 20 October 1972, 1050 UNTS 16 (entered into force 15 July 1977) (COLREGS). COLREGS Rule 3(g)(i).

128 COLREGS Rule 27.

129 COLREGS Rule 18.

130 1972 Convention on the Prevention of Marine Pollution by Dumping of Wastes and Other Matter, 29 November 1972, 1046 UNTS 120 (entered into force 30 August 1975).

131 1996 Protocol to the 1972 Convention on the Prevention of Marine Pollution by Dumping of Wastes and Other Matter, 7 November 1996, 2006 ATS 11 (entered into force 24 March 2006).

and other competing uses of the sea. As international conventions are political as well as legal documents, a level of compromise is inevitably involved in the wording of the provisions. In this regard, it should be noted that international law on treaty interpretation stipulates that a "treaty shall be interpreted in good faith in accordance with the ordinary meaning to be given to the terms of the treaty in their context and in the light of its object and purpose".[132] This is salient advice to governments, policy-makers, lawyers and the industry when interpreting the sometimes ambiguous provisions in UNCLOS it is critical to give the provisions on submarine cables a "good faith" interpretation, bearing in mind the object and purpose of UNCLOS.

The question is whether the international legal regime described above can meet the challenges faced today and whether it continues to be an effective framework for the governance of submarine cables. As will be discussed throughout the course of this Handbook, State practice, particularly with regard to coastal State rights to regulate cable operations[133] and the obligations of States to protect cables,[134] is at variance with UNCLOS provisions on submarine cables, arguably undermining the effectiveness of the regime. UNCLOS also does not address certain issues, particularly in relation to the protection of cableships from interference with other activities[135] and the protection of cables from intentional damage.[136] The objective of the remaining chapters of this Handbook is to examine the law and policy issues that confront the cable industry and governments alike and propose recommendations on how to move forward.

---

132 Vienna Convention on the Law of Treaties, adopted 23 May 1969, 1155 UNTS 331 (entered into force 27 January 1980) [1969 VCLT], Art 31.

133 Please refer to Part III of the Handbook on 'Cable Operations—Law and Practice'.

134 Please refer to Part IV of the Handbook on 'The Protection of Submarine Cables'.

135 This will be addressed in Chapter 9 of the Handbook on Protecting Cableships Engaged in Cable Operations.

136 This will be addressed in Chapter 12 of the Handbook on Protecting Submarine Cables from Intentional Damage: The Security Gap.

PART III

CABLE OPERATIONS—LAW AND PRACTICE

CHAPTER FOUR

THE PLANNING AND SURVEYING OF SUBMARINE CABLE ROUTES

Graham Evans and Monique Page

Introduction

Since the late 1980s there have been unprecedented levels of activity in the construction of submarine telecommunication cable infrastructure. In parallel with the construction of systems, submarine telecommunications technology has advanced from offering system capacities of 280 Mega bits per second (TAT 8, 1988) to system capacities of multi Terra bits per second. This phenomenal increase in submarine telecommunications infrastructure and system capacity has elevated submarine telecommunications networks to global critical infrastructure status. The criticality of submarine telecommunications infrastructure has focused attention on the need for planning, with system security as a primary goal.

Part IV of this Handbook addresses the protection of submarine cables, with Chapters 11 and 12 discussing the causes of cable faults and illustrating that most system outages occur as a result of post installation damage caused to wet plant by third parties. It is therefore essential that the planning and selection of the submarine cable route, together with the most appropriate cable protection, installation and maintenance solutions, be viewed as vital components of system security.

The cable route survey and associated activities described in this Chapter are essential ingredients in planning for cable route security. Part I of the Chapter gives a brief background and history of cable route surveys and Part II discusses pre-survey activities, such as preliminary route planning, landing site selection and the desktop study. Part III examines in detail the actual cable route survey and Part IV offers a brief summary of operational planning and schedules for cable route surveys. Parts V and VI address the international law governing cable route surveys and law and policy challenges respectively, and Part VII sets out some recommendations on how these challenges can be addressed.

I. Background and History

When the first submarine cable was laid between Dover and Calais in 1850[1] virtually no information on seabed topography or physical conditions were known. The charts available at the time were the result of depth information collected by lead-line and seabed sediment information from material adhering to the tallow wax coating the lead weight. For a cable across the English Channel, even the sparse depth information available would indicate minimal risk posed by the seabed profile and physical seabed conditions. Just seven years later, when the first attempt to lay a trans-Atlantic cable was undertaken, the situation was completely different. Whilst the same sparse bathymetric data was available in coastal waters, nothing was known about conditions at the bottom of the Atlantic Ocean, which today we know is divided by a submarine mountain chain of similar topographic relief to that of the European alpine regions. Laying the first Atlantic cable would therefore have been equivalent to laying a rope across a mountain chain from a helicopter through dense cloud cover.

A lack of accurate bathymetric data prevailed until the development of the first echo sounders towards the end of World War 1. It was not until the 1960s and 1970s that the true complexity of the floors of the world's oceans were revealed as a result of research into plate tectonics using geophysical techniques, including side scan sonar, which enabled surface features to be mapped, and seismic methods, which provided information on subsurface geology and sediments.

In parallel with the development of techniques and tools that provided information of the seafloor and geology of the oceans, developments in accurate horizontal positioning were also taking place. The earliest cables were laid by applying well practiced celestial navigation methods using sextants, a method that continued to be used until the advent of radio navigation, which has today been supplanted by the use of highly accurate global positioning system (GPS) satellite navigation.

Today, whilst accurately measured bathymetric tracks across the oceans are still sparse, the use of satellite gravity derived bathymetry is available worldwide and this information is routinely used as the basis of preliminary cable route planning.

II. Pre-Survey Activities

Preliminary Route Planning

Preliminary planning and route selection are critical elements in planning for route security and total cost of ownership of the system. It should (but does not

---

[1] For an overview of the history and development of submarine cables, see Chapter 1.

always) form part of the feasibility study and conceptual design phase of the planning process. During this route development phase, it is important for the design team to identify, assess and evaluate the risks and hazards to which the cable system may be exposed during its design life. In this way system planners are able to design a route that either avoids potential hazardous areas or, where this is not feasible, estimate the costs of risk avoidance and prevention engineering. These considerations are particularly important where systems are to be installed in areas where traffic restoration possibilities are restricted and where system owners may have limited access to maintenance facilities, for example geographically isolated areas such as small island communities with low population density and limited budget to cover system operational and maintenance costs. During the preliminary planning stage the potential need for specialized non-standard project specific installation procedures should be identified and costed together with the system life cycle maintenance budget.

Preliminary planning will typically include, but not be limited to:

- Preliminary examination of available charts, satellite gravity bathymetric data (ETOPO) and literature pertaining to the landing sites and possible cable routes between the landing sites.
- Determination of political constraints to cable routing, for example international boundaries and disputed territorial claims.
- Consideration of maritime jurisdiction issues including permitting requirements for survey, installation and maintenance related permit procurement lead-times, permit interdependencies and the responsibilities of all stakeholders in the permitting process.
- Determination of physical constraints to cable routing and installation, for example rock and coral outcrops, seismic activity, excessive seabed slopes, seabed slope stability potential for submarine sediment flows, sand waves, coastline stability etc.
- Determination of cultural constraints to cable routing and associated risks, for example fishing activities, offshore mining, hydrocarbon exploration and production, coastal developments, offshore dumping grounds, marine parks etc.
- Determination of wayleaves[2] and rights of way at the landing sites and the availability of land for terminal station construction.
- Identification of viable cable landing site locations.

[2] A wayleave refers to a right of use over the property of another party. In the current context it refers to the right to conduct activity and or lay and maintain a cable in a particular area of water or across land from the cable landing site to the terminal station.

Landing Site Selection

Landing site selection should be finalized during the feasibility study and initial conceptual design phase of the system planning process. However, technical viability can often only be proven during the desk study (cable route study) landing site visits. A prime objective of the system planners is to avoid the need to conduct costly re-surveys at landing sites that require relocation following the completion of the route survey. Key factors in selecting the system landing sites are:

- The availability and establishment of the necessary permits, rights of way and wayleaves for the route approaches, manhole position and cable landing station site prior to commencing the route survey.
- Land ownership issues related to the beach manhole location and the landing site terminal station location.
- Landing site technical viability taking account of:
  - Design life of the system.
  - Protection and burial requirements.
  - Constraints imposed by cultural activities.
  - Constraints imposed by other cables sharing the same landing site.
  - Constraints imposed by the physical environment.

Desktop Study (Cable Route Study)

The importance of the desktop study (also commonly referred to as the cable route study) cannot be overstated when planning submarine cable systems. Errors and omissions during the desktop study phase of planning can have extremely costly and far-reaching consequences during the later phases of the project. The minimum requirements of desktop studies are set out in Recommendation 9 of the International Cable Protection Committee (ICPC).[3]

The desktop study should form a logical, more in-depth continuation of the processes performed during the feasibility and initial conceptual system design phases of the project. The desktop study should address the following key objectives:

- Confirm system feasibility and enable system budgets to be refined.
- Enable preliminary system design and configuration parameters to be confirmed and refined.
- Identify and fully define cultural activities and physical conditions along the preliminary cable route that may pose risks to the installed cable and/or impact system design.

---

[3] 'Minimum Technical Requirements for a Desktop Study' Recommendation 9, International Cable Protection Committee. The document is available upon request from the ICPC, see http://www.iscpc.org/.

- Identify and fully define political and environmental constraints along the preliminary cable route that may impact system design.
- Identify and/or confirm the often complex permitting requirements governing all aspects of the construction and operation of a submarine cable system; and, within the context of this Chapter, ensure that particular attention is given to the permitting requirements governing marine survey activities.
- Enable the route survey scope and procedures to be correctly defined.

The purpose of the desktop study in the route selection process is to examine existing literature and information held in the public domain and, where available, other less accessible databases. In developing regions this information may be sparse, but nonetheless it is important to source all the information available.

Close attention to system security and survival potential is fundamental to the route planning and selection process performed at the desktop study phase. Key areas of focus during the desktop study are the evaluation of physical conditions prevailing along the provisional route developed during the feasibility and conceptual design phase of route planning, and cultural activities along and close to the selected route. The latter have been shown to be a far greater cause of cable system outages than the impact of the physical environment (see Part I, Chapter 11). A primary function of the desktop study is therefore to identify and notify all parties who may potentially have a conflicting interest in the area of the cable owner's intention to install a cable. In some cases it will be necessary to negotiate a mutually acceptable route with the various stakeholders, including cable and pipeline crossing agreements or arrangements, routing through hydrocarbon lease blocks, routing through or around seabed mining tenements, and the complexities of routing through or proximal to offshore renewable energy developments. This negotiation process should ideally be concluded prior to the commencement of survey operations.

Under the 1982 United Nations Convention on the Law of the Sea (UNCLOS) there are no requirements for crossing agreements and there is no basis under international law to compel a party to agree to an unreasonable crossing agreement. This makes sense because, as discussed in Chapter 3, all States enjoy the freedom to lay cables, subject only to the requirement that if the first laid cable or pipeline is injured in the crossing, the crossing party must indemnify the first laid cable or pipeline for the cost of repairs. Cable companies approach crossing arrangements or agreements with a view to minimizing crossing risks by exchanging data and establishing procedural protocols and physical protections that allow both systems to exist without conflict.

Typically, the most common routing conflicts occur with offshore hydrocarbon developments, offshore renewable energy developments or with seabed mining operations. In developing regions such operations may be absent, in the early stages of development or be playing an important role in the economic development of the region. Other likely routing conflicts may occur as a result of coastal fishing activities, marine conservation areas and coastal tourism developments.

As submarine cables typically land at and/or link main coastal population centres, it is important for the desktop study to identify routing conflicts that may arise as a result of existing or planned future coastal construction projects, for example ports and harbours, coastal power stations and chemical plants.

During the desktop study a clear understanding must be gained of political issues that can impact on cable route planning and selection, installation and long term maintenance of the cable system. Such issues include permitting (which can seriously impact system implementation programs), international boundary crossings and territorial claims of both landing and non-landing parties.

As discussed above, visits to all of the landing sites should form an integral part of the desktop study. The site visits should be used to test the technical viability of the site and to gather information from local government offices and other relevant authorities administering the regions in which the landing sites are located.

An important function of the desktop study is to research accurate as-laid position lists of existing cables, repeater locations and histories of previous faults. This information is critical when developing new submarine cable routes to ensure that normally expected separation, crossing angle and minimum crossing distances from repeater criteria are met. This information can be sourced from system owners, commercial databases, for example the Global Marine GeoCable™ database and databases held by suppliers, maintenance authorities, installation contractors and survey companies.

The output from the desktop study forms an indispensable part of the overall process of route planning for security and is also an important element when considering total cost of cable system ownership.

The output from the desktop study should include:

- The provisional cable route in the form of a physical description, Route Position Lists and Straight Line Diagrams that have been developed during the initial conceptual system design and desktop study phases of system planning.
- A series of system planning charts depicting the route, showing all route alter course points and landing site details.
- Definition of provisional cable quantities and cable engineering, including provisional cable armoring schemes.
- Full and detailed descriptions of the system landing sites.
- Full details of route permitting issues and procedures, including the status of routing negotiations, permit lead-times and impacts on survey operational sequencing.
- Definition of detailed route survey procedures and scope of work, based on the most appropriate technical approach which addresses the prevailing physical conditions of the route and the cable protection and installation strategies.

Definition of Pre-survey Route

The pre-survey cable route will be one of the outputs from the desktop study which may have been subsequently updated prior to survey operations. The route will be defined as a series of *Route Position Lists* and *Strait Line Diagrams*. The Route Position List defines the route in terms of a tabulated list of coordinates, route distances, water depth, cable distances (including cable slack allowances), cable burial requirements and cable engineering parameters. Strait Line Diagrams show a graphical representation of the route and depict distances by cable type, water depth at cable type transition points, cable burial distances and target burial depth.

III. The Cable Route Survey

Survey Technology

Submarine cable route surveys employ a range of technologies that address specific objectives and requirements. The technologies typically employed include:

- *Multibeam Echo Sounders*
  Multibeam echo sounders are acoustic sonar devices comprising an array of focused individual narrow angle beams typically operating within the frequency range 12 kHz to 300 kHz. The array is mounted to the hull of the survey vessel for deep water operations or, in the case of shallow water surveys, may be mounted to the side of the hull. Multibeam echo sounders are a fundamental tool in cable route surveys and enable a swath of seabed to be mapped in a

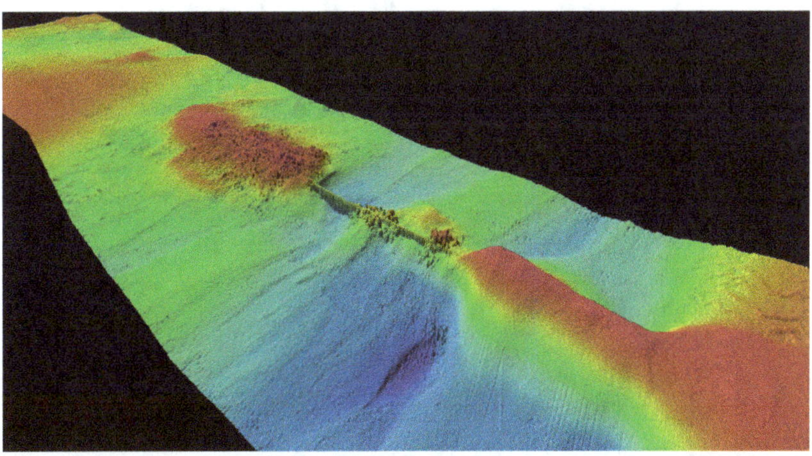

Figure 4.1 Exposed submarine pipeline identified from high resolution multibeam echo sounder data. (Image courtesy of EGS International Ltd)

single pass of the survey vessel (a typical swath width is from 2.5 to 7 times water depth). Multibeam echo sounders, depending on their operating frequency, can operate from very shallow water to full ocean depth.

- *Side Scan Sonar*
  Side scan sonar systems are acoustic sonar devices that transmit horizontal beams within the frequency range 100 kHz to 500 kHz. The systems are normally towed behind the survey vessel and enable seabed surface features (rock, coral, shipwrecks etc.) to be accurately mapped beneath and on each side of the survey track line. Side scan sonar is typically used within the water depth range where cable burial is required.

- *Sub-bottom Profilers*
  Sub-bottom profilers may be towed or hull mounted acoustic devices. They transmit a vertical beam of acoustic energy typically operating within the frequency range 1.7 kHz to 7 kHz and provide a continuous sub-seabed profile of sediments and shallow geology along the cable route. Sub-bottom profiling is typically carried out within the water depth range where cable burial is required.

- *Magnetometers*
  Magnetometers are typically used to map and confirm the position of existing cables and pipelines that cross the cable route being surveyed. They can also be used to detect ferrous metal objects that could present a hazard or obstruction

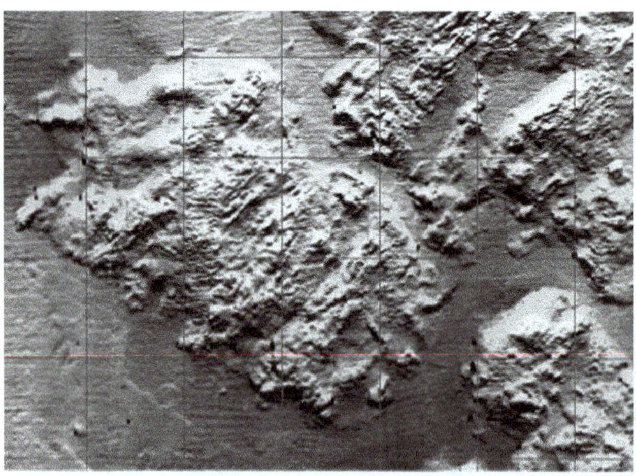

Figure 4.2 Regional seabed geology identified from side scan sonar mosaic imagery depicting areas of rock outcrop. (Image courtesy of EGS (Asia) Ltd)

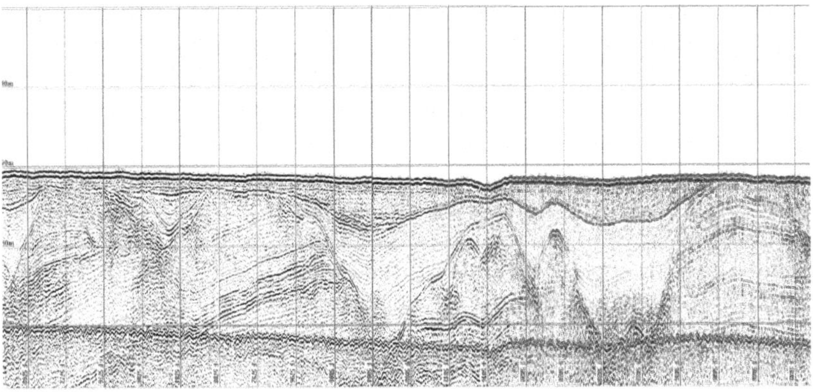

Figure 4.3 Shallow sub seabed sediments identified from sub-bottom profiling data. Interpretation of such data, with the help of gravity cores and Mini Cone Penetration Test measurements, assist in assessing cable burial conditions. (Image courtesy of EGS (Asia) Ltd)

to cable laying operations, such as unexploded ordnance. This technology is used only within the water depth range where cable burial is required.

- *Gravity Corers*
  Gravity coring equipment is used to obtain sediment cores, typically up to 3 m long, within the water depth range where cable burial is required. Samples taken are used for 'ground truthing' (i.e. to enable the geophysicists to validate their interpretation of the seabed soils obtained from the sub-bottom profiling data).

- *Burial Assessment Technology—Mini Cone Penetration Test (MCPT)*
  Where cable burial is required, a burial assessment survey (BAS) is typically required. Where a BAS is carried out it is usual to employ an MCPT, which normally comprises a coiled rod fitted with a cone tip which has a surface area of 2 cm^2 which is pushed into the seabed at a controlled rate to the point of refusal, or typically 3–6 m below seabed. The standard MCPT system measures the tip and side wall resistance during the push, the ratio of which provides a measure of the mechanical properties of the soil (the shear strength or the relative density) which can then be translated into burial performance of the cable burial equipment. If fitted with a piezocone, the equipment will also measure sediment pore pressure. BAS operations are typically carried out within the water depth range 15 m to the maximum water depth where cable burial is required.

- *Vessel Navigation and Positioning*
  Survey vessels use Differential Global Positioning Systems (DGPS) to provide navigation from inshore operations to transoceanic operations. DGPS provides positional accuracy typically within the +/–1 meter range. The DGPS

equipment is interfaced to a vessel navigation computer system that provides the survey team and helmsman with a real time graphical display of the vessel position relative to the pre-planned survey line. For the positioning of sensors that are towed behind the survey vessel, acoustic positioning is used. All positioning data is inputted to the data acquisition and processing software.

- *Data Acquisition, Management and Processing Systems*
  All data acquired by the survey systems in use is typically managed by an integrated data acquisition, processing and management system. This system facilitates the rapid processing and interpretation of data to enable preliminary survey output to be available within 24 hours of its acquisition. This rapid turnaround of data enables routing decisions to be made quickly thus avoiding the potential need to transit long distances to carry out re-route surveys.

- *Communications*
  Most survey vessels carry V-Sat broadband satellite communication systems. This facilitates the transfer of survey data, and enables survey deliverables and route engineering information to be transmitted to shore based clients or contractor offices as the survey progresses.

Survey Scope of Work and the Cable Route Survey

The degree to which the route survey can successfully achieve its primary objectives will be considerably influenced by the survey scope of work. One of the key outputs from the information gathered during the desk study (described in Part II above) is the definition of the field investigation procedures and survey scope of work. It is an essential requirement of the scope of work that the survey fully address the predicted conditions along the route in order to fulfil the requirements of the system supplier to confirm or amend preliminary cable engineering and cable type selection. It also provides the installer with all the information needed to achieve installation objectives. The cable route survey is a key building block when considering total cost of ownership for the system.

Route survey objectives include the following:

- Provide information required to confirm or amend the preliminary pre-survey desktop study route;
- Define and document the final selected route;
- Enable final cable engineering to be defined;
- Provide the system installer with data required to finalize installation procedures; and
- Identify potential post installation residual risks and hazards during the system design life.

The survey scope of work typically falls into the following zones:

- Landing site survey—a zone centred on the beach manhole location extending a specified distance behind and each side of the beach manhole to the low water mark;
- Inshore survey—a specified corridor centred on the pre-survey defined route from the low water mark to a water depth typically between 15–20 m;
- Shallow water survey—a specified corridor width centred on the pre-survey defined route from the limit of the inshore survey to a specified water depth which is typically, but not always, the limit where cable burial is required; and
- Deep water survey—all other sections of the pre-survey defined route with a specified corridor width typically defined as a multiple of water depth.

Survey data is thus collected along a narrow strip of seabed with the width of the survey corridor varying for each zone. Corridor widths typically range from 500 to 1000 m in shallow water and 3 × water depth in deep water.

The most fundamental data components of a cable route investigation are:

- Bathymetry—using multibeam echo sounders;
- Seabed imagery—using side scan sonar;
- High resolution seismic reflection profiling—using sub-bottom profilers;
- Seabed soils data—using gravity corers;
- Submarine geology;
- Electronic burial and plow assessment; and
- Oceanography.

Bathymetry

The collection of bathymetric data is fundamental in any marine survey operation. Using multibeam echo sounders, accurate water depth data, co-located back-scatter data and a digital terrain model of the seabed topography within the survey corridor is derived. This swath mapping technology can be rapidly processed onboard the survey vessel, enabling routing decisions to be made in near real-time when adverse bathymetric terrain is encountered.

Seabed Imagery

Side scan sonar imagery provides a plan view of the seabed surface along the corridor centred on the track of the survey vessel. This data not only provides valuable information on seabed topography that may be correlated with the bathymetric data, it also yields information on the characteristics of the seabed surface sediments through seabed surface back-scatter. In this way outcropping rock and coral and non-geological obstructions such as shipwrecks showing positive relief above the seabed occurring off the track of the survey vessel may be identified and mapped.

High Resolution Seismic Reflection Profiling

Seismic reflection profiling is used to provide a continuous along route record of the shallow sub-seabed soils profile and underlying geology. Seismic reflection profiling allows significant geological features such as submarine landslides, faults and rock outcrops to be immediately identified and mapped. Where cable burial is required, route deviations can be implemented to avoid terrain where burial may be difficult or impossible.

Seabed Soils Data

Shallow (near surface) seabed soils data is required for 'ground truthing' the seismic data collected during cable route surveys and for assessing the burial performance of cable burial equipment. Obtaining this data traditionally uses gravity coring,[4] or more rarely vibrocoring.[5] As cable burial is the preferred method for protecting cables laid in the continental shelf sections of cable systems, the installer requires a detailed knowledge of the soil profile to the full depth where burial will be required. The burial assessment aspect of the survey is a holistic process, integrating geophysical data with soils data from gravity coring and from remotely operated lightweight mini cone penetrometers (MCPTs). It is important that during the cable route survey sufficient ground truthing data is collected that adequately represents the variations in soil conditions identified along the route. In addition to the collection of the raw ground truthing data, onboard testing procedures must enable provisional burial and plow assessment to be made so that rapid re-routing decisions can be made while the survey vessel is in the problem area.

Submarine Geology

In the case of cable route surveys it is unusual to carry out a programme of borings deeper than the typical 3 m cores obtained by gravity coring. The prevailing geological conditions along the cable route are therefore most commonly

[4] A gravity corer is a simple and reliable instrument for collecting sediment cores from coastal and deepwater locations for sample analysis. The corer uses the pull of gravity to penetrate the seabed with a steel core barrel within which a replaceable core liner is housed. Gravity corers can collect samples typically 3–6 m in length.

[5] Vibrocoring is a technique used for collecting samples of unconsolidated saturated sediments. A core tube is attached to a source of mechanical vibration (the power head) and lowered into the sediment. The vibrations provide energy for rearranging the particles within the sediment in such a way that the core tube penetrates under the static weight of the vibrocoring apparatus.

derived from a literature search during the desktop study phase of the investigation, from interpretation of the seismic data and from the drop cores collected during the survey. An understanding of the route geology is important where the route passes through seismically active areas, areas of active faults such as may occur at the continental shelf edge and continental slope, areas of submarine volcanic activity and in areas where active hydrocarbon gas venting occurs.[6] Other geological information that may need to be understood within the context of cable system design includes seabed temperatures and radioactivity.

Oceanography

Oceanographic data, particularly data on seafloor currents, may be important in some marine environments, for example in areas where seabed scouring[7] could affect the long term security of the cable system, in areas of high current flow near beach landings where shore end installation could be adversely affected, in areas of sand waves where cable re-exposure may be a significant risk and in

Figure 4.4 Deep water ridge and trough submarine topography identified from high resolution multibeam echo sounder data in water depth range 500–2500 m over an area of 25 km × 20 km. (Image courtesy of EGS (Asia) Ltd)

---

6 For a discussion of the relationship between the geological marine environment and submarine cables, refer to Chapter 7.

7 Seabed scouring refers to the erosion of seabed sediments due to hydrodynamic forces, for example water currents. Seabed scouring may occur around seabed structures or other objects on the seabed that interfere with current flow. Scouring may also occur as a result of an iceberg scraping along the seabed which is capable of crushing and ripping cables.

areas where high current velocities combined with a rough seabed could result in excessive wear on a surface laid cable. In many cases, oceanographic information will be gathered during the desktop study. This information may, in a benign environment, be considered adequate for the purposes of installation planning and system risk analysis. However, seismic, imagery and bathymetric data collected during the survey may show submarine evidence indicative of high current velocities along the cable route. In these cases there may need to be sufficient flexibility in the survey procedures to accommodate the need to acquire additional current data.

Onboard Survey Team

To achieve reduced survey lead-times, route survey practices include procedures for at-sea development of a comprehensive understanding of the often complex physical along-route conditions and the impact on cable engineering and installation. To meet this challenge the survey vessel must have the capability and facilities to process, integrate and evaluate the large volumes of multi-parameter data collected throughout the entire data acquisition phase of the survey. This practice facilitates the making of objective routing refinements at sea which may be necessary to meet the requirements for cable protection, optimum system engineering and installation.

In support of these practices and to meet the objective of achieving final route selection, engineering and the production of final system Route Position Lists and Straight Line Diagrams concurrently with survey operations, the onboard survey team, including representatives from purchasers and suppliers, are required to

Figure 4.5 Offshore cable route survey vessel *RV Ridley Thomas*. (Photograph courtesy EGS Survey Group)

have the necessary experience and authority (or have ready access to the necessary authorizing parties) to approve final route selection at sea. To this end the survey vessel must not only have the capability to communicate with the on-shore offices of the purchasers and suppliers, but also have the capability to transmit large data files that may be needed to support routing decisions. For this reason most offshore survey vessels are equipped with V-Sat broadband satellite communication systems.

The Survey Output

The output from the survey is presented as a series of deliverable products. It is normal practice for the offshore survey team to include full reporting and charting capabilities, utilizing integrated data management software systems. This facilitates data turnaround from data acquisition, processing and interpretation to onboard delivery of preliminary key survey outputs within 24 hours. The delivery of the products is therefore progressive as survey operations advance along the route, which enables routing issues to be identified and re-routing decisions to be made without delay. It also enables route development decisions to be made in a timely manner and minimizes the risk of costly re-survey transits. The preliminary survey outputs made available onboard the survey vessels include the following:

- Preliminary pre-final Route Position Lists and Straight Line Diagrams;
- Preliminary charts;
- Preliminary geophysical interpretation;
- Preliminary route description and risk assessment;
- Preliminary assessment of cable burial performance; and
- Recommendations for route development.

In addition to discrete delivery products, data output is available to the supplier's onboard route engineer in digital format as a direct input to route planning software, such as Makai, enabling route engineering decisions to be made as the survey progresses.

Upon completion of the survey operations a final report will be issued which will provide a fully documented record of the survey activities. The final report preparation will typically be office based and will have undergone rigorous quality control checking and review by the supplier and system owner prior to release. The final report will include the following minimum deliverables:

- Final pre-installation Route Position Lists and Straight Line Diagrams for the entire system;
- Final cable engineering;
- Complete set of cable route charts including north-up charts for the entire system and alignment charts for sections of the route where cable burial is required;

- Detailed narrative text describing all survey activities, including a diary of events;
- Detailed route description on a chart by chart basis;
- Burial assessment report;
- Definition of residual risks to the cable; and
- Cable protection recommendations.

IV. Operational Planning and Schedules

The cable route survey and operational planning is influenced to a considerable degree by permitting lead-times and seasonal weather patterns. Survey operations should always schedule marine activities to be performed during the time of year when minimal delays due to bad weather can be anticipated. Survey operations performed during bad weather not only impact on operational safety, they also compromise data quality and will result in expensive operational delays.

Permits and permitting lead-times are discussed in more detail below, but it should be noted that when setting survey deliverable milestones, issues related to permitting have increasingly become a critical path activity. A detailed understanding of the impact of permitting lead-times on operational planning and schedules is therefore fundamentally important.

V. International Law Governing Survey Activities

As is evident from the above discussion, cable route surveys form an integral part of the activities associated with laying and maintaining submarine cables. When a new cable is proposed the route for the cable must be mapped in a desktop study and a survey of the route conducted. Survey data is collected along the proposed cable route through the use of geophysical survey methods and extraction of core samples. These steps are essential components of the cable laying process as they identify the safest route for the cable in order to preserve its lifespan, minimize interference with other marine uses in the area and identify potential obstacles and hazards. The survey also serves to determine final cable lengths and cable engineering criteria and to confirm or amend preliminary cable protection strategies. In addition, it provides essential data and documentation to support cable installation and a database framework for future maintenance of the cable system.

A comprehensive discussion of the international legal regime governing the surveying, planning, laying and repair of submarine cables in the territorial sea, EEZ, continental shelf and high seas is provided in Chapter 3. The present Chapter summarizes the relevant principles of international law relating to cable route surveys and supplements where necessary.

Generally speaking, as with other activities, coastal State jurisdiction over cable route surveys will depend on the type of activity involved and the maritime zone in which the activity is being conducted.[8] Notwithstanding the limits to jurisdiction set out in UNCLOS, some coastal States adopt the view that cable route surveys are *marine scientific research* or that they otherwise present a threat to the State's security or economic interests and seek to monitor and regulate the activities. Security concerns may be intimately connected with the determination of coastal States to define and protect perceived entitlements to maritime spaces,[9] and as a result some States have demonstrated a willingness to exercise jurisdiction in excess of that afforded to them by international law. A serious question for the cable industry, therefore, is how to allay the suspicions of the small but important number of coastal States who place onerous and time consuming restrictions on cable route surveys in a manner inconsistent with UNCLOS.

Prior to addressing the limits of coastal State jurisdiction, it is necessary to first understand how activities relating to cable route surveys are characterized in the context of UNCLOS.

Cable Route Surveys and Marine Data Collection

At the outset, it is important to note that a cable route survey is a form of marine data collection[10] undertaken for the specific purpose of preparing for cable installation. There are several categories of marine data collection, these include: (1) marine scientific research; (2) surveys (including hydrographic surveys for navigational purposes and military surveys for military purposes); (3) surveys for the exploration and exploitation of resources; (4) operational oceanography (including ocean state estimation, weather forecasting, climate prediction) and (5) cable route surveys.[11]

As noted above, coastal State jurisdiction over these categories of data collection will depend on the specific type of activity involved and on the maritime zone in which it is conducted.[12] However, determining whether a particular marine data collection activity may be regulated under coastal State jurisdiction is not as straightforward as it seems and has been the subject of numerous debates between States and academics.[13]

---

8 J.A. Roach and R.W. Smith, *Excessive Maritime Claims* (3rd ed, Martinus Nijhoff Publishers, 2012) at 413.

9 N. Klein, *Maritime Security and the Law of the Sea* (Oxford University Press, 2011) at 7.

10 Note that UNCLOS does not use the term 'marine data collection'. As noted by Roach and Smith 'marine data collection' is used as a generic term without legal content, by way of providing an umbrella under which the various collection activities can be considered, see Roach and Smith, *supra* note 8 at 413.

11 Roach and Smith *supra* note 8 at 449.

12 *Ibid.*

13 See for example R. Pedrozo, "Preserving Navigational Rights and Freedoms: The Right to Conduct Military Activities in China's Exclusive Economic Zone" (2010) 9 *Chinese Journal of International Law* 9–29; Z. Haiwen, "Is it Safeguarding the Freedom of Navigation or Maritime Hegemony of the United States? Comments on Raul (Pete) Pedrozo's

The difficulty in distinguishing different types of marine data collection and survey activities is partly attributable to the lack of definitions provided for these terms and activities in UNCLOS. Of the five categories described above, only marine scientific research,[14] and surveys and hydrographic surveys (terms which are used interchangeably)[15] are expressly mentioned in UNCLOS. Even then, there is no comprehensive definition provided for these terms. The remaining categories, i.e. surveys for the exploration and exploitation of resources, operational oceanography, military surveys and cable route surveys, are not expressly referred to in UNCLOS.

The lack of definition of these key terms must be seen in the context of the negotiations for UNCLOS. The convention was exhaustively negotiated over a period of nine years and although it is a *legal* document, the final version of the treaty that States were prepared to sign and ratify contains numerous *political* compromises. There were many economic, scientific, technological, military, social and political factors that influenced these compromises. As a result, key terms affecting cable route surveys, such as *marine scientific research* and *hydrographic surveying*, appear to have little to do with how scientists and industry may conceive of these activities and a great deal to do with compromises that lawyers and diplomats could settle upon.[16]

Another factor that has led to difficulty ascertaining whether particular marine data collection activities are subject to coastal State jurisdiction is the fact that many of the activities utilize similar methods of data collection[17] and in some cases the data collected may be the same.[18] This is perhaps best exemplified by the cable route survey which employs a variety of invasive and non-invasive techniques to gather data. The surveys obtain bathymetric data through a hydrographic survey conducted by a multibeam echo sounder. The bathymetric data provides information on the morphology of the seabed and the seabed slopes. Sonar data provides a three-dimensional picture of the seabed surface and enables geophysicists to distinguish between different types of sediment. Gravity coring and cone penetrometer tests are also conducted to obtain data on seabed sediment, using methods similar to those utilized in surveys for exploration and exploitation of resources. In addition the surveys also gather oceanographic data in a comparable manner to the methods used in operational oceanography. However, as will be explained below, the **parameters** of data collected and

Article on Military Activities in the EEZ?" (2010) 9 *Chinese Journal of International Law* 31–47, 43.

14 UNCLOS Part XIII, Arts 238–265.

15 UNCLOS Arts 19, 21, 40 and 54.

16 A.H. Soons, *Marine Scientific Research* (Kluwer Law and Taxation Publishers, 1982) at 5.

17 S. Bateman, "Hydrographic Surveying in the EEZ: Differences and Overlaps with Marine Scientific Research" (2005) 29 *Marine Policy* 163–174.

18 Roach and Smith, *supra* note 8 at 450.

their **intended use**[19] distinguish cable route surveys from the other categories of marine data collection and prevent them from being a threat to the economic interests of the coastal State.

Cable Route Surveys and Hydrographic Surveys

It is the opinion of the authors that a cable route survey cannot accurately be described as a 'hydrographic survey'. Hydrographic surveys are commonly understood as surveys used for navigational purposes, i.e. for making navigational charts and ensuring safety of navigation.[20] They "include the determination of the depth of water, the configuration and nature of the seafloor, the direction and force of currents, heights and times of tides and water stages, and hazards for navigation".[21] While there is no doubt that a hydrographic survey is one component of a cable route survey and is used to gather critical data on bathymetry, akin to hydrographic surveys used for navigation, the collection of bathymetric data is for a fundamentally different purpose directly related to the laying of cables.

The survey technology used in cable route surveys not only collects bathymetric data used for navigational purposes referred to above, it also employs acoustic (seismic) energy to penetrate the upper layers of the seabed soil in order to provide shallow sub-seabed information regarding the initial few meters of seabed sediment. The primary use of this technology is to help understand the physical properties and composition of the sediment so as to plan for the type of cable burial that will be required and the tools that will be appropriate for completing it (for example, plows or jetting tools). Plow shares are only capable of penetrating soft sediment such as silt, sand, clay, chalk, marl and so forth, and testing of sediment helps identify areas in which plows can be used. The bathymetric data collected from the survey helps derive a picture of the seabed topography and enables cable engineers to accurately calculate cable lengths and cable slack values, to identify areas of steep slopes where cable plow operations could be compromised and where potential cable suspensions could affect the integrity of the installed cable system. Given the distinct scope and purpose of these survey activities it is clear that they are not hydrographic surveys.

Cable Route Surveys and Surveys for the Exploration and Exploitation of Resources

Similarly, cable route surveys should not be equated with surveys used for the exploration and exploitation of resources. It is correct to state that the activities of sub-bottom profiling and gravity coring essentially involve an examination of the seabed in a comparable manner to the surveys related to the discovery and/

---

19 *Ibid.*

20 *Ibid.*, at 416.

21 *Ibid.*

Figure 4.6 'Mowing the lawn'. Survey vessel using multibeam side scan sonar to delineate cable route along a 1000 m swath on the desktop study route position list. (Image courtesy of NIWA)

or evaluation or exploitation of economic resources. However, there are two critical differences.

The first difference is the *purpose* for which these activities are conducted. The technology used for sub-bottom profiling, which is a non-invasive data collection technique, only penetrates the upper layers of the seabed soil. The energy levels used during the surveys are not sufficient to penetrate more than a few meters of soft sediment and have little or no capacity to penetrate hard seabed materials. Given that data is only retrieved from the shallow sub-seabed to a depth of a few meters of sediment, this technology is not used, and in fact would be inappropriate, for the exploration of economic resources. The data collected during these surveys would therefore have no value for oil and gas prospectors.

Similarly, with regard to gravity coring, which is an invasive technique that typically penetrates the seabed to depths of up to 6 m, the primary purpose of extracting the core samples is to 'ground truth'. When undertaking gravity coring, sediment samples are brought onto the survey vessel and tested in order to derive engineering parameters. The sediments are then typically discarded and are not tested for hydrocarbon potential. It should be noted that neither explosives nor air-guns are used during cable route surveys.

The second essential difference between cable survey route survey activities and surveys conducted for mineral resources is the breadth of area in which the activities are carried out. Data collection for submarine cable planning, whether it is bathymetric data, sub-bottom profiling or gravity coring, takes place along the very narrow strip of seabed that represents the proposed route of the cable. The strips of seabed are typically between 500–1000 m wide in areas anticipating cable burial to a maximum three times water depth or 10 km in deep water areas beyond the continental shelf. Within the industry, cable route surveys are colloquially referred to as *long thin surveys* and they bear no relationship to the wide-scale block or area surveys typically employed for mapping economic resources. Again, the data retrieved from the survey activities would be of no value to oil and gas prospectors.

Cable Route Surveys and Marine Scientific Research

A further important distinction should also be made between cable route surveys and *marine scientific research* (MSR). No definition of MSR is provided in UNCLOS. During treaty negotiations, various proposals were made for a definition, with the Chairman of the Committee charged with drafting the relevant provisions noting that "[m]arine scientific research means any study or related experimental work designed to increase man's knowledge of the marine environment".[22] The definition was soon abandoned and a consensus appears to have been reached that it was not necessary to include a definition, as the substantive provisions of the convention clearly established the meaning intended for MSR.[23]

Although the term is not defined, UNCLOS has created a very comprehensive regime for MSR in Part XIII. Within Part XIII, Article 245 confers upon coastal States exclusive rights to regulate, authorize and conduct MSR in their territorial sea and Article 246 provides that coastal States, in the exercise of their jurisdiction, have the right to regulate, authorize and conduct MSR in the EEZ and on their continental shelf. In these zones MSR shall be conducted with the consent of the coastal State.[24]

It is clear however, that cable route surveys are not MSR and should not be subjected to this regime. First, surveys and MSR are distinguished in UNCLOS. Although MSR may appear to incorporate activities such as hydrographic surveying and research, the latter are in fact separately provided for in UNCLOS. For example, Article 19(2)(j) refers to "research or survey activities", Article 21(1)(g) to "marine scientific research and hydrographic surveys", and Article 40 to "marine scientific research ships and hydrographic survey ships" and "research or survey activities". The separate treatment afforded to these activities indicates that surveying and MSR were viewed as distinct during drafting of the convention.[25] Second, and perhaps most compelling, the intended purposes for cable route surveys and MSR are fundamentally different, with data collected for cable route surveys being used solely for cable installation whereas data collected through marine scientific research is used either for the benefit of humankind or for

22 The initial single negotiating text, part III, document number A/CONF.62/WP.8/ drafted by the Chairman of the Third Committee, Conference for the Law of the Sea; *see* Part II, Article 1, at page 177, available online at http://untreaty.un.org/cod/diplomatic conferences/lawofthesea-1982/docs/vol_IV/a_conf-62_wp-8_part-3.pdf (last visited 7 June 2013).

23 Soons, *supra* note 16, at 122.

24 UNCLOS Art 246(2).

25 Soons, *supra* note 16, at 125.

resource related research. Other than requests by the coastal State, cable route surveys are not published or shared with third parties.

What is a Cable Route Survey?

It is evident that a cable route survey cannot be characterized as a hydrographic survey, a survey for the exploration and exploitation of resources or as MSR. In the view of the authors, a cable route survey is a separate category of marine data collection essential for the laying of cables and hence the resilience and integrity of the world's telecommunications systems. With this in mind, the following two sections will provide a brief overview of the applicable international law.[26]

Cable Route Survey Activities in Territorial Sea/Archipelagic Waters

In the territorial sea/archipelagic waters the coastal State has sovereignty over the water column, the airspace above it and the seabed and subsoil below it.[27] As the coastal State has sovereignty over this maritime zone, it has a large degree of jurisdictional competence to prescribe and enforce its laws. The only restriction on the authority of the coastal State is that its sovereignty must be exercised "subject to this Convention and to other rules of international law".[28] Principally this means that the coastal State must allow the ships of all States the right of innocent passage.[29]

Similarly, an archipelagic State, as defined in Article 46 of UNCLOS, has sovereignty over the waters enclosed by its archipelagic baselines.[30] In these waters, referred to as archipelagic waters, the archipelagic State must exercise its sovereignty subject to Part IV of UNCLOS.[31] The ships of all States enjoy innocent passage through archipelagic waters.[32]

Pursuant to this sovereignty, coastal States and archipelagic States have the right to regulate cable route survey vessels. For example, in the territorial seas and archipelagic waters, ships carrying out "survey activities" are not engaged in innocent passage[33] and archipelagic States and coastal States can adopt regulations on innocent passage relating to hydrographic surveys within these maritime zones.[34] Foreign ships (including hydrographic survey ships) undertaking

---

26 An expansive discussion of coastal State authority in different maritime zones is provided in Chapter 3.
27 UNCLOS Art 2.
28 UNCLOS Art 2(3).
29 UNCLOS Art 17.
30 UNCLOS Art 49.
31 UNCLOS Art 49.
32 UNCLOS Art 52. It should be noted that archipelagic States are permitted to designate sea lanes to ensure that the right of innocent passage is exercised in a safe manner, see Art 53.
33 UNCLOS Art 19(2)(j).
34 UNCLOS Art 21(1)(g).

passage in archipelagic sea lanes may not carry out any survey activities without the prior authorization of the archipelagic state.[35] A similar prohibition applies to foreign ships exercising the right of transit passage in straits used for international navigation.[36] While the terms *survey activities* and *hydrographic surveys* are used interchangeably, it is reasonably clear that cable route surveys would fall within the definition of *survey activities*.

Freedom of All States to Conduct Cable Route Surveys in the EEZ

Article 58(1) of UNCLOS provides that in the EEZ all States enjoy the freedom to lay submarine cables and pipelines and "other internationally lawful uses of the sea related to these freedoms, such as those associated with the operation of... submarine cables and pipelines". There are no express provisions in UNCLOS relating to cable route *survey* activities conducted in the EEZ and continental shelf. However, given that cables cannot be installed or operated without pre-laying survey activities, it is reasonable to submit that cable route surveys are a lawful use of the sea related to the freedom to lay submarine cables pursuant to Article 58(1). All States therefore enjoy the freedom to conduct cable route surveys in the EEZ.

This freedom is not, however, unconditional. As discussed in Chapter 3, when States are exercising their right to conduct these surveys, they have certain obligations that they must fulfil, including the obligation to have due regard to cables and pipelines already in position,[37] to have due regard to the rights and duties of the coastal State and to comply with the laws and regulations adopted by the coastal State in accordance with UNCLOS.[38] In particular, due regard should be given to the sovereign rights that the coastal State enjoys for the purpose of exploring and exploiting, conserving and managing its natural resources in the EEZ.[39] Arguably this due regard obligation would include the need for parties conducting cable route surveys to notify the coastal State of their intended activities and provide details of the nature and timing of the activities. This would alleviate concerns that the surveys are being conducted for the purpose of collecting data related to hydrocarbon resources or otherwise prejudicing the resource-related interests of the coastal State. Such notification would afford coastal States the opportunity to advise of potentially conflicting interests in the area, such as intended gas and oil exploitation activities, and also to assist in ensuring the safety of the cable survey vessel from competing activities when conducting the surveys. Exchanges of information and/or consultation are clearly useful to all parties and are part of the due regard obligation.

---

[35] UNCLOS Art 54.
[36] UNCLOS Art 40.
[37] UNCLOS Art 79(5).
[38] UNCLOS Art 58(3).
[39] UNCLOS Art 56(1).

Rights and Obligations of Coastal States with Regard to Cable Route Surveys

Pursuant to Article 79(2) coastal States have an obligation not to impede the laying or maintenance of submarine cables on the continental shelf. As noted above, a cable route survey is a lawful use of the sea related to the freedom of States to lay and maintain submarine cables. The coastal State therefore has an obligation not to impede cable route surveys. However, Article 79(2) also provides that the obligation not to impede cable route surveys is subject to the right of the coastal State to take *reasonable measures* for the exploration of the continental shelf and the exploitation of its natural resources. What is deemed a *reasonable measure* is not clear.[40] Given that the interests the coastal State has in these maritime zones are resource-related, a reasonable measure would need to be related solely to resource interests and not to other concerns of the State, such as security concerns pertaining to crew and vessels or the imposition of taxes and customs duties. Seeking information about the cable route survey activities and requesting that an official of the coastal State be allowed on to the survey vessel would arguably be reasonable measures.

The coastal State is also required to have due regard to the rights and duties of other States, to act in a manner compatible with the provisions of UNCLOS[41] and, on the continental shelf, refrain from exercizing its rights in a manner that will infringe or result in "any unjustifiable interference" with the rights and freedoms of other States as provided for in UNCLOS.[42]

It should be noted that the coastal State may also take reasonable measures to prevent, reduce and control pollution from pipelines and to delineate the course of the laying of pipelines.[43] However, these measures only apply with respect to pipelines. There are no comparable provisions with respect to submarine cables.

High Seas and Deep Seabed

The high seas and deep seabed are areas beyond the national jurisdiction of any State. All States are free to conduct survey activities in these zones, subject to the requirement to show due regard for the interests of other States in the exercise

---

[40] While it is not clear what is meant by *reasonable* "no more definite criterion than that of reasonableness could be established for the measures which coastal States may take, for the reason that it was impossible to foresee all situations that might arise in the application of this article": Statement by the US Representative during the Eighth Session of the International Law Commission cited in M. Whiteman, "Conference on the Law of the Sea: Convention on the Continental Shelf" (1958) 52 *American Journal of International Law* at 642.

[41] UNCLOS Art 56(2).

[42] UNCLOS Art 78(2).

[43] UNCLOS Arts 79(2) and 79(3). UNCLOS Art 79(2) distinguishes between pipelines and cables. It is only in respect of pipelines that a coastal State is permitted to impose reasonable measures for (1) the exploration of the continental shelf; (2) the exploitation of its natural resources; and (3) the prevention, reduction and contol of pollution.

of the freedom of the high seas.[44] Further discussion of this maritime area is set out in Chapter 3.

VI. Law and Policy Challenges for Cable Route Surveys

With increasing numbers of submarine telecommunications cables being planned and installed, the infrastructure supporting these developments are coming under increasing pressure. There are demands on providers to reduce project schedules during the construction phases, with related demands being imposed on system planners and supporting service providers to examine ways to expedite the various planning and installation processes. The pressure to reduce lead-times in the planning process places system planners in a dilemma. Many of the time constraints imposed during the planning process fall outside the control of the system planning team, primary examples being permitting and the finalization of routing negotiations with other seabed users. As survey activities are the first tangible evidence of the progress of a new submarine cable system, it is this activity that is most severely targeted for a reduction in the program schedule.

Excessive Coastal State Regulation of Cable Route Surveys within the Territorial Sea/Archipelagic Waters

The operational permits required for performing a submarine cable route survey are essentially the same as they are for any other survey activity conducted within the territorial sea/archipelagic waters. Agencies that regulate the issuing and/or authorization of operational permits typically include:

- Port authorities for routes passing through a gazetted port authority's jurisdiction;
- Hydrographic office and/or navy having responsibility for charting and navigation within the State's maritime jurisdiction; and
- Government ministries, which may include those responsible for security, defence, foreign and home affairs, transport, communications, environment, energy and fisheries.

Information requested as part of the operational permitting process typically includes details of the survey vessel/s (including all mandatory vessel certificates), a list of marine crew and survey team members, details of supplier and purchaser representatives and a description of survey activities including the provision of charts showing the cable route to be surveyed and survey schedules.

The agencies responsible for authorising survey operational permits may link various conditions to the granting of permits. Such conditions may include the following:

---

44 UNCLOS Art 87(2).

- Requirement for all vessel crew and survey team members to undergo security checks;
- Requirement for survey operations to be witnessed by security officers;
- Requirement for copies of survey data and survey reports to be made available to the coastal State upon completion of the survey;
- Requirements that survey operations be carried out by national research institutions;
- Compliance with coastal State cabotage regulations requiring surveys be conducted by vessels flagged by the coastal State;
- The imposition of restrictions on certain nationalities within the vessel crew and survey team.

Additional permitting requirements may also arise to address environmental regulations or requirements set out by development funding agencies such as the World Bank. Such requirements may not impact survey permit lead-times but may impact the survey scope of work, landing site location and the pre-survey route, particularly where cable burial is planned. As submarine cables terminate at the beach manhole it may be that ordinances governing the near shore and foreshore will be in force. Such ordinances may require that the regulator gazette details of the intended installation and maintain such a gazettal for a prescribed period of time, which may range from two months to as long as six months. This process will typically require provision of the system Route Position Lists and overall project information. Where cable routes pass through areas regulated by local marine departments responsible for policing vessel movements and ensuring safety and navigation, any requirements directed by these departments would typically be included in this process. In addition to any gazettal for the marine portion of the route, additional permission may be required from the local lands department and/or the local landowner for the land portion of the route. This would be in the form of a wayleave application to occupy the land. Depending on the area concerned for the proposed land portion of the route or where a new cable landing station is to be constructed, this application may also need to be submitted to the local planning controls office or its equivalent. In addition, if any trees or ecologically sensitive areas are to be affected, or where routing passes through coastal recreational areas and/or marine parks, then permits will be required from the relevant regulating authority.

Where cable protection by burial is required, this would typically extend from the beach manhole to a predetermined water depth based on the perceived threat to the cable from activities such as fishing and anchoring. Embedding cables into the seabed may be regulated under an Environmental Impact Assessment Ordinance (EIAO) requiring that an Environmental Permit (EP) be approved and issued prior to construction. The EP may require that the application forms part of a mandated EIA or, in certain cases, a direct application for an EP may be entertained by the regulating authority, which avoids the need for the application to be subject to the typically more lengthy EIA process. The direct approach may be based on submitting a Project Profile, which incorporates a general environmental assessment. Part of this process may require meetings with all interested stakeholders, comprising government bodies (including environmental protection agencies) and agencies having interests in agriculture, fisheries and conservation as well as local communities perceived to be impacted by the project. Depending on the route, there may also be a requirement to undertake ecological studies, water quality monitoring and modelling studies and a marine archaeological investigation (MAI) may be requested.

It should be noted that any significant change to the route or methodology during the permitting process might require the EP to be re-issued if the route installation impacts any sensitive area. Where changes are deemed to present a minor impact, a variation to the original submission to qualify changes may be allowed. The lead-time for these processes may vary from two to six months. Where a variation is required, this lead-time may well be extended by between one and three months. Where significant changes to the route are made after the permitting process has started, it is likely that the gazettal will be impacted. If changes exceed certain published (sometimes unpublished) criteria this may severely impact the gazettal schedule and such changes will have to be reincorporated into the gazettal with both the government and public consultation periods restarted.

Variations in permitting lead-times may be attributed to:

- Failure by the regulating agencies to understand the scope of the cable route survey;
- Lack of established process within regulatory agencies;
- The number of systems following similar routes overwhelming existing processes;
- Regulatory inefficiency;
- Competing systems landing in the same country serving the same market; and
- Actual failure by permit applicants to fully comply with regulatory processes.

Excessive Coastal State Regulation of Cable Route Surveys in the EEZ/Continental Shelf

As with permitting regimes for the territorial sea, coastal State regulations for cable route surveys in the EEZ and continental shelf may also be lengthy, complex and lacking in transparency. Some coastal States require cable system owners to submit the same permit applications for surveys conducted in the EEZ/continental shelf as they do for the territorial sea. Notwithstanding the policy of such States, attempts to subject cable route surveys in the EEZ/continental shelf to domestic permitting requirements are in fact contrary to UNCLOS. This is because UNCLOS recognizes cable route surveys as an internationally lawful use of the sea related to the operation of submarine cables and does not provide authority for coastal States to regulate these activities.

It appears that the right to conduct cable route surveys in the EEZ has become a casualty of the long-standing debate on the permissibility of surveys in the EEZ. Many coastal States consider surveys, such as hydrographic surveys and military surveys, to be part of the "other internationally lawful uses of the sea related to [high seas] freedoms, ... associated with the operation of ... submarine cables and pipelines"[45] afforded to all other States in their EEZ and hence are not subject to coastal State consent. Conversely, other States have argued that any type of survey, including hydrographic surveys and military surveys, are a form of *marine scientific research* and are therefore subject to coastal State consent in Article 246 of UNCLOS (as discussed above).

However, as formerly noted, a cable route survey is neither a hydrographic survey nor marine scientific research but rather is a right associated with the operation of cables. Accordingly, these permits are *prima facie* inconsistent with UNCLOS. Of course, cable route surveys can be subject to "the reasonable measures" of States relating to exploration and exploitation to ensure that survey activities do not interfere with coastal State rights and jurisdiction within the EEZ/continental shelf, but this does not extend to an excuse for coastal States to demand permits for cable route surveys.

Implications of Excessive Coastal State Regulations on Cable Route Surveys

Securing permissions to carry out survey operations vary from coastal State to coastal State and range from the straightforward to the highly complex, with lead-times measured in days or multiple months. The impact of variations in permitting lead-times on project planning is minimal when dealing with domestic national cable systems but extremely complex when planning for long haul international cable systems that traverse waters adjacent to States with highly variable national jurisdictional requirements.

Underestimating the time required to address permitting issues is an increasing and frequent cause of project delays, missed project milestones and cost overruns. This can, and increasingly does, result in project implementation phase dislocations. It is therefore critical that during the earliest stages of project planning and development for the implementation schedule of a new submarine cable system, permit interdependencies are identified and the associated processing responsibilities addressed. It is not uncommon for system owners to abrogate or fail to recognize their responsibility in the permitting process by passing responsibility and permit accountability to the supplier or even the survey contractor. It is also not uncommon for suppliers to accept such responsibilities; an inevitable consequence of an increasingly competitive supply environment. There are also States whose regulatory agencies will only consider a full project brief from the system owner as a prerequisite to initiating the permitting process.

---

45 UNCLOS Art 58(1).

A thorough understanding of the permitting requirements pertinent to the construction of the system is therefore imperative, and certainly prior to committing to *Ready for Service* (RFS) dates. There is no doubt that past examples of serious project delays and cost overruns has focused project funding institution audits to pay particular attention to how well permitting has been addressed and lead-times anticipated in the Business Plan.

VII. The Way Forward

In territorial seas/archipelagic waters the permitting regimes, despite being complex, expensive and cumbersome, are nonetheless largely consistent with UNCLOS provisions that allow coastal States/archipelagic States to regulate cable route surveys. The balancing of the interests that coastal States have in preserving their security and economic interests and the freedoms that other States have to undertake survey activities is clearly tilted in favour of coastal States in this maritime zone. However, as States gradually come to recognize that submarine cables are critical infrastructure underpinning their security interests and economic development, it is hoped that their policies will be reviewed to ameliorate the effects of onerous and lengthy permitting requirements required for cable route survey activities.

In the EEZ and continental shelf, coastal States should also review their permitting regimes and recognize that cable route surveys should not be subject to coastal State regulation. They should also acknowledge that cable route hydrographic surveys are not marine scientific research and should not be subject to the marine scientific research regimes provided for in UNCLOS.

It is recommended that Coastal States balance the interests of both cable route survey vessels and the wider community interest in telecommunications and give weight to the need to make permitting regimes more transparent and less time consuming for companies seeking to conduct survey activities. Helpful steps that coastal States could adopt include nominating a lead agency responsible for:

(i) publishing procedures/requirements for obtaining permits in territorial sea/archipelagic waters;
(ii) streamlining the application processes for permits required in territorial sea/archipelagic waters, for example ensuring that in circumstances where permits require the consent of multiple government agencies, a single permit application can be filed and the lead agency will co-ordinate with the other agencies to circulate the application and secure consent; and
(iii) work to remove permitting obligations that do not serve a useful and legitimate purpose, such as the need to undertake environmental impact assessments prior to undertaking survey activities.

As this Chapter demonstrates, the interests of the cable industry have been compromised by the creeping jurisdiction of a small but important number of coastal States. Notwithstanding the permissibility or otherwise of the regulatory regimes these States have put in place, at a practical level the cable industry seeks to mitigate the effects of the regimes as far as possible in order to continue its work. Those representing the cable industry continue to engage with States through workshops, lobbying groups and information sessions, to exchange views with State and policy officials and to seek compromises to allay suspicions that survey activities present a threat to coastal States' security and economic interests.

Given that there are few objective means by which a coastal State can confirm that cable route survey vessels are engaging in innocuous activities, it has been the practice of many in the cable industry to share extensive information about their survey activities with coastal States. In keeping with this pragmatic approach the following recommendations are made with respect to the cable industry:

(i) when conducting cable route surveys in the EEZ/continental shelf cable companies may wish to notify the coastal State of the purpose, route and timing of the survey pursuant to its obligation to give due regard to the rights and duties of the coastal State.[46] Such notification would provide assurance to the coastal State that the survey activities are part of the cable planning process and are not prejudicial to its sovereign rights to explore and exploit the natural resources in this maritime zone.[47] Notification of intended surveys would also serve to avoid interference with other users of maritime space and ensure safety for the vessels and crew. From the perspective of the cable industry, this notification is typically given (particularly when cableships will be working near busy shipping lanes and other areas subject to intense use), but in some instances it is not as it may trigger interference by some coastal States;

(ii) Cable companies could allow national observers on board the survey vessels (this is commonly done where there is a request by the coastal State);

(iii) Cable companies may share the results of their survey, including sediment data, with the coastal State, subject to confidentiality rules (again, this is commonly done where there is a request by the coastal State).

By voluntarily adopting these types of notification and reporting measures, cable companies can demonstrate that their cable survey activities pose no threat to States seeking to safeguard security and economic interests. As coastal States become more aware of the importance of submarine cables, it is hoped that States with excessive regulatory regimes will re-balance their policies to give more weight to the interests of cables companies and appreciate the benefits that cables bring. States that persist with oppressive and illegitimate regulatory regimes may find themselves increasingly left out of the technology loop.

46 Roach and Smith, *supra* note 8 at 459.

47 *Ibid.*

CHAPTER FIVE

THE MANUFACTURE AND LAYING OF SUBMARINE CABLES

Keith Ford-Ramsden and Tara Davenport

Introduction

This Chapter provides an overview of the detailed planning and installation of the telecommunications cables that operate to transmit voice, data and internet communications around the world. The work that goes into achieving these installations is undertaken by a small group of experts of many nationalities based around the globe. Their expertise ensures that submarine cables can be installed anywhere from the deepest parts of the oceans to diverse landings on all continents. Cable system installation entails five major stages: the planning and surveying stage, applying for permits, the manufacturing stage, ship loading stage, and the cable laying stage. The planning and surveying stage has been addressed in Chapter 4. This Chapter will provide an overview of the application for permits (Part I), cable manufacture (Part II), ship loading (Part III) and laying operations (Part IV). It will then discuss the law and policy challenges for laying operations (Part V) and conclude with some recommendations on how these challenges can be addressed (Part VI).

I. Applying for Permits

Telecommunications cables are by their very nature designed to connect many States. The permitting and legal regulations that need to be observed may be wide-ranging. It is necessary to have a thorough understanding of the jurisdiction applicable to each portion of the cable in order to ensure that the correct permits are sought and issued prior to cable surveys or installation being undertaken. This is critical for ensuring that the cable can be installed in the first instance and that there will be no delays in the cable installation. The permitting process is often the critical path of any submarine cable project.

Accordingly, after a cable route has been identified, the next stage is to identify the permits required for each segment or section of the cable route and to enter into discussions and negotiations with the relevant permitting authorities at the earliest opportunity. The first permit required, to ensure the cable can be installed in the planned Cable Landing Stations, is the Telecoms Operations License or Landing License. These are issued by the national governments of the countries where the cable is landing. Once this license is in place, the Permits in Principle (or System Permits) and the Operational Permits can be applied for. The Permits in Principle are the system owner's submissions to, and/or permissions from, governmental agencies that authorize the cable to be placed and remain on the seabed for the duration of the cable system's life. They also apply to the land sections of the cable route. The Operational Permits are the permits required by the cable system provider and/or the installer to carry out all of the activities needed to install the cable on land and on/under the seabed (including access for cable route survey vessels and cable laying vessels etc).

Permits for the Land Section of the Cable

Each cable landing has a 'Landing Party'. This is generally a telecommunications company based within the country where the cable is landed.[1] These companies are critical in assisting with and obtaining the permits for the Cable Landing Station, land based fronthaul cable route and beach manhole selection and installation. The Landing Party is also likely to be a key contact for obtaining the marine permits and liaising with stakeholders in the vicinity of the marine portion of the cable landing.

Permits in Principle/Operational Permits

When the Telecoms Operating License has been approved, the Permits in Principle must be obtained for the route permits and/or approvals from the various governmental stakeholders at national and local levels. These stakeholders may include Ministries of Communications, Environment, Defence, Transport, National Security, Coast Guard agencies, and the Hydrographic Office in each country where the cable is landing. Permits in Principle must also be obtained from countries whose territorial waters the cable passes through, but does not land. A number of States also require permits for transiting the EEZ or continental shelf even though the cable never enters into the territorial waters, especially if the EEZ includes offshore exploration areas. This will be dealt with below.

[1] The Submarine Cable Almanac produced by the Submarine Telecoms Forum provides details of telecommunications cables worldwide, including the Landing Parties in each country. See www.subtelforum.com/articles/2012/submarine-cable-almanac-issue-4/ (last visited 7 June 2013).

Environmental Permits

In addition a separate Environmental Permit may also be required. This will only be issued after an Environmental Impact Study (EIS) has been conducted. The Permits in Principle or Letter of No Objection from the various departments may run in parallel or be sequential, so the estimated timelines must be predicted and accommodated into the overall Project Plan.

Consideration of Other Stakeholders and Interests

Oil and gas and renewable energy concession crossings may also require permits and/or agreements and these will need to be obtained either from State authorities or private companies. Appropriate permits or agreements may also be obtained from other seabed users and maritime interests that will have been identified in the Desktop Study.

The International Cable Protection Committee (ICPC) has issued a recommendation that be can used as the basis for cable crossing agreements between the new cable system and other telecommunications cables, pipelines and power cables. ICPC has also issued a recommendation for cable routing in the vicinity of other cables and offshore wind farms.[2]

When the requisite permits and agreements are in place the marine survey of the intended route can be undertaken (as described in Chapter 4). If the survey identifies changes that are required to the provisional route, the necessary amendments will be made in order to produce the final 'as built route position list'.

II. Cable Manufacture

The survey of the cable system route provides the information that is needed to refine the planned route and to determine the type of subsea plant and cable lengths to be manufactured. The manufacturing specification is described in the Straight Line Diagram (SLD) and the Route Position List (RPL), two key documents used in planning, manufacturing and installing a cable system. These documents set out the requirements with respect to the following:

i. number and position of optical repeaters, equalizers and branching units;
ii. type of cable armoring required in different parts of the system, typically lightweight for deep water and a range of armors for shallower water, plus the location of the transitions required to change one type of cable to another.

---

[2] Refer to International Cable Protection Committee, 'Recommendations' at http://www.iscpc.org/. ICPC Recommendations No. 2 (Cable Routing and Reporting Criteria), No. 3 (Telecommunications Cable and Oil Pipelines/Power Cables Crossing Criteria) and No. 13 (Proximity of Wind Farm Developments and Submarine Cables).

Note that burial expectations, determined from the burial assessment survey, will also influence armor decisions. Heavier armor may be considered in areas where the seabed does not allow burial;

iii. accurate positions of the crossings over other cables and pipelines; and
iv. the order in which the cable factories manufacture the lightweight and armored cable, insert the repeaters and equalizers into the cable and the order in which the cable system is to be laid.

III. Ship Loading

It is the role of the System Provider's Project Management Team (SPMT) to match the production of the various parts of the cable system with the freighting and installation vessels available, having due regard to the location of the cable system with respect to the cable factories and the size, seagoing capability and cable carrying capacity of the installation vessels (both volume and weight), as well as the installation methodology and estimated completion times.

Small single segment systems may be able to be loaded onto and then laid by a single vessel. Large multi-landing and/or trans-oceanic cable systems may require a number of different types of vessels to undertake a variety of roles. These tasks may include transporting the cable and plant from the factory to a port close to the area of operation for transfer to the installation vessels.

The manufacture of the cable, repeaters, equalizers and branching units, plus the system segment(s) assembly and testing are normally completed prior to their being freighted to the installation site. These processes form the initial part of the overall Project Installation Plan (PIP), which includes the planned durations for each part of the Project. This PIP allows the SPMT to ensure that continuity between the production and installation can be achieved, because any deviation from the PIP may seriously affect the timing of the installation and may result in a delay in bringing the system into service for the owners.

IV. Laying Operations

Preparation

Cable Station, Fronthaul and Ocean Ground Bed

The main components for a telecommunications cable landing are shown in the diagram below, Figure 5.1.

The 'backhaul' is the cable that connects the telecom operator's network to the marine telecommunications Cable Landing Station. The data signal is prepared and transmitted from the Cable Landing Station through the marine cable after passing through the land section of the marine system (fronthaul) and the joint between the fronthaul and marine cable at the beach manhole.

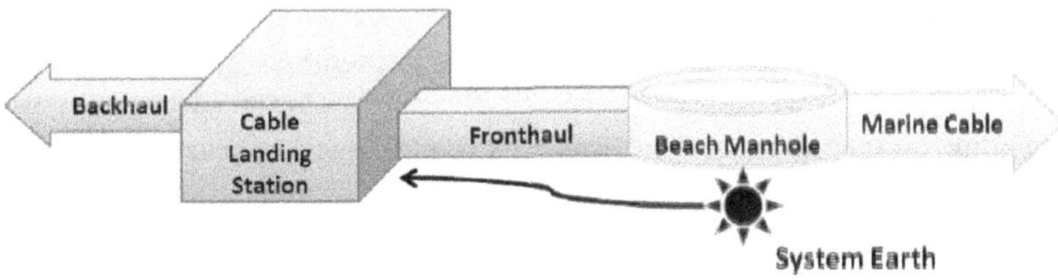

Figure 5.1 Structure of telecommunications cable landing.

All long-haul telecommunication cables have repeaters to regenerate the optical signal after distances of approximately 60–90 km. Power Feed Equipment (PFE), sited in the Cable Landing Stations, supplies up to a maximum of 15,000 volts at a constant current which can be up to 1.5 amps (depending on the number of optical amplifiers in each repeater). The PFE provides the power to the repeaters in order to generate the optical signals that carry the data over the trans-oceanic distances. The Power Feed Systems require an electrical earth at each Cable Landing Station. This is called a System Earth. The System Earth can be a land earth in the grounds of the Cable Landing Station on the beach or an electrode laid in the ocean.

Before the installation of the marine cables begins the infrastructure must be put in place. This includes the beach manhole, the fronthaul ducting and cabling, System Earth and the Cable Landing Station. Existing cable landing facilities and/or new constructions are subject to coastal State regulation and require local permits. The lead-time required in obtaining the necessary permits and complying with local regulations needs to be factored into the overall Project Plan.

Advance Notices to Fishermen and Mariners, Other Cables and Pipelines, Broadcasts from Ship

During the route planning and survey phases of the project the cable purchaser and installer will need to engage in dialogue with other sea and seabed users who may be impacted by the preparations for, as well as the actual installation of, the cable and any post cable installation operations. Other seabed users may include:

- fishermen;
- oil and gas companies, where the cable(s) pass through exploration blocks or over pipelines;
- seabed mining and dredging companies, where the cable(s) pass through mineral extraction or dredging blocks;
- cable owners from telecoms and power companies;
- renewable energy providers that generate and export energy from wind farms, tidal turbines and wave generators;

- fish/shellfish farm operators; and
- owners of out-of-service cables, if they can be located.[3]

Depending on the nature of the discussions and the relevant coastal State law, a written agreement may be entered into between the affected parties to cover just the survey and installations operations or, alternatively, a longer term agreement could be negotiated that covers the lifespan of the cable system. As a matter of good practice the installer of the new cable(s) should provide prior notice to any affected parties advising of the nature and timing of various operations associated with installation of the cable. This notice is usually given by the cable installer, either by directly advising the affected party and/or by providing a Notice to Mariners through the local coast guard agencies. The notification process continues throughout the cable laying operations via broadcasts from the laying vessels.

Route Clearance and Pre-Lay Grapnel Runs

Before laying operations commence, it is necessary to ensure that there are no obstructions along the selected cable route through Route Clearance (RC) and Pre-Lay Grapnel Runs (PLGR).

The Desktop Study and marine survey of the cable route will have identified any known objects that are on the selected cable route(s) where cable burial by plow or remotely operated vehicle (ROV) is planned.[4] Precise locations are determined during the marine survey with side-scan sonar and magnetometers or from databases and crossing agreements available to the survey and installation companies. The objects that lie close to or on the proposed cable route will be objects such as out-of-service cables that are buried or on the seabed surface and/or a wide variety of objects and debris that have been accidently or purposely disposed of in the oceans.

First, items such as out-of-service telecommunications cables that may foul or damage the plow during installation are cleared from the route by route clearance vessels. In special cases, unexploded ordnances that may hazard the cableship, the plow or the ROV's operations have to be cleared by specialized companies. This requires a specialized survey in order to identify such munitions prior to clearance.

Second, a separate operation is carried out to clear any surface debris that lies on or just beneath the surface of the seabed prior to the main cable installation

[3] ICPC, Recommendation No. 1 on Recovery of Out-of-Service Cables covers the method of clearing out-of-service cables, see http://www.iscpc.org/. See also Chapter 8 on Out-of-Service Submarine Cables.

[4] In areas where the cable is surface laid RC and PLGR are not normally carried out as the object is to prevent damage to plows and ROVs. In surface laid sections the cable is normally routed around the hazard, if this is not possible the armoring is increased or additional protection is provided over the object being crossed or onto the cable itself.

in shallower waters where cable burial is planned. This operation is known as the Pre-Lay Grapnel Run and involves a set of grapnels being towed along the seabed and recovered at regular intervals to remove any debris from the grapnels. Once the PLGR operation has cleared the route, it is ready for the installation and burial of the cable below the seabed. The ship's master carrying out the PGLR is responsible for this operation.

Requirements and Characteristics of Installation Vessels

The installation of cables is undertaken by a variety of specialized vessels, each chosen to suit the particular requirements of the project phases. Consideration must also be given to the cabotage regulations of some coastal States, which may require use of a locally flagged vessel (this is discussed further in Part V below).

Barges and Ships Used for Pre-Laid Shore Ends

For the pre-laid shore ends of cable systems, barges or shallow draft vessels are used. These may be required to sit on the ground when the tide goes out and bury the cable with a vertical injector or sled when the tide comes back in. Due to the very shallow areas that these vessels operate in they use anchors and spud poles to hold their position and move, until they are in sufficient water depth to use their thrusters (if fitted).

In order to operate in extended very shallow water areas (i.e. from the beach to approximately 10–15 m water depth), the shore end vessel must have a shallow draft and hence is limited in the amount of cable that it can carry, especially as the cable used in these areas are generally the heavier double or single armored cable types that are required to provide the requisite level of protection.

Cableships

Most cable installation is carried out by cableships that have been specifically built or converted to carry and install the long lengths of cable required to connect countries and continents. Their crews are highly trained and specialized.

There are a limited number of specialized cableships available worldwide and they are required to be able to safely install cable and withstand the severe weather encountered across the world's oceans and seas.

The laying of trans-oceanic or festoon systems may require the cableship to remain at sea for extended periods. Most cableships are capable of carrying sufficient fuel, food and provisions, water and personnel to work 24 hours a day for two months.

Typically the vessels are 100–140 m in length, over 20 m beam[5] and are able to transit at a speed of at least 12 knots. Vessels of this size are capable of carrying

[5] The beam is the width of the vessel at its widest point.

Figure 5.2 Surface laying in rough weather.
(Photograph courtesy of Keith Ford-Ramsden)

4000–6000 tonnes of cable, which may be sufficient for a single trans-Atlantic lay, depending on cable types.

The ships are fitted out with cable tanks to store many thousands of kilometers of cable. (Figure 9.3.) The internal cone of the cable tanks must have a radius that is greater than the minimum bending[6] diameter of the cables being loaded (typically 3 m) and the outside diameter is dependent on the vessel's beam. The cables exit from the top of the cable-tanks and are guided via trackways of rollers known as cable highways to the Linear Cable Engine that controls (holds back) the pay out speed and/or tension of the cable over the ship's stern.

During the laying of the cable it is necessary to confirm that no damage has taken place during the installation process. In order to do this the cable is powered up from the ship and the fibers are monitored to ensure there are no faults. The powering of the cable creates a potentially lethal hazard to those working close to the cable onboard the ship. Power Safety Officers control and restrict access around the system whilst it is powered, as well as monitoring the system for faults. The cable system is depowered during any operations that require the cable, repeaters or any other bodies, such as equalizers, to be handled.

The laser light carrying the data in the fiber optic cable needs to be amplified every 60–80 km by repeaters. These repeaters are designed to operate on the seabed, and specialized temperature controlled repeater storage stacks are required to ensure they do not overheat during the period they are onboard prior to deployment.

6 The Minimum Bending Diameter is the minimum diameter the cable should have without the probability of damage to the optical fibers.

Cableships use two methods for installing cable:

i. plow burial, where the cable is simultaneously laid and buried at slow speed with the cable pay out being controlled so as to lay the cable on the seabed in front of the plow with minimal residual tension; and
ii. surface lay, where the cable is directly laid from the cableship onto the seabed.

The cables have to be laid at a speed that can exactly match the required pay out speed (slow speed for plow burial and surface laying of armored cable, i.e. 1–4 km/hour or ½–2 knots, and up to 11–15 km/hour or 6–8 knots for lightweight cable surface laying). The cable handling machinery used to lay and repair cable consists of a combination of a Linear Cable Engine (LCE) and one or two powered cable drums. The LCE, which is normally used for laying operations, uses up to 21 pairs of wheels mounted above and below the cable that can grip cable of varying diameters. The wheels rotate in the direction required to lay or recover the cable and have the ability to lay cable at up to 8 knots.

The tires fitted to the LCE have a limited holding power for higher tensions experienced during cable recovery and a cable drum is the alternative method for laying and/or recovery of cable. The cable drums are in the order of 4 m in diameter and have the capability of exerting 30–40 tonnes lifting tension. The cable laying machinery has to be able to react rapidly and accurately to changes in speed and tension requirements when installing a cable.

In order for a cableship to install the cable on the permitted and surveyed route they are fitted with Dynamic Positioning (DP) systems that automatically control the vessel's position, speed and heading by using the ship's rudder and powerful propellers and thrusters. The ship's position is accurately determined by Differential Global Position Systems and, together with inputs from sensors that measure the vessel's pitch and roll, wind speed/direction and the ship's heading, the DP enables the cableship to operate in various modes in order to 'hover' in one position, pull a plow with a tow force of up to 100 tonnes or move at speeds up to 11–15 km/hr (6–8 knots); all of these modes are required during a typical cable installation. The accuracy of this position keeping is in the order of a few meters and ensures that the cable is laid accurately on the planned route, with the correct amount of slack and residual tension. Note, however, that other factors, such as ocean surface and sub-surface currents, will influence how accurately the cable can be placed on the seabed.

The cable may, at various stages in an installation, need to be jointed to other sections of cable. This is a highly specialized discipline that requires exacting standards in order to prepare the cable ends, fusion splice the fibers together, terminate the elements of the cable and mechanically assemble the cable joint in a clean environment. The joint is then encapsulated in polyethylene by using specialized molding equipment to complete the lightweight joint and, if the cable is armored, an armored protection kit is fitted to ensure that 90 per cent of the tensile strength of the parent cable is maintained and to protect from crush forces in the area of the joint. At 6000 m the hydrostatic pressure is 600 times the atmospheric pressure that is experienced at the surface. Joints are x-rayed to detect any imperfections in the polyethylene mold.

Installation Operations Onshore or Near-shore

Cable Station, Land Cable/Fronthaul

The Landing Party will fit out the cable station to a standard required to receive the Supplier's Power Feed and Optical Transmission Equipment. Once the cable station has been accepted by the supplier, the shipment and installation of the Power Feed Equipment, Terminal Transmission and System Monitoring Equipment will commence.

The installation of the fronthaul ducting or route from the cable station to beach manhole needs to be prepared for the installation of the cables required for the power, ocean ground bed and fiber cables. In many cases, timely attention must be paid to coastal State importation, customs and tax requirements for components installed in the coastal State's territory. As explained in Chapter 3 and elaborated below, under the 1982 United Nations Convention on the Law of the Sea (UNCLOS), no such requirements should be imposed for cables outside of territorial waters.

Shore End Installation and Burial

The area from the beach manhole to the position where the cable vessel commences the main (majority) portion of the cable installation is known as the Shore End. The bathymetry, geology, weather and permits will be considered by the cable installer prior to deciding what type of Shore End operation will be required. There are three main types of cable landing as described below.

(i) *Direct Landing*
The cableship sets up as close to the beach manhole as is safe and practical in order to land the cable at either the start or the end of the lay. As with most procedures in cable laying there are a number of recognized methods to achieve the desired results. For direct landings the transfer of cable from the vessel to the shore can be achieved by one or a combination of the following:

(a) A direct pull with a rope from the shore using a winch or back-hoe/bull-dozer;
(b) Pulling the cable ashore with a work boat/small tugboat;
(c) Pulling the cable ashore with a rope that is passed around a sheave secured on the beach and back to the vessel.

During the pull in floats are tied to the cable so that it is floated ashore. When sufficient cable has been pulled ashore, it is secured and divers cut off the floats and the cable drops to the seabed. The divers then check that the cable conforms to the seabed profile.

The next task is for the installer to provide protection for the cable commensurate with the risk of damage that the cable faces in the area. The protection is usually a combination of additional armoring with articulated split metal pipes and/or Post Lay Burial (PLB) with high pressure water jetting by divers or ROVs.

Figure 5.3 Direct landing from a cableship.
(Photograph courtesy of Keith Ford-Ramsden)

(ii) *Directionally Drilled*

In some situations the landing permit may prohibit the disturbance of the beach and/or near-shore seabed, or the beach/near-beach topography may not be suitable for a standard direct landing. In these circumstances *horizontally directionally drilled* (HDD) or thrust-bored pipe may be used as an alternative and will need to be installed some distance inland to a position on the seabed that may be in excess of 1 km from the seabed offshore.

This type of cable landing operation will require different permits prior to carrying out the pull in. The operation will require the cableship to hold stationary within a very small footprint so that the cable can be pulled through the HDD pipe, whilst being assisted and monitored by divers. This is a very costly operation and is only resorted to if the direct or pre-laid options are not feasible.

(iii) *Pre-laid Shore End*

There are a number of circumstances where it may be necessary to lay a section of cable to a designated position offshore, where the main lay vessel will joint on to the pre-laid cable end. These circumstances include landings where:

(a) Shallow water precludes the main lay vessel getting close enough to the shore to carry out a direct landing and a shallow draft vessel or barge is used;
(b) Specialized burial requirements such as deep burial (8–15 m) by a barge with a vertical injector when the cable has to be routed through anchor-

Figure 5.4 Rocksaw ready for deployment off Singapore.
(Photograph courtesy of Global Marine Systems Ltd)

ages (such as off Hong Kong) or cutting of a trench by a rock saw/barge in hard seabeds, followed by installation of the cable in the trench;

(c) Permitting requirements may restrict access to landing sites during periods when the main-lay vessel is not available (e.g. to address environmental issues such as breeding seasons for fish, birds or animals and/or tourist seasons on particular beaches).

Installation Operations Offshore

Burial of Cables

The majority of damage to cables is caused by external aggression on the continental shelf.[7] This is because bottom fishing and ship anchors cause the majority of faults to subsea cables in water depths less than 200 m. (Figure 11.3.) The most effective method for protecting submarine cables from these hazards is to bury the cables below the seabed using sea plows that are deployed from cable vessels. Sea plows can weigh between 15 and 30 tonnes and are towed from a cableship in water depths from 10–1500 m. The cable passes from the stern of the cableship through the water column to the plow, where it is fed through into the bottom of a narrow furrow or trench cut by the plowshare.

The cable engine driver who operates the LCE or cable drum that lays the cable is instructed to pay out the cable at a specific tension that is slightly more than

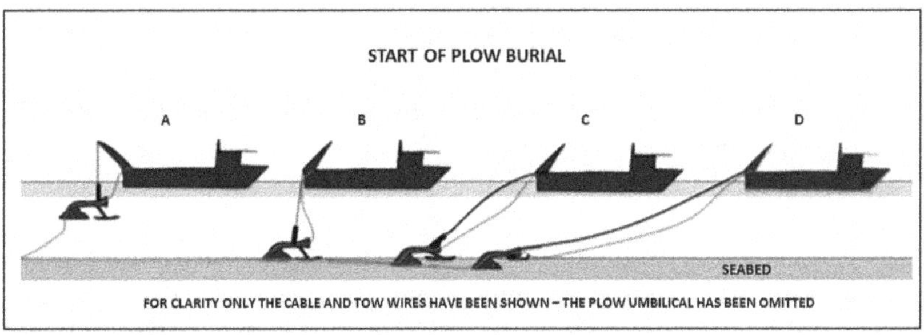

Figure 5.5 The initial deployment of the plow and start of burial.
A. Plow lowered with A frame outboard
B. A frame inboarded and plow landed on seabed
C. Grading in of plow starts
D. Catenaries set and plow to desired burial depth starts

[7] L. Carter *et al.*, "Submarine Cables and the Oceans: Connecting the World" (2009) Report of the United Nations Environment Program and the International Cable Protection Committee (UNEP-WCMC-ICPC) at 34, available at http://www.iscpc.org/publications/ICPC-UNEP_Report.pdf (last accessed 11 May 2013). Also see the discussion in Chapter 11 on "Protecting Submarine Cables from Competing Uses."

the weight of the cable between the ship and the plow. The aim is to install the cable in the bottom furrow with the minimal amount of residual tension, otherwise the cable may be pulled out of the trench as the cable is buried in an undulating seabed. If too much cable is paid out excess cable will be laid on the seabed in front of the plow and the plow skids will run over and damage the cable.

The usual target for burial of cable to protect against bottom fishing and trawling is 1.0–1.5 m, but this may be increased in softer seabeds to 3 m because of the type of fishing activities that take place in the area. Waters off China and South Korea where stow net fishing is prevalent require this deep burial. (Figure 6.8.)

The cable route is carefully selected to try and maximize the protection for the cable, where necessary, by plowing. However, plows are not capable of burying the cables in all types of seabeds, such as bedrock and rugged rocky areas. In these circumstances the plow is either recovered to deck or lifted off the seabed and 'flown' over the hazard and the cable is surface laid on the seabed. The plow will be landed back on the seabed and plow burial will be resumed as and when it is practicable.

The plow will also be lifted and 'flown' over or recovered to deck for the crossing of other cables and pipelines and re-deployed at a predetermined distance past the crossing. As part of the crossing agreement between the cable owner and pipeline owner it may be necessary to have additional protection laid over the pipeline at the crossing and/or to apply polyurethane half shells to the cable to protect the pipeline from the cables armor wires. The exact type of protection and distances that the plow is flown will be determined by the pipeline or cable crossing agreements entered into by the parties.

If the cable has been surface laid in areas where burial is required then Post Installation Burial will be carried out. This may occur in planned locations such as crossings, or unplanned locations for a variety of other reasons. The Post Installation Burial will be carried out either by divers in very shallow water or by a tethered ROV using high-pressure water jetting swords.

As the cableship nears the end of the area that requires burial to protect against potential external aggression, the cable system is depowered and the plow recovered to deck. The armored cable is removed from the plow and the plow is moved clear in preparation for surface laying.

Surface Laying of Cables

The route planning and survey will have identified and recommended areas in which cables should be protected, by either plow[8] or ROV burial.[9] The other sections of the cable route to a maximum water depth of 2000 m will be surface

[8] A plow can achieve burial down to a maximum water depth of 1000–1500 m, providing the seabed slopes on the continental shelf are not excessive.

[9] ROVs can operate in water depths of around 10 m down to 2000–3000 m.

laid with armored cable which provides protection commensurate with the risk to the cable in that area. An ROV is used if burial protection is required below the plow's maximum operational depth or limits. (Figure 7.4.) Areas such as the Red Sea have little plow burial, but long sections of surface laid armored cable. The surface laying of armored cable is carried out at relatively slow speeds and under tension control so as to allow the cable to conform to the seabed contours. In this way if PLB is required there will be sufficient slack to bury the cable, however, excessive slack is not desirable as this can lead to the forming of loops in the cable during the lay.

When the protection afforded by the single armor (SA) cable is no longer required there would be a transition to Lightweight Protected (LWP), Lightweight Screened (LWS) or Special Applications (SPA) cables. (Figure 1.4.) These are lightweight (LW) cables with an additional layer of high-density polyethylene sheath and, in the case of LWS and SPA, an additional metallic screen to provide protection from fishing hooks, abrasion and fishbites from sharks. These cable types are used in water depths of 1000–3000 m. LWP is normally surface laid in areas where the medium density polyethylene outer of the LW cable does not afford adequate protection. The positions of all transitions, repeaters and joints will have been inputted into the Lay Plan and Route Position Lists and the installer and cable owner will have verified that the seabed at the planned positions is suitable.

A further transition is required to change from LWP to LW cable, which has some limited resistance to abrasion and is predominantly used for the long haul distances across the benign ocean floor in depths ranging from 2000–8000 m.

The method utilized for surface laying of LW cable over long distances and higher speeds is to manage the slack. The slack is the difference between the distance travelled and the amount of cable laid. The correct amount of slack will ensure that the cable conforms to the seabed contours with minimal residual tension, no loops and will minimize the risk of any cable suspensions.

The ideal steady state for the surface laying of a cable in 6000 m of water onto a flat benign seabed is 12 km/hour (6.5 knots). Because the cable can only sink at its 'terminal velocity' it will take approximately 4 hours for the LW to touch down on the seabed at a distance of some 48 km from the stern of the cableship. The seabed is not flat but has minor undulations, therefore around 1–2 per cent of slack is allowed for to ensure that the cable conforms to the seabed contours and there are minimal suspensions.

As can be seen from Figure 5.6, the cable is at a steady state shallow angle to the seabed at this speed (ship A) and the cable would be in suspension if this steady state were to be maintained when approaching a seabed rise, as shown by the dotted-line in Figure 5.6. By slowing the vessel's speed, the angle between the seabed and cable increases until the desired angle that is a few degrees greater than the slope is achieved (position B) and the cable is laid on the seabed conforming to the upward slope until the top of the slope (position C).

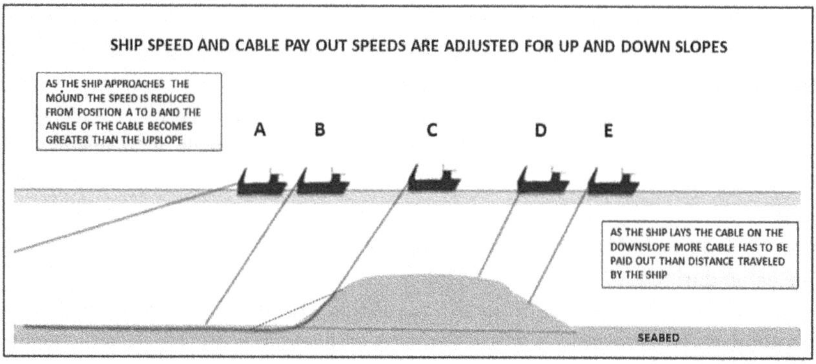

Figure 5.6 Adjustments to surface lay over subsea mounds.

On the downward slope from position D the speed is maintained and additional cable has to be paid out to impart more slack to fill the down-slope.

The processes depicted above usually apply to the laying of LW, SPA or LWP cable, however the cable system will also include repeaters, joints and equalizers that are heavy bodies and these will change the angle of the cable at the cable-ship once they enter the water column. Hence the only way to effectively lay a cable system is to pre-plan the whole of the surface lay with a Slack Plan which takes account of the slack required for each section of the lay, the positioning of the repeaters and other bodies, and the ship and cable pay out speeds. The plan must allow for surface laying of armored cable which is often carried out in 'tension' mode to avoid loops in the cable. This cable lay engineering function used to be undertaken with manual calculations but is now undertaken with computer models and full simulations of the lay which identify any potential issues for the cable engineers. Expert cable engineering knowledge is still required to input the data into the cable lay computer, monitor the lay and make the necessary adjustments during the lay, as and when required.

Branching Units

The traditional design of systems used to be that of a point to point, similar to a festoon system. This meant that all cable landing stations, except at either end of the system, had a double cable landing.

This is an expensive option, with two cable landings, extra cable and repeaters; it also means that all traffic that is sent from the initial Cable Landing Station to the final Cable Landing Station has to pass through every Cable Landing Station in the system. This increases latency and gives rise to data security issues.

Branching units (BUs) provide the system designer with an alternative. With repeatered systems offering a maximum of four or eight fiber pairs it is possible to have express traffic between any combinations of the cable stations. The fibers in the BUs can be either physically routed or multiplexed to add and/or drop wavelengths, depending on the customer's needs.

Figure 5.7 Branching unit deployment.
(Photograph courtesy of Keith Ford-Ramsden)

The deployment of the BUs is a complex operation that involves laying one of the three cables on the seabed and buoying it off, then laying the second cable and maneuvering and picking up the cable buoy and cable from the first cable. The cableship and cables are adjusted so they form the top end of the Y, the cables are then cut to length and jointed to the BU. The BU is then lowered to the seabed as the third end of the BU is laid and the installation of the third leg of the cable is commenced.

Post Lay Burial and Inspection

After the completion of the main lay on the continental shelf there will be a number of planned and unplanned sections of cable that have not been buried.

The planned areas will include cable and pipeline crossings, and a Dynamic Positioning vessel fitted with an ROV will undertake burial of the cable in these areas. The ROV is fitted with a set of high pressure jetting swords that either cut a trench or fluidize the seabed so that the cable drops to the bottom of the trench.

A certain percentage of the cable will be inspected by the ROV to confirm whether the target burial depth has been achieved and other areas where burial was not achieved will be Post Lay Buried by the ROV and the revised burial depth will be inspected.

Weather and Currents

Humans have been able to populate some very harsh climatic environments in the world and many areas with large populations now require voice and data traffic. If these populations are close to the sea or ocean a submarine telecommunications cable will probably be necessary. The cable installer must have the capability to meet this demand and be able to install cable in extremes of temperature and high local sea currents close to shore.

As with all marine operations the weather conditions are always monitored and a close eye is kept on the weather forecasts to ensure there is an adequate weather window to carry out the installation related work. Although the modern cableship is very seaworthy, the extreme weather and sea conditions caused by tropical revolving storms or other extreme weather may well result in the cableship cutting the cable and seeking shelter. In this matter the role of the ship's captain has not changed in hundreds of years and he or she always has the final say on the safety of the ship and its personnel.

V. Law and Policy Challenges with Respect to Laying Operations

UNCLOS governs cable laying operations and an overview of the relevant provisions is provided in Chapter 3.[10] For the present purposes of this Chapter, it suffices to say that the extent to which UNCLOS allows the coastal State to regulate cable laying operations will depend on where the operations take place, namely whether they take place in territorial seas/archipelagic waters (territorial waters), the exclusive economic zone (EEZ)/continental shelf or high seas/deep seabed. The following sections address the various issues relating to coastal State regulation of cable laying operations in the different maritime zones.

Coastal State Regulation within Territorial Waters

Pursuant to its sovereignty over its territorial sea and/or archipelagic waters,[11] coastal States and archipelagic States clearly have extensive authority to regulate ships engaged in laying operations. However, some coastal States have imposed excessive regulations on such operations which have resulted in significant delays

---

[10] The 1884 Convention for the Protection of Submarine Telegraph Cables does not address the laying of cables.

[11] UNCLOS Art 2 (territorial sea) and Art 49 (archipelagic waters).

and costs to the cable owner/operator thereby undermining the integrity and reliability of global and regional telecommunications systems. Some examples of such regulations are found below.

Lengthy, Costly and Unpredictable Permitting Processes

As noted above, coastal States require permits or licenses before a cable can land in their territory and before laying operations can take place within territorial waters, even if the cable is transiting territorial waters and not landing (See Part I on Applying for Permits). The Telecommunications Operations License or Landing License issued by the relevant national telecommunications authority can itself take several months.[12] Before a Permit in Principle or Operational Permit for laying is granted, other permits before may also be required, "including defense or national security authorizations, environmental permits and permits for construction and land use".[13] Further, it is often difficult to ascertain the relevant procedures for the application of permits for the laying of cables and such information is not publicly available.[14]

The United States regulation of submarine cables is a good illustration of how permitting processes can be lengthy and complex. Generally speaking, there are a number of different permits and licenses required, depending on where the cable lands and whether it passes through environmentally sensitive areas.[15] First, all submarine cable operators must be granted a cable landing license from the Federal Communications Commission (FCC) for the installation and operation of any undersea cable in US Territory, pursuant to the Cable Landing License Act of 1921.[16] The FCC must seek the views of the US Department of State, the US Department of Commerce's National Telecommunications and Information Administration and the Defense Information System Agency.[17] An additional

---

12 A.D. Lipman and N.T. Vu, "Building a Submarine Cable: Navigating the Regulatory Waters of Licensing and Permitting", (March 2011) 56 *Submarine Telecoms Forum*, 21–24, at 21 available at http://www.bingham.com/Publications/Files/2011/04/Building-a-Submarine-Cable-Navigating-the-Regulatory-Waters-of-Licensing-and-Permitting.

13 *Ibid.*

14 For example, in Hong Kong, it has been observed that "the cable industry may find it difficult to get hold of necessary information in respect of the application procedures and statutory approvals for landing a new submarine cable in Hong Kong". See Hong Kong, Legislative Council Panel on Information Technology and Broadcasting, "Landing of Submarine Cables in Hong Kong" 8 March 2010, LC Paper No CB(1) 1289/09-10(04), which also observed that "there was a need to increase the transparency of the application process".

15 Lipman and Vu, *supra* note 12, at 21.

16 An Act Relating to the Landing and Operation of Submarine Cables in the United States codified at 47 USC Sections 34–39; Executive Order 10, 530, reprinted in 3 USC Section 301; 46 CFR Section 1.767.

17 *Ibid.*, 47 CFR Section 1.767 (j).

authorization may also be needed if the submarine cable system is operated on a common carrier basis.[18]

Second, for undersea cables connecting the US with foreign points or with significant foreign ownership, the cable application has to go through a review by 'Team Telecom' which consists of the US Department of Defense, Homeland Security and Justice and the Federal Bureau of Investigations.[19] The Team Telecom review asks a series of questions relating to the storage and security of call data and other information and will ask the FCC to defer granting the license application until Team Telecom has finished its review.[20] They also often request the FCC to impose security-related conditions in the cable landing license in order to assure both infrastructure security and information security.[21] Given that the FCC will not grant a landing license until Team Telecom has given its approval, it takes as long as six months for a license to be issued.[22]

Third, in addition to the FCC cable license, submarine cable systems must also obtain a federal permit from the Army Corps of Engineers (ACOE), pursuant to the Clean Water Act[23] and the Rivers and Harbor Act.[24] In some cases, these cables are authorized under the ACOE's Nationwide Permit Program.[25] In other cases, individual permits are required, and the ACOE must complete an environmental review under the National Environmental Policy Act before issuing the permit.[26] The ACOE is also required to consult with the applicable federal resource agencies under the Endangered Species Act if the project affects protected species.[27] Further, if the cable falls in an area over which a state also has jurisdiction, a consistency determination must also be issued by the state which stipulates that the activities authorized under the permit are consistent with the state's Coastal Zone Management Plan.[28]

Lastly, even if all the Federal permits have been obtained, it may also be necessary to obtain state and local permits, depending on where the cable lands. For example, according to Lipman and Vu, California has some of the most onerous

---

18 See Section 214 of the Communications Act of 1934; Lipman and Vu *supra* note 12 at 22.

19 See Comments of the North American Submarine Cable Association Before the Bureau of Ocean Energy Management, US Department of the Interior In the Matter of Atlantic OCS Proposed Geological Services, Mid-Atlantic and South Atlantic Planning Areas and Draft Programmatic Environmental Impact Statement, OCS EIS/EA BOEM 2012-005, 30 May 2012, at 13.

20 Lipman and Vu *supra* note 12 at 23.

21 Comments of the North American Submarine Cable Association, *supra* note 19.

22 Lipman and Vu *supra* note 12 at 23.

23 Section 404, Clean Waters Act.

24 Section 10, Rivers and Harbor Act of 1899.

25 Comments of the North American Submarine Cable Association, *supra* note 19.

26 Lipman and Vu *supra* note 12 at 23.

27 *Ibid.*

28 Section 307(c)(1) of the Coastal Zone Management Act.

permitting requirements of any state in America and getting an Environmental Impact Report is a process that can take several years, particularly if there is local opposition.[29]

It is evident from the above that there are many steps that have to be taken to install a cable in the US, a country which has a developed infrastructure and regulatory regime. One can imagine that obtaining licenses and permits from developing countries where there are less developed regulatory regimes, less transparency and more bureaucratic red-tape, can be infinitely more difficult.[30] India, for example, has one of the most complicated licensing regimes for the installation of submarine cables. One of the reasons for this is that India perceives telecommunications as a source of security threats.[31] Another reason is that "India sees a vibrant telecommunications services sector operating largely with equipment manufactured and tested outside of India using technologies developed outside of India, inflaming sensitivities about the need for self-sufficiency and about the comparative success of China in encouraging domestic innovation and manufacturing of electronic equipment".[32] To address these concerns, the Indian government has amended the regulations governing the licenses granted to Indian telecommunications providers and internet service providers "imposing new telecom equipment security requirements and proposing a variety of measures to encourage or require development, manufacturing and testing of equipment in India".[33] For example, it is presently a requirement that the equipment vendor employs an Indian national with a security clearance as its security contact.[34]

Aside from these licensing requirements, what is more troubling for the cable industry is the number of permits needed to lay and repair a cable within India's maritime zones. It is reported that no less than seven different permits are required from various agencies before a vessel can conduct laying or repair operations in Indian territorial waters and EEZ.[35] These include: (1) a Research, Survey, Exploration and Exploitation Permit from the Ministry of Defense; (2) a Ministry

29 Lipman and Vu *supra* note 12 at 24.

30 *Ibid.*, at 21.

31 This is attributable to firstly, India's fear of terrorism in India's maritime zones (for example, the 2008 terrorism attacks in Mumbai originated from the sea) and secondly, the terrorists in that attack used mobile phones and VoIP to direct the attacks. See K. Bressie and M. Findley, "Coping with India's New Telecom: Equipment Security Requirements and Indigenous Innovation" (March 2102) 62 *Submarine Telecoms Forum* 15–19 at 16 available online at http://www.subtelforum.com/articles/wp-content/STF-62.pdf and R. Rapp *et al.*, "India's Critical Role in the Resilience of the Global Undersea Communications Cable Infrastructure" (May–June 2012) 36(3) *Strategic Analysis*, 375–383 at 378–379.

32 See Bressie and Findley *supra* note 31 at 16.

33 *Ibid.*

34 *Ibid.*, at 18.

35 Rapp *et al.*, *supra* note 31 at 380–381.

of Home Affairs Clearance for all personnel and crew involved; (3) a Specified Period License from the Director-General of Shipping; (4) clearance from the Indian National Shipowners Association; (5) naval clearance from the Flag Officer Offshore Defence Advisory Group; (6) Crew Visas from the Immigration Office and (7) Vessel Importation from the Customs Office.

'Localization' Requirements

Adding to the lengthy and costly permitting requirements, is a requirement imposed by some States that the crew and/or the vessel carrying out cable operations have the same nationality of the coastal State. India, for example (apart from the already onerous permitting requirements described above) has imposed a requirement that a minimum of 1/3 of the crew need to be of Indian nationality on foreign vessels granted a license to operate in Indian coastal waters beyond 30 days and a minimum of ½ of the crew need to be of Indian nationality if the foreign vessel is operating in Indian coastal waters beyond 90 days.[36]

Indonesia has imposed similar requirements. In 2008, in a bid to boost its shipping industry, Indonesia introduced Cabotage Regulations, which allow only Indonesian-flagged vessels to carry cargo or passengers from one Indonesian port to another.[37] These regulations applied to vessels carrying out cable operations.[38] While the cable industry obtained a short reprieve through an exemption from the full application of the Regulations,[39] foreign cableship owners must meet several conditions before they can operate in Indonesian waters.[40] This includes

---

36 See Shipping Development Circular No. 1 of 2013 on "Conditions for grant of license to foreign flag ships under sections 406 and 407 of the Merchant Shipping Act 1958, laying down norms for engaging Indian crew and trainees on board these ships engaged in shipping and related activities in Indian coastal waters" issued on 18 January 2013 by the Directorate General of Shipping, Ministry of Shipping, Government of India.

37 K. Hasan, "Indonesia Tells Owners to Shape up or Ship Out" (28 September 2011) *Lloyds List.*

38 See generally, K. Bressie and M. Findley, "Indonesia's 2008 Shipping Law: Unintended Harms to Undersea Cable Installation and Maintenance" (May 2011) 5 *Submarine Telecoms*, 28–32 available online at http://www.wiltshiregrannis.com/sitefiles/news/f82af38b23b5f8b61aaf27bba8575103.pdf (last visited 7 June 2013).

39 The concerns of the oil and gas industry about the lack of Indonesian vessels able to carry out the highly specialized operations of hydrocarbon exploration and exploitation and concern on the potential impact on the Indonesian economy, resulted in an amendment to the 2008 Cabotage Regulations which allows foreign vessels to carry out specified activities in Indonesian waters provided they comply with certain conditions: see Hasan *supra* note 37; R. Kusuma, "Enforcing the Cabotage Law Could Cost the Country Billions" (2 March 2011) *Jakarta Globe.*

40 See Article 5 of Regulation of the Minister of Transportation No. 48 Year 2011 Regarding the Procedures and Requirements to Grant on the Use of Foreign Vessels for Other Activities that do not Include Activities of Transporting Passenger and/or Goods in the Domestic Sea Freight Activities, 18 April 2011.

having to prove that: (1) a minimum one-time effort has been made to procure an Indonesian-flagged vessel or if it is unable to do so, to enter into a charter with a national shipping company[41] and (2) that the owners of the foreign vessel have submitted certain documents, including a work plan and a schedule of activities, a charter party between national sea transport companies and foreign shipowners (if applicable) and the relevant certificates of the vessel.[42] Once the permit is granted, it lasts for a period of only three months, although this can be extended by the director-general of sea communications of the Ministry of Transportation.[43] In any event, the exemption will lapse in December 2013, by which time only Indonesian vessels or crew will be allowed to conduct cable operations in the territorial waters of Indonesia.[44]

Given that Indonesia does not presently have any of these highly specialized cableships under its flag, cabotage regulations may have a significant impact on laying operations in Indonesia's waters. Even if Indonesia is successful in compelling a single ship to fly its flag, it will be inadequate to deal with the high number of faults experienced in Indonesia archipelagic waters. Cable companies and neighboring States have every reason to be concerned, given that Indonesian archipelagic waters are home to numerous cables that serve Indonesia, Australia, India, Malaysia, Myanmar, Singapore, South Africa and Thailand.[45] Moreover, this is a negative precedent for the reliability and cooperation that is characteristic of the modern international cable infrastructure. If every nation insists on only allowing cableships flying its flag to repair cables landing in that country or in its waters, the current efficient, flexible and proven cable maintenance zone agreements described in Chapter 6 which allow many cable systems to share the cost of cableships will be undermined.

Implications of Excessive Coastal State Regulations in Territorial Waters

The above-mentioned regulations are not inconsistent *per se* with the rights of the coastal State in territorial waters and the desire of coastal States to regulate activities in areas near their coasts is understandable. Indeed, UNCLOS recognizes that in territorial waters, coastal States have a right to take measures to protect the marine environment and their security interests, as well as to ensure minimum interference with competing activities under their jurisdiction.[46]

---

41 Bressie and Findley *supra* note 38 at 30.

42 *Ibid.*

43 *Ibid.*

44 *Ibid.*

45 *Ibid.*, at 31–32.

46 For example, UNCLOS Art 19 recognizes that certain activities of a foreign ship in the territorial sea are prejudicial to the peace, good order and security of the coastal State including any act of propaganda aimed at affecting the defence or security of the

That said, the number of permits that need to be obtained, coupled with the fact that in many States there is no lead agency in charge of coordinating application procedures, can lead to significant delays and increased costs to cable owners/operators in cable installation. As explained above, while cable owners/operators are well aware of the importance of determining the applicable permitting processes as early as possible and often build-in the amount of lead time in the overall project for permitting processes, there is no doubt that construction of new submarine cables can be hindered by onerous permitting requirements which stand as "roadblocks to the rapid deployment of new international submarine cable systems".[47] It is a fact that in deciding cable routes, the difficulty, unpredictability and costs of permitting are key factors taken into consideration by cable owners. In some cases, unreasonable permitting delays, uncertainty and costs have led cable owners to by-pass one State in favor of a coastal State with predictable and reasonable permitting requirements. This is certainly the case for the state of California in the United States[48] and recently with India. For example, operators of the SEA-ME-WE-5 Cable System decided not to land in India due to a decision by the Indian government to impose an extensive bond before foreign submarine cables can land in India, a decision which will no doubt also have a detrimental impact on route diversity and connectivity of less developed countries.[49]

Coastal State Regulation of Submarine Cables in the EEZ/Continental Shelf

The Freedom to Lay, Repair and Maintain Cables under UNCLOS

In contrast to territorial waters, UNCLOS recognizes that all States have the freedom to lay submarine cables in the EEZ[50] and on the continental shelf.[51] It has been noted that the right to lay submarine cables given to "all States" should not be read restrictively as "in practice many submarine cables and pipelines are privately owned and are laid by corporations or other private entities. The term therefore refers to the right of States or their nationals to lay cables and pipelines".[52]

---

coastal State, any act of wilful and serious pollution contrary to the Convention and any fishing activities.

47 Lipman and Vu *supra* note 12 at 21.

48 See the North American Submarine Cable Association's Recommended Changes to "State Planning and Evaluation Guidelines for Submarine Fiber Optic Cables for Hawaii", 11 October 2002, at 2 [personal copy with authors].

49 See S. Tagare, "SEA-ME-WE-5 not to land in India" 17 February 2012 available online at http://blog.buysellbandwidth.com/sea-me-we-5-not-to-land-in-india/ (last accessed 7 June 2013).

50 UNCLOS Arts 58 and 87(1).

51 UNCLOS Art 79(1).

52 M. Nordquist *et al.*, eds, *United Nations Convention on the Law of the Sea 1982: A Commentary*, Volume III (Martinus Nijhoff, 1995) at 264.

States or companies that do wish to conduct laying operations in the EEZ or continental shelf of another State have certain obligations under UNCLOS. First, such States or companies must have due regard to the cables and pipelines already in position and must not prejudice the possibilities of repairing existing cables or pipelines.[53] Second, States or companies exercising the right to lay cables in the EEZ (and in the continental shelf to the extent it overlaps with the EEZ) must have "due regard" to the rights and duties of the coastal States recognized in the EEZ.[54] The "rights and duties" of the coastal State refer to the rights and duties in Article 56, namely rights over the exploration and exploitation of living resources; non-living resources; other economic resources such as the production of energy from the water, currents and winds as well as jurisdiction over artificial islands, installations and structures, marine scientific research and the marine environment. Third, States or companies "shall comply with the laws and regulations adopted by the coastal State in accordance with the provisions of [UNCLOS] and other rules of international law in so far as they are not incompatible with this Part [on the EEZ]".[55] The question is what "laws and regulations" can the coastal State impose on cable laying operations.

The Right of Coastal States to Regulate Cable Operations under UNCLOS

First, Article 79(2) of UNCLOS makes clear that a coastal State may subject cable operations only to its right to take reasonable measures for the exploration of the continental shelf and the exploitation of its natural resources, and **not to** reasonable measures for the prevention, reduction and control of pollution from pipelines.[56] Second, coastal States cannot adopt regulations on the delineation of the cable route. Article 79(3) provides that "the delineation of the course for the laying of such pipelines on the continental shelf is subject to the consent of the coastal State" which suggests that coastal State consent for the delineation of a cable is not required.[57] This distinction between a pipeline and a cable reflects the benign environmental impact of a cable fault.[58] Unlike the case of a leaking or

53 UNCLOS Art 79(5).
54 UNCLOS Art 58(2).
55 UNCLOS Art 58(3).
56 Article 79(2) distinguishes between pipelines and cables. It is only in respect of pipelines that a coastal State is permitted to impose reasonable measures for: (1) the exploration of the continental shelf; (2) the exploitation of its natural resources and (3) the prevention, reduction and control of pollution from pipelines.
57 This is supported by the legislative history of this provision: See M. Nordquist *et al.*, eds, *The United Nations Convention on the Law of the Sea 1982: A Commentary*, Volume II (Martinus Nijhoff, 1993) at 915. See also Chapter 3 Overview of the International Legal Regime Governing Submarine Cables.
58 See Chapter 7 on the Relationship between Submarine Cables and the Marine Environment.

ruptured oil pipeline, there is no environmental impact when a cable is damaged, only a loss of communications or power.

UNCLOS also imposes certain procedural requirements on the coastal State when imposing resource-related measures on the laying of cables. First, the measures must be reasonable.[59] Second, in the EEZ, the coastal State must have due regard to the rights and duties of other States and shall act in a manner compatible with the provisions of UNCLOS.[60] Third, on the continental shelf a coastal State must not exercise its rights in a manner that will infringe or result in "any unjustifiable interference" with the rights and freedoms of other States as provided for in UNCLOS.[61]

The major issue for laying operations in the EEZ/continental shelf is that some coastal States have adopted laws and regulations which are arguably inconsistent with UNCLOS. These regulations can delay laying operations and consequently may seriously undermine the connectivity of the world's telecommunications systems. The following paragraphs set out some examples of laws and regulations that are not consistent with UNCLOS.

Coastal State Consent for the Delineation of Cable Routes

A surprising number of States have treated cables and pipelines in the same way and have adopted regulations which require the consent of the coastal State for the delineation of cable routes along with pipelines.[62] Coastal States may of course argue that the delineation of cable routes is a "reasonable measure related to their resource exploration and exploitation rights" under UNCLOS as

---

59 While it is not clear what is meant by reasonable, "no more definite criterion than that of reasonableness could be established for the measures which coastal States may take, for the reason that it was impossible to foresee all situations that might arise in the application of this article": Statement by the US Representative during the Eighth Session of the International Law Commission cited in M. Whiteman, "Conference on the Law of the Sea: Convention on the Continental Shelf" (1958) 52 *American Journal of International Law* at 642.

60 UNCLOS Art 58(2).

61 UNCLOS Art 78(2).

62 Examples of States that subject the delineation of cable routes to their consent include China, Provisions Governing the Laying of Submarine Cables and Pipelines, Article 4, Decree No 27 of the State Council of the People's Republic of China, 15 February 1989, available at Law Info China Web site (for subscribers only); India, Indian Territorial Waters, Continental Shelf, Exclusive Economic Zone and other Maritime Zones Act, 1976, Act No 80 of 28 May 1976, Article 7(8), UN Division of Ocean Affairs and Law of the Sea (DOALOS), available at www.un.org/Depts/los/LEGISLATIONANDTREATIES; Malaysia, Exclusive Economic Zone Act 1984, Act No 311, Section 22, DOALOS, available at www.un.org/Depts/los/LEGISLATIONANDTREATIES; Saint Lucia, Maritime Areas Act, Act No 8 of July 18, 1984, Section 13(2), DOALOS, available at www.un.org/Depts/los/LEGISLATIONANDTREATIES; and Uruguay, Act No 17,033, 20 November 1998, Article 12, DOALOS, available at www.un.org/Depts/los/LEGISLATIONANDTREATIES.

this requirement ensures that there is minimum interference with those activities. As explained above, however, this is clearly contrary to UNCLOS which only imposes coastal State consent on the delineation of pipelines.[63] Further, it should be borne in mind that during the desktop and cable route surveys done prior to laying a cable, every effort is made to avoid areas where there are intense competing uses, such as fishing and anchoring or other environmental considerations.[64] This should assuage coastal State concerns on the possible interference a cable may pose to other competing activities.

Permit Requirements

Another type of regulation arguably inconsistent with UNCLOS is the requirement that cable laying ships obtain permits before laying operations can take place in the EEZ and/or continental shelf, contrary to the freedom to lay cables recognized in these zones. States that require permits or their consent for the laying of submarine cables include China, Cyprus, Guyana, India, Malaysia, Mauritius, Pakistan, Portugal, Russia, Saint Lucia, the United States and Uruguay.[65] Not only are these permits *prima facie* inconsistent with UNCLOS, the permitting processes in States such as India are also complex and lack transparency, which causes further delays to laying operations.[66]

Coastal States may argue that permits are "reasonable measures" for the exploration of the continental shelf and the exploitation of its natural resources allowed under Article 79(2) of UNCLOS. This is on the basis that permitting requirements are necessary to ascertain that (a) foreign cable laying ships are not engaging in exploration or exploitation activities and (b) to ensure that cable laying activities do not interfere with fishing, hydrocarbon exploration, or exploitation activities and *vice versa*.[67] However, there are also weaknesses in this argument. First, on the continental shelf/EEZ, coastal States can only regulate exploration/exploitation activities, and permits for cable laying fall outside their jurisdiction. Second, it is undeniable that such permit requirements are *prima facie* inconsistent with the freedom to lay cables in the EEZ/continental shelf recognized in UNCLOS. Third, such permitting processes, particularly those that are lengthy and complicated, arguably do not meet the procedural obligations that coastal States have when imposing such measures in the sense they are not reasonable, are contrary to the obligation of exercising due regard to the rights of other States

---

63 UNCLOS Art 79(3). However, because submarine cables and pipelines are dealt with together in this article, coastal States may be under the misapprehension that cable route delineation is also subject to their consent.

64 See Chapter 4 on the Planning and Surveying of Submarine Cable Routes.

65 J.A. Roach and R.W. Smith, *Excessive Maritime Claims* (3rd ed, Leiden, 2012) at 461.

66 See Rapp et al., *supra* note 31.

67 T. Davenport, "Submarine Communications Cables and Law of the Sea: Problems in Law and Practice" (2012) 43 *Ocean Development and International Law* 201–242 at 212.

in the EEZ and also constitute unjustifiable interference with the rights and freedoms other States under UNCLOS.[68]

Taxes on Submarine Cables/Cable Operations in the EEZ/Continental Shelf

Some States, such as Malta, charge an annual license fee or tax on submarine cables which do not enter into Malta's territorial waters or land but do transit waters outside territorial waters.[69] For example, the Europe India Gateway (EIG) cable system is a 15,000 km submarine telecommunications system co-owned by a consortium of 18 companies and links the United Kingdom with Gibraltar, Portugal, Monaco, France, Libya, Egypt, Saudi Arabia, Djibouti, Oman, United Arab Emirates and India.[70] Segments of the cable pass through Malta's continental shelf without entering Malta's territorial sea or contiguous zone. As a term of the license to lay cables on the continental shelf (a license that is already inconsistent with UNCLOS), Malta has demanded an annual fee for the life of the cable.[71]

Similarly, India has "imposed excessive customs duties and taxes on undersea operators and their suppliers and maintenance providers" which apply in India's EEZ.[72] For example, India's Central Board of Excise and Customs have assessed customs duties on all goods imported in the EEZ (even temporarily), as well as all services provided within the EEZ, which apply to cable vessels conducting laying operations within the EEZ.[73] Further, Indian Customs have also claimed an Indian Services Tax (in addition to the customs duties mentioned above) on the value of services provided within the EEZ and, since 2009, has assessed its service tax on installations, structures and vessels in its EEZ.[74]

There is nothing in UNCLOS that allows coastal States to require licenses, let alone impose fees or taxes for cables which transit the continental shelf, or for cable operations in the EEZ/continental shelf. The imposition of a tax or duties is clearly unrelated to any resource rights that a coastal State has over its EEZ/continental shelf.[75] If other coastal States followed the example of India and Malta, particularly for international cables that do not enter territorial seas but merely

---

68 *Ibid.*

69 Roach and Smith, *Excessive Maritime Claims, supra* note 65 at 462.

70 See EIG Website available online at europeindiagateway.com/webclient/common/html/aboutus.html (last accessed 2 April 2013).

71 Apparently, Malta relies on Section 8 of its Continental Shelf Act of 29 July 1966 as amended to justify the imposition of such fees. However, Section 8 of its Continental Shelf Act only provides that no person shall lay or maintain any submarine cables in the high seas without a license and imposes a fine for any contravention of this provision. There is no clear reference to an annual fee.

72 Bressie and Findley, *supra* note 31 at 16.

73 *Ibid.*, at 17.

74 *Ibid.*

75 Supreme Court (Contentious-Administrative Decision, 5th Chamber) Ruling of 16 June 2008, JUR 2008/211246, *Telefónica de España S.A. v Ministry of the Environment.*

transit the EEZ or continental shelf, the freedom to lay and maintain international cables would be effectively nullified.

Implications of Excessive Coastal State Regulations in the EEZ and Continental Shelf

Unlike coastal State regulations in respect of submarine cables in territorial waters, the above-mentioned regulations are not consistent with the rights of the coastal State in the EEZ and the continental shelf. From a practical perspective, such regulations can cause significant delay and/or costs to cable owners/operators during cable installation. As explained above, while cable owners/operators are usually cognizant of the regulations of a particular coastal State in the EEZ/continental shelf and the time and costs are generally built into project costing and schedules, the construction of new submarine cables can be hindered by onerous permitting requirements and may result in some cable owners/operators bypassing these States.

Apart from the practical consequences of excessive permitting requirements for the laying of cables, there are also wider implications. Such regulations on cable operations in areas outside of territorial sovereignty can be seen as part of a wider trend of what has been described as the 'territorialization' or 'creeping jurisdiction' of coastal States in the EEZ.[76] This trend is hardly surprising or unpredictable. The EEZ has always been perceived in "quasi-territorial terms".[77] However, the overarching objective of UNCLOS was to limit the expanding nature of coastal State jurisdiction through a series of carefully constructed compromises. These compromises took into account coastal State interests, the interests of other States as well as the interests of the wider community. Regulations which are contrary to UNCLOS threaten the substantive balance UNCLOS sought to achieve and put the entire legal order of the sea at risk.

Permit Requirements in Disputed Areas or Areas where Boundaries are Undefined

Another issue that often confronts the cable industry is that of overlapping or disputed maritime areas. UNCLOS allows States to claim maritime zones where they are given certain rights and jurisdiction, particularly over resources. In view of these numerous benefits to coastal States, particularly in terms of control over fisheries and hydrocarbon resources in the waters and seabed, it is unsurprising that coastal States frequently make maritime claims to ocean space that maximize their maritime entitlements. In certain regions, this has resulted in a multitude

76 See generally B. Oxman, "The Territorial Temptation: A Siren Song at Sea" (2006) 100 *American Journal of International Law* 830–851.

77 *Ibid.*, at 839.

of overlapping claims where two or more States claim either sovereignty or sovereign rights and jurisdiction in the same area. While UNCLOS contains provisions on the delimitation of maritime boundaries,[78] there is still a considerable amount of uncertainty surrounding the principles and rules governing maritime delimitation. As noted by two prominent scholars, there is much room "for radically differing interpretations as to which factors and methods of delimitation are appropriate to a particular case, and therefore potential for dispute and deadlock in delimitation negotiations".[79]

When a cable needs to be laid (or repaired) in an area in which two or more States claim sovereignty or sovereign rights, it is difficult for the cable owner/cableship owner/operator to ascertain the correct permitting authority. In practice, the cable industry errs on the side of caution and seeks permits from all States that may have a claim in the area, with the inevitable addition in the cost and time factor that this entails. Cableship operators and cable system owners have no desire to be dragged into maritime boundary disputes. They are neutral with regard to boundary and sovereignty disputes and their activities should be recognized as such by all States. One solution is the development of an international consensus that cable laying and repair activities should be regarded as 'without prejudice' to any State's maritime boundary claims.

Competing Activities in the High Seas/Deep Seabed

The high seas and deep seabed are areas beyond the national jurisdiction of any State. The latter is described as "the Area" under UNCLOS and is defined as the "seabed and ocean floor and subsoil thereof, beyond the limits of national jurisdiction".[80] UNCLOS established the International Seabed Authority (ISA) to regulate the exploration and exploitation activities that occur in the Area.

The water above the Area is considered high seas and is governed by Part VII of UNCLOS. Accordingly, Article 87 freedoms apply, including the freedom to lay submarine cables. Article 112(1) of UNCLOS recognizes that States are entitled to lay submarine cables on the bed of the high seas beyond the continental shelf, which refers to the Area. However, under Article 112(2), cable owners/operators must have due regard to cables already in position and not prejudice the possibility of repairing existing cables or pipelines. Further, Article 87(2) requires that the freedom to lay submarine cables be exercised with due regard for the interests of others States in their exercise of high seas freedoms and also with due regard for the rights under UNCLOS with respect to activities in the Area.

---

[78] UNCLOS Arts 15, 74 and 83.

[79] V. Prescott and C. Schofield, *The Maritime Political Boundaries of the World* (2nd ed, Martinus Nijhoff Publishers, 2005) at 246.

[80] UNCLOS Art 1(1).

There is a potential for conflict between the laying and repair of cables and ocean mineral extraction in the Area.[81] While the ISA does not have the authority to regulate submarine cables, as it is unconnected with the exploitation of seabed resources,[82] both the ICPC and ISA have recognized the need for practical cooperation in the use of the Area and have signed Memorandums of Understanding to this effect.[83]

VI. The Way Forward—Recommendations

The above discussions demonstrate that there are serious issues with regard to laying operations. The recommendations described below are intended to provide constructive guidance to governments, the cable industry and other relevant stakeholders on how these issues can be addressed.

Appointment of Lead Agency to Coordinate National Policy on Submarine Cables

One of the major problems with coastal State regulations relating to submarine cables is the absence of a lead agency. National telecommunications agencies frequently only address telecommunications standardization, licensing and competition issues and may not be familiar with maritime issues. Similarly, maritime agencies may not be aware of the critical nature of submarine cables. Such a lead agency, if appointed, could act as a focal point in the approval process for laying operations or assist in the coordination of the activities of all relevant government agencies that deal with submarine cables permits. It could also take the lead in formulating a cohesive national policy on submarine cables, including liaising with the cable industry and raising awareness on the importance of submarine cables.

Streamlining of Permitting Processes in the Territorial Sea

While governments have every right to regulate laying operations within the territorial seas, given the importance of submarine cables and the fact that cable laying vessels do not pose a threat to the security of the coastal State, governments should consider streamlining and simplifying their permitting processes within

81 See S. Coffen-Smout and G.J. Herbert, "Submarine Cables: A Challenge for Ocean Management" (2000) 24 *Marine Policy* 441–448, at 444.

82 R. Churchill and A.V. Lowe, *The Law of the Sea* (3rd ed, Juris Publishing, 1999) at 240.

83 Memorandum of Understanding between the International Cable Protection Committee and the International Seabed Authority signed on 15 December 2009, Annex to Note by the Secretariat at the 16th Session, 26 April to 7 May 2010, available online at http://www.isa.org.jm/files/documents/EN/Regs/MOU-ICPC.pdf (last accessed 7 June 2013).

the territorial sea. As a matter of public policy when imposing such regulations, they should examine: (1) the coastal State interests that such regulations seek to protect; (2) whether the regulations genuinely protect this interest and (3) the impact of such regulations on common interests or goods such as international communications.

One possible suggestion is for coastal States in a region to standardize the information that is needed before cable laying can take place, including standard forms for cableship operators to provide information on the background of crew members.

Good Faith Exchange of Information Instead of Permit Requirements in the EEZ/Continental Shelf

As discussed above, permit or consent requirements in the EEZ/continental shelf are not consistent with the freedom to lay cables in the EEZ. The imposition of such requirements decreases the trust and confidence between governments and the cable industry. One alternative would be for coastal States to remove any permit requirements or regulations contrary to UNCLOS that are presently contained in their national legislation/regulations and for the cable industry, as a matter of 'best practice', to exchange information with the relevant government agency in relation to its planned laying activities. Similarly, coastal States should also inform cable companies of activities under their jurisdiction and control which may impact cable operations. This would minimize the risk of interference between cable laying activities and other activities within the EEZ. Such an exchange of information and/or consultation is consistent with the mutual obligations of coastal States and cable companies to give due regard to each other's rights and duties. Of course, such a system would be most effective if there was a lead agency responsible for overall coordination.

Enhance Mechanisms for Dialogue, Consultation and Cooperation between States and Cable Companies

The problems described above are arguably symptomatic of a lack of communication and consultation between the cable industry and governments. This could be attributable to the fact that governments have not appreciated the full importance of submarine cables (although there are signs that this is changing). Accordingly, it is critical that the cable industry and governments work together to create formal and informal mechanisms/fora for dialogue, consultation and cooperation through workshops, meetings and other confidence-building measures.

CHAPTER SIX

SUBMARINE CABLE REPAIR AND MAINTENANCE

Keith Ford-Ramsden and Douglas Burnett

Introduction

As described in Chapters 4 and 5, cable owners go to great lengths to minimize the risk of damage to cables in the planning, construction and installation phases of a cable project. There are, however, other seabed users and natural events that can damage cables. In order to keep these interruptions to a minimum, cable system owners enter into contracts with marine maintenance companies that have cable and equipment storage depots and cableships strategically positioned throughout the world, available on a 24/7 standby basis to repair cable faults.

Cable faults receive immediate attention not only because of the impact of lost service, but because every cable down is one less cable system available to restore traffic from other cable systems. For these reasons cable owners should consider a cable repair to be an emergency action, so as to directly meet the obligations they owe to their customers and indirectly to restore the back-up capability that other cable systems depend upon. To the surprise of some, however, cables are repaired not at the direction of a national government, but rather by contract.

I. Maintenance Agreements

There are a limited number of companies that are able to provide the specialized ships, equipment and trained crews, splicers and engineers required to undertake cable repairs worldwide.

There are two types of maintenance agreements that offer 24/7 responses, 365 days of the year, in the manner demanded by cable owners. The first is a *Consortium Agreement*, whereby the cable owners work together to set up and enter their cables into regional maintenance agreements. These regional or *Zone* agreements divide the oceans and seas of the world into areas that are serviced by regional strategically based vessels and supported by cable depots where the system spares are stored. Normally, Zone agreements require the cableship to sail within 24 hours of being mobilized, ready to carry out an emergency repair. The Zone agreements efficiently distribute cable vessels and depots on a shared cost effective basis. The second type of agreement is a *Private Maintenance Agreement*. Private maintenance agreements are provided by recognized marine service providers who offer similar services to those provided in Zone agreements, but on an individual contract basis. It is up to the cable owners to decide which service agreement best suits their needs, based on response times, costs and other factors.

Cable maintenance agreements (both Zone agreements and Private maintenance agreements) usually have a two tier payment system in which a *standing charge* is paid to ensure the ship operator has sufficient income to cover the costs of keeping the cable maintenance vessel operationally ready, at its base port, to sail within 24 hours of notification of a repair. The annual standing charges would include, but not be limited to, the cableship's general maintenance, berthing fees, fuel (bunkers) costs for port operations, crew, in-port insurance, and company overheads. The second set of charges would commence when a request has been

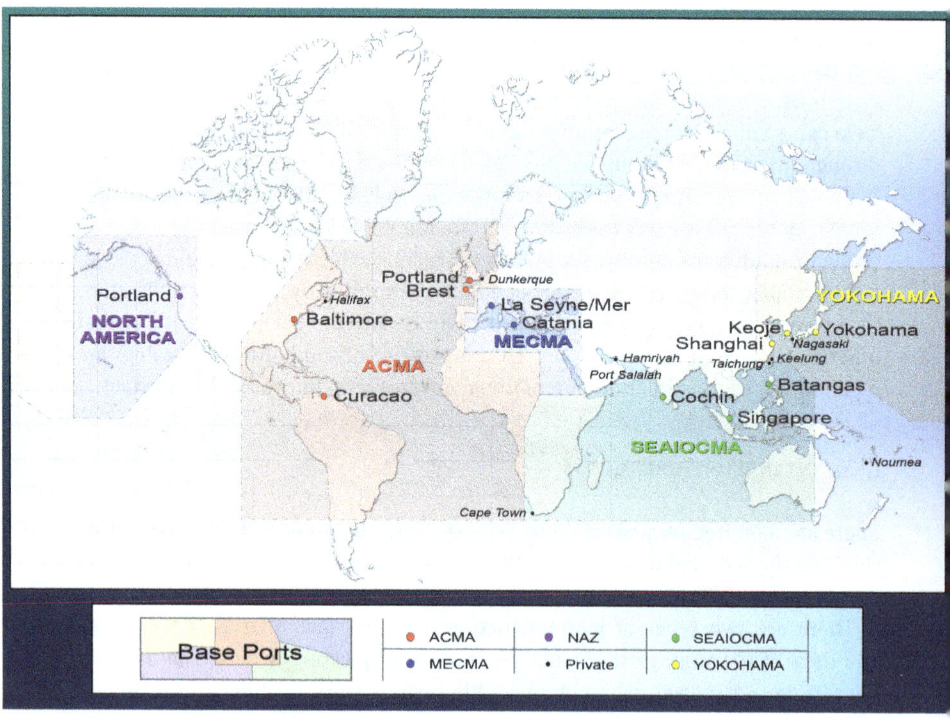

Figure 6.1 Worldwide Zone maintenance agreement areas and base ports. The map is current as of 12 April 2012. (Map courtesy of D. Burnett and Squire Sanders (US) LLP)

made for the cableship to prepare for a repair operation, after which time all costs associated with the cable repair operation would be charged to the owner of the cable system requiring the repair. These charges are called *running costs* and include incremental at-sea insurance costs, fuel (bunkers) consumed at sea, additional at-sea personnel such as splicers, pilot fees and other costs that arise when the ship is at sea and engaged in cable operations (Figure 2.1).

Large cable systems may span a number of these maintenance agreements and hence the owner will need to ensure that sufficient spare cable, jointing piece parts, branching units, equalizers and repeaters are stored in each Zone or agreement area to enable repairs to be carried out by the designated repair vessels. As part of the cable system procurement process, spares are estimated and purchased at the time when the system is ordered from the manufacturer. The rationale is to obtain the lowest cost for the spares and to minimize the risk of increased costs and delays in ordering spares from factories which may subsequently be unavailable.

The spares required to maintain various cable systems may be too large to be stowed on a cableship, therefore a number of cable depots have been constructed around the world, close to the maintenance vessels' base ports. These cable depots store the large volumes of spares required for multiple cable systems. The cost of this storage, loading, discharge and disposal is borne by the cable owners, under a separate depot agreement, for the life of the cable system.

II. Cableships and Equipment

The cableships used for the repair and maintenance of subsea telecommunication systems worldwide are required to have the same ocean going capability as the vessels used to install the cables. In some cases the same class of vessel is used for installation and maintenance, and in other cases the vessels have been specifically built or converted for maintenance operations. Vessels that have been specifically built or converted for maintenance do not need to carry the large amounts of cable necessary for installation, as the total quantity of spare cable required for maintenance is usually stored in purpose-built cable depots strategically located around the world, and only a small quantity of the appropriate cable type is taken out of the depot for each repair.

It is necessary for a cableship that operates in a repair and maintenance role to have a minimum of one cable drum, as this enables the recovery of lightweight cable from deep water or armored cable that is buried, under tension, without the risk of the cable being damaged by wheel slippage, as may be the case with Linear Cable Engines (LCE).

Maintenance vessels are also fitted out with tethered remotely operated vehicles (ROVs) that are capable of operating to water depths in excess of 2000 m.

Figure 6.2 Cable maintenance vessel, *Cable Retriever*.
(Photograph courtesy of Global Marine Systems Ltd)

III. Fault Detection, Location and Restoration

The operators of telecommunication cables use Network Operations Centers (NOCs) to monitor the traffic flow through their networks on a 24/7 basis and are able to immediately identify any interruption to the traffic or a change in the normal operating conditions of the marine portion of their network.

If a customer experiences an interruption of traffic, the NOC operators will, providing the customer has signed up for the service, restore the customer's traffic as soon as possible onto an alternative routing. Normally, this alternative routing or restoration takes place on other cable systems pursuant to mutual restoration agreements between different cable systems, or through purchasing restoration capacity on other systems. Each cable system employs a Restorations Liaison Officer (RLO) who is charged with planning, executing and exercising restoration plans to minimize disruptions in service.

In cases of multiple simultaneous failures there may be delays in this restoration process if the other cables used to restore traffic are also damaged. This is more likely to occur in situations involving major seismic or weather-related events. In other instances the fault may not affect the traffic immediately but may develop into a fault that causes loss of traffic, so maintenance will be required at some stage to rectify the damage.

The fault location needs to be identified as accurately as possible in order to ensure that the cableship undertakes the cable repair in the correct section of cable, optimizes the timescale and cost of the repair, and minimizes the traffic interruption.

Fiber Breaks

In the case of a loss of traffic due to fiber damage or a fiber break, the Cable Landing Stations (CLS) at either end of the section of damaged cable commence procedures to determine the location of the damage. A broken or damaged fiber can occur with or without damage to, or a break in, the cable.

When a cable is broken, the two halves of the system can still be powered from the CLS to the earth grounds either side of the cable fault. The powering allows the repeater's internal supervisory system to be interrogated by each CLS. The repeaters are accessed from each shore terminal up until the cable break and not beyond. Using the supervisory system the cable break can be localized to a point between two repeaters. A repeater span will typically be 60–80 km so the supervisory system cannot provide a sufficiently accurate fault location for a repair operation. The Power Feed Equipment (PFE) voltage required from each CLS can be used to calculate a more accurate fault location, but for a variety of reasons this calculation can only reduce the detection of the fault location to a position accurate to approximately 10 km either side of the fault. The most accurate fault location is achieved by low current, direct current (DC) tests, which can negate the effect of end resistance. These can provide a good approximation of the distance of the fault from each CLS and, therefore, a cable distance from the repeaters on either side of the fault. After referring to the cable as laid Route Position List (RPL), the location and potential extent of the damaged cable can be determined. In the case of external aggression by an anchor or fishing gear this may be small, but in the case of a major natural event (such as an earthquake) large amounts of cable and even a number of repeaters may not be able to be seen from the CLSs.

Fiber faults within repeater spans can be located through the use of Coherent Optical Time-Domain Reflectometers (COTDR) from each CLS. COTDRs can see through repeaters and give accurate fault locations for fiber breaks, but they are expensive pieces of equipment and may not always be readily available.

Shunt Faults

The DC electrical power supply to the cable (which is necessary to energize the repeaters that amplify the data signals every 60–80 km) is normally fed from the CLS located at each end of the cable. Constant current is fed into the system, with one CLS applying positive voltage and the other negative voltage. The PFE that supplies this power can vary depending on the length of system and the number of fibers in the cable. The largest systems can have PFE equipment capable of a maximum output of up to 15,000 volts and 1.5 amps. Feeding positive and negative voltages from each end of the system creates a virtual earth that is roughly half way between the CLS. When the cable is damaged it can induce an actual earth indication at the point of damage. This is called a *shunt fault* and can occur without damage to the fibers. The PFE systems may be able to automatically adjust the voltages' feed to line to balance the system. If the fibers are not damaged and the PFEs on either side of the fault are capable of powering the repeaters the data transmissions on the fibers may not be affected. If this is the case the service may remain on the cable until the operator decides to undertake the repair operation.

Locating a shunt fault relies on assessing PFE voltages to the earth and low current DC tests. This requires considerable skill and knowledge of the system in order to provide an accurate fault location. In some instances the power is ramped up from the PFEs so as to try and create a 'blowout' and damage the fibers at the fault location to provide additional information.

Once the location of a fault has been determined and the owner has decided that a repair is required he/she will advise the marine maintenance provider to make the necessary preparations.

IV. Mobilization of the Repair Ship and Loading Plant and Spares

Having determined the approximate location of the fault a detailed Repair Plan identifying the type of repair operation will be produced. The location of the fault is checked against the 'as laid' cable data provided by the cable installer. This information allows the cable owner, in conjunction with the maintenance provider, to determine:

- whether the cable is buried and requires mobilization of the vessel's tethered ROV for fault location, recovery and reburial of the cable; if the cable is surface laid it may not require the services of the ROV;
- method of cable recovery and whether specialized grapnels are necessary;
- the sections of spare cable that are to be loaded, taking into consideration the armoring and fiber characteristics required for the repair;
- whether spare repeaters and other equipment, such as branching units and equalizers are required;
- the correct type and number of jointing piece part kits for the repair;
- whether the necessary calibrated test equipment and jointing tools and equipment are onboard the repair vessel;
- the amount of ship's stores and fuel required;
- the need to obtain special permitting requirements or notifications;
- whether any additional cable protection such as Uraduct™ is required;
- whether coordination with other cableship operators is required. This is especially relevant when multiple faults have occurred in an area so as to prevent any interference with each other's repairs. (The International Cable Protection Committee (ICPC) has issued a recommendation to coordinate the scheduling

and operations in circumstances such as these where repairs are required for a number of cables with faults.)[1]

V. Repair Operations

The Repair Plan is specific to each fault and depends on the location of the fault and the original protection afforded to the cable.

Lightweight Cable Faults

Lightweight cable is laid on ocean floors at depths ranging from 1000–8000 m in areas where there is minimal risk of external aggression or damage to the cable from strong seabed currents that may cause damage by abrasion. Protected versions of lightweight cable, such as LWS (lightweight screened), LWP (lightweight protected) and SPA (single protection armored), are available and afford greater protection from abrasion and damage from fish bites or fishing hooks over the same range of depths. (Figure 1.4.)

The lightweight cable is laid with sufficient slack to allow the cable to conform to the contours of the seabed and the normal slack allowance is in the order of 1–2 per cent, which is not sufficient to allow it to be recovered from the seabed to the surface. Even if the fibers are all broken, there is no way of determining for certain whether the cable has been severed by the fault. The first task of the cableship when it arrives on the repair site is to cut the cable close to the calculated cable fault position. This is achieved by deploying a cutting grapnel about two to three times water depth away from the cable route and then dragging it along the seabed, perpendicular to the line of the cable, until it engages the cable. Once the cable is engaged there will be a steady rise in tension and this continues to rise until the steel knife-edge in the grapnel cuts through the cable and a rapid drop in tension is noted on the grapnel rope that is trailing approximately twice the depth of water behind the cableship.

After the cable has been cut the grapnel is recovered to the cableship and changed for a holding grapnel in order to begin the process of cable recovery. The method for recovering the cable has changed little since the first cables were laid and repaired in the 1860s. The cableship then repositions to conduct the first *holding drive* at a distance roughly 1–1.5 times water depth from the position where the cable was cut. This ensures that when the cable is recovered to the surface, there is sufficient weight on the free side so that the end does not slide off the grapnel.

[1] International Cable Protection Committee Recommendation No 4—Recommended Co-ordination Procedures for Repair Operations near In-Service Cable Systems. See http://www.iscpc.org/ via the 'Publications' link.

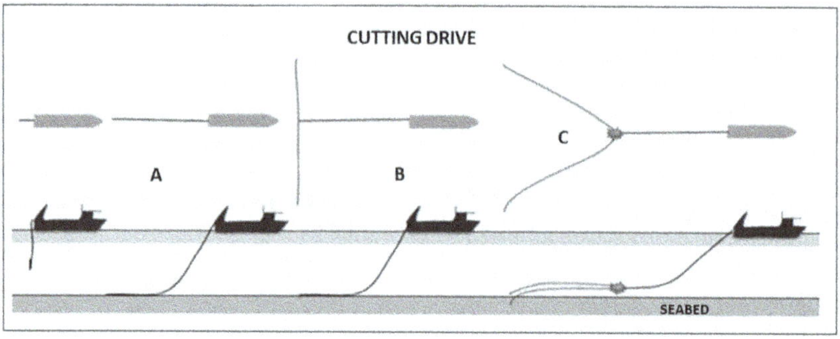

Figure 6.3 The process of the cutting drive.

A. The grapnels are lowered as the cableship moves slowly towards the cable until they are on the seabed. The cableship continues to move slowly ahead until the appropriate amount of grapnel rope is paid out and continues towards the cable line.
B. The cutting grapnel hooks the cable and the cableship sees a gradual rise in tension.
C. The grapnel cuts the cable, a rapid drop in tension is noted and the two ends fall to the seabed.

Figure 6.4 Armored cable recovered by Rennie grapnels during a holding drive. (Photograph courtesy of Keith Ford-Ramsden)

When the cable is brought on to the deck of the cableship the cable on both sides of the grapnel is stoppered off and the cable is cut. The stray end that leads to the cut end of the cable is recovered to the cableship for later disposal. The other cable end is recovered and placed in position for testing.

The onboard testing personnel prepare the cable end, and testing of the fibers and electrical conductor is commenced. If the fibers or electrical continuity do not test satisfactorily more cable is recovered, cut and re-tested until the tests show there is no further damage in the cable. When the cableship testers are satisfied that the cable is good, the cable end is sealed and lowered onto the seabed.

The cable seal is secured to a ground rope and anchor that are lowered to the seabed by the riser ropes followed by an orange or yellow cable repair buoy which is attached on the surface for recovery at a later stage of the repair operation.

After the cable buoy has been released the cableship moves to conduct a second *holding grapnel drive* to recover the other end of the cable in a similar manner to the first end. Once the cable has been recovered, cut and tested to the satisfaction of the onboard testers, a suitable section of replacement cable is selected and jointed on to the cable end. The type of cable used for the repair may require additional protection and therefore a more robust cable type may be selected. The insertion of more cable, especially in deeper waters, will affect the optical characteristic of the system and this may require correction with specialized types of fiber being inserted into the repair section of cable.

The jointers may take up to 24 hours to complete the initial joint between the installed cable and the new cable section. The initial joint is then deployed onto the seabed as the cableship moves towards the cable buoy whilst paying out the repair cable. The cable buoy, buoy rope and first cable end are recovered onto the cableship and both ends of the cable are placed in the cableship's jointing area. The cableship is manoeuvered into the correct position, whilst adjusting the cable so as to have the correct catenaries. Once in position the cables are cut to length and the final joint, that joins the two cable ends together, is started. On completion of the final splice the cableship moves perpendicular to the cable route and pays out the final joint and cable bight over the ship's cable sheaves. At a suitable height above the seabed the final bight is released and the cable sinks and comes to rest on the seabed. The cable station personnel carry out a final set of tests before restoring the customer's traffic back on to the repaired cable.

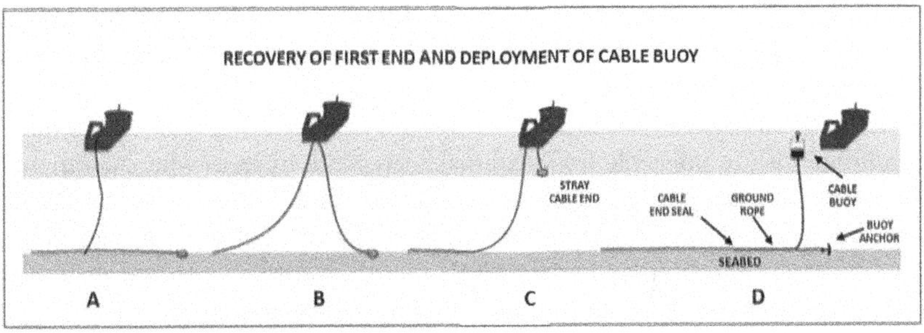

Figure 6.5 The cable recovery process.

A. The holding grapnels are dragged perpendicular to the cut cable at a distance of approximately one and a half times water depth from the cut end.
B. The grapnels are recovered to the deck of the cableship without the cut cable end sliding off.
C. The cable is secured onboard and cut and tested. If the test is successful the good end is sealed and the stray end is recovered.
D. The sealed cable end is attached to the ground rope. The ground rope, anchor, buoy moorings and buoy are deployed.

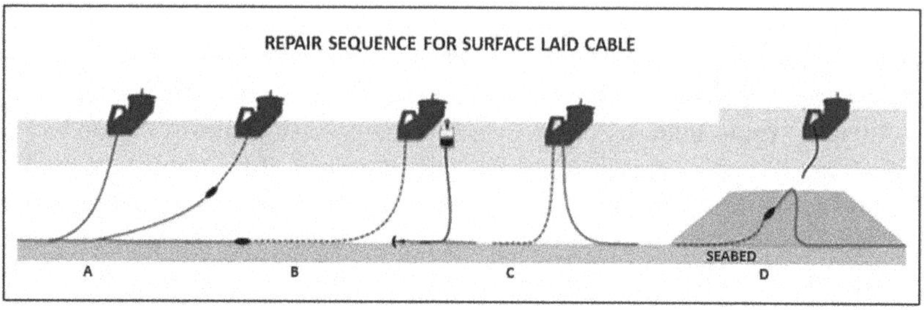

Figure 6.6 The repair sequence for a surface laid cable.

A. The second end is recovered and the initial joint connects the repair section of cable to the original good cable.
B. The new repair cable section is paid out as the cableship moves towards the cable buoy and the initial joint is lowered on to the seabed.
C. The cable buoy, moorings and first end are recovered back to the cableship and the repair section on the cable is paid out whilst the first cable end is recovered to the jointing area.
D. The final splice is completed in the jointing area on the cableship and the vessel is manoeuvered so as to lower the final splice bight of the cable onto the seabed without causing any non-conforming bends in the cable. It is then released to complete the repair to the surface laid section of cable.

Cable Faults in Armored Cable

Armored cable is used in water depths of less than 2000 m where there is a greater need to protect the cable from damage caused by human or natural external aggression. Single Armor (SA) consists of a single layer of galvanized steel wire wrapped around the lightweight cable core and is used down to water depths of 2000 m. A further layer of wire armoring is wrapped around the SA cable to produce Double Armor (DA) cable that can provide far greater protection. DA cable can be used to depths of 500 m, however it is normally only used to water depths of 200 m.

Both DA and SA can be surface laid when it has been determined that there is minimal risk to the cable from external aggression. In areas where additional protection is required the cables can be buried below the seabed. SA cable is normally selected either for simultaneous lay and burial (by plow or injector) or for surface lay and post lay burial. Where there is a specific need for increased armoring DA is selected, with post lay burial being undertaken by an ROV.

When a fault occurs in armored cable the cable stations employ the same process used for locating faults in lightweight cable. Further refinements are available to the maintenance provider to localize the fault to a greater degree of accuracy.

The CLSs are able to inject a low frequency alternating current (AC) signal, known as a 25 hertz electroding tone into the cable. This tone may be detected hundreds of kilometers from the cable station along the cable by electrodes that are trailed behind a cableship or by a detecting system fitted to a tethered ROV. The trailed electrodes will detect the electroding tone on the cable line, prior to the fault, at positions 1, 2 and 3 in Figure 6.7 below, but the tone will not be detected when the electrodes cross beyond the fault at position 4, because the electroding tone will have gone to earth at the fault. At this time the cableship will turn back towards the CLS and cross the cable again, but will still find no signal at position 5. The electroding tone will be detected at the next two crossings, these being positions 6 and 7. This indicates that the location of the fault can be narrowed to a point between positions 5 and 7.

An alternative method for determining the fault location, which may be used independently of or in conjunction with the trailed electrodes, is to deploy a tethered ROV with tone detectors onto the seabed to determine the fault location both electronically and visually.

ROV's are fitted with active and passive systems used to detect the cable. The cable can be detected at a far greater range if the 25 hertz tone is detected by the active system rather than by the passive system which relies on detection of an anomaly in the Earth's magnetic field.

The ROV is also fitted with a sonar that can detect a cable protruding from the seabed or an object, such as an anchor scar, where the cable has been fouled. The tethered ROV is capable of being used to depths in excess of 2000 m and can be fitted with tracks to enable it to manoeuver along the seabed or with skids so that it can fly above the seabed. Decisions regarding configuration depend on prevailing currents and seabed conditions.

The ROV is deployed from and attached to the cableship by a tether. The tether carries the power and telemetry to enable the ROV to move, operate manipulators and high pressure water swords and to power the high pressure water pumps. The position of the ROV is determined through the use of *hydroacoustic position reference* beacons that are attached to the vehicle and are monitored from the cableship. These beacons allow the cableship to track the ROV and to follow and accurately identify its position. With this information and the images and data transmitted by the ROV, the fault location is determined.

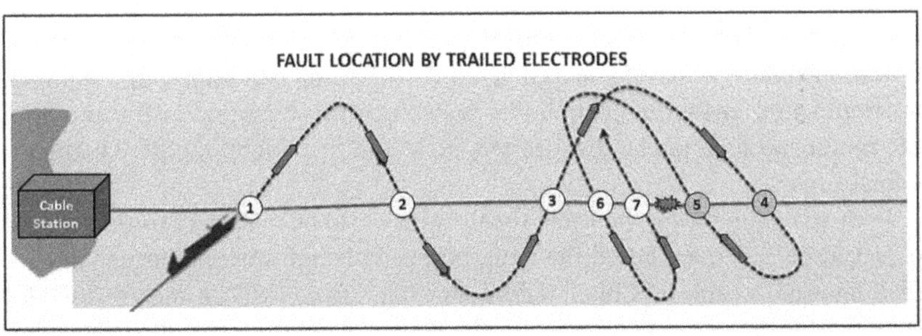

Figure 6.7 How trailed electrodes can be used to detect a cable fault.

After the fault location has been determined the cable repair can be carried out in a similar manner used for lightweight cable and surface laid armored cable.

Where possible, it is prudent to utilize the ROV to cut and recover the cable ends in shallow water, as this minimizes cable damage during recovery and reduces the amount of cable to be inserted during the repair.

After the cable is cut the ROV is recovered and a cable gripper and recovery line will be attached to one end of the cable for recovery to the cableship. The other end of the cable will be recovered in a similar manner.

In areas of strong currents the use of the ROV may not be possible and the armored cable will be cut and recovered through the use of a set of grapnels.

To recover cables from buried sections where the cable is not exposed on the surface of the seabed, specialized de-trenching grapnels will be used to bring the cable to the surface. The de-trenching grapnels are specifically designed to penetrate the seabed to engage and recover the cable from buried depths of 0.8 to 2.0 m. Alternatively an ROV may be used to de-bury the cable, with the ROV or grapnels being used to cut and recover the cable.

From the time that the cable ends have been recovered onboard the cableship through to the deployment and laying down of the final bight of cable on to the seabed, the process is the same for both lightweight and armored cable. The repair plan will specify whether reburial of the repaired section of cable is required. Reburial is a standard requirement for repaired sections of cable buried during installation. The cable owners may also require burial in areas previously surface laid so as to provide additional protection for cables.

After the final bight of the repaired cable has been lowered to the seabed the ROV is deployed from the cableship to conduct a survey of the cable, using either the passive or active tracking system. The survey is conducted to identify the route of the newly inserted cable and the positions of the initial and final joints.

After the survey has been completed the ROV positions to the cable and deploys the burial swords, with one sword on either side of the cable. The onboard ROV high pressure water pumps are started and the swords are gradually lowered into the seabed. High pressure water is injected through the jet nozzles on the ROV burial swords and the water cuts a trench and/or fluidizes the seabed underneath the cable so that it falls into the trench created by the ROV. It may take a number of passes along the cable to ensure that it is buried to the required depth or to the maximum achievable depth given the local soil conditions. The minimal environmental impact of cable burial is described in Chapter 7. After burial of the initial and final joints, the inserted cable and final bight, the ROV conducts a final survey.

Prior to commencing the cable repair it may also be necessary to remove the object or objects that caused the fault, for example, the stow net fishing anchor that damaged a cable off China in 1999 shown in Figure 6.8. The anchors used for stow net fishing can penetrate to depths of over 2 m into soft seabeds. Another repair off Hong Kong required the removal of a 20 foot container that had fallen

Figure 6.8 Stow net fishing anchor recovered during a repair operation. (Photograph courtesy of Keith Ford-Ramsden)

off a ship and had been swept along the seabed by strong currents until it caught on and damaged a communications cable. These impacts and other human impacts on cables are described in more detail in Chapter 11.

Power Safety

Throughout the duration of the repair operation the cableship designates a Power Safety Officer (PSO) who is responsible for ensuring that in circumstances where repeatered system repairs are involved the correct electrical power configurations are applied at the correct phases of the operation. For unrepeatered systems, the PSO need only address optical power safety. The repair may take place in a system whereby a number of cable stations provide the electrical power and the laser signals that enable the cable to carry the data traffic. The laser light and electrical power must be rigidly controlled to protect the personnel onboard the ship from electrical shock or damage to their eyes from high powered laser light in the fibers. This is especially important with respect to the jointers and testers who spend a great deal of time handling and manipulating the bare cable ends during the testing and jointing phases of the repair.

All written instructions sent by the PSO must receive written confirmation from the relevant cable stations that his or her instructions have been carried out. These cable stations may be located hundreds or thousands of kilometers from the repair.

Only after the PSO has confirmed that he or she is satisfied that the cable is safe to handle will repair operations commence. It is also the responsibility of the PSO to ensure that personnel are clear of the cable when any testing of the cable or joints is carried out onboard the cableship.

Jointing

It is not only the fiber optic cable that has to withstand the extreme pressures exerted when they are laid on the ocean floor at depths of up to 8000 m. Other components, such as repeaters, equalizers and branching units that are connected to the cable, must also be able to resist the ingress of water. The manufacturers will have developed their own jointing technologies, joint kits and methodologies to join the various types of cable and components in order to produce the owner's system.

Cable owners may own or be partners in a large number of cable systems and therefore have the option to have the cable supplier(s) provide the jointing kits, piece parts and equipment needed to assemble the kits for the systems. This may require the purchase of specific equipment for each cable system and require the maintenance provider to retain a large amount of equipment in order to maintain all of their cable systems. The alternative is to have a set of common components capable of being used on all cables, with interchangeable piece parts that specifically fit the owner's cable irrespective of the manufacturer or cable type. It is for the cable owner to decide upon the preferred option.

Each cableship has a dedicated dry and clean area where the various processes required for the jointing of the cables and components can take place; this area is known as the Jointing Space. The specialized personnel who undertake the jointing of subsea fiber optic cables are known as jointers and they undergo rigorous training and testing at regular intervals to ensure they have the skill set and aptitude to successfully complete the construction of a cable joint.

After the two cable ends have been placed in the jointing space the jointers prepare the ends of the lightweight portion of the cable. After the ends have been prepared the assembly of the joint commences. The colored coating of the fibers is removed and the ends of the fiber are cleanly cut. The fibers are then placed in a fusion splicer that automatically lines up the two fibers and fuses them together with an arc to prevent reflections or distortions in the splice. The splice is protected with a sleeve and the spliced fibers are placed into the main body of the joint. The mechanical construction of the joint is completed and the fibers are tested.

For repeatered systems the joint is then placed into a mold so that the joint is encapsulated in polyethylene. The encapsulated joint is then x-rayed to ensure that there are no metallic inclusions or void spaces in the molded section that could cause electrical breakdown or implode under hydrostatic pressure on the seabed. For unrepeatered systems high voltage performance is not necessary so molding is replaced by a heat shrink system that makes the joint watertight.

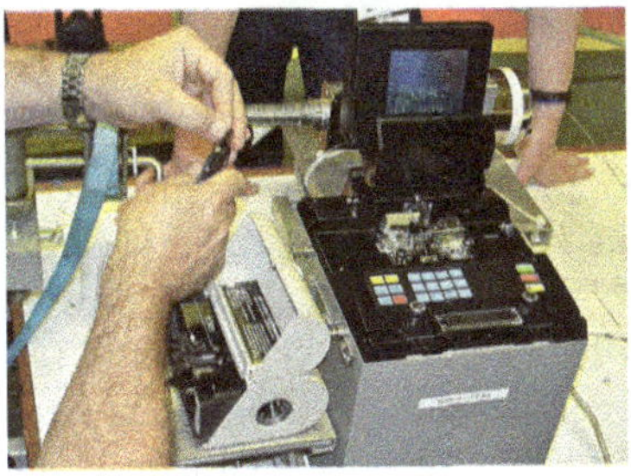

Figure 6.9 Fibers being prepared for fusion splicing. (Photograph courtesy of Global Marine Systems Ltd)

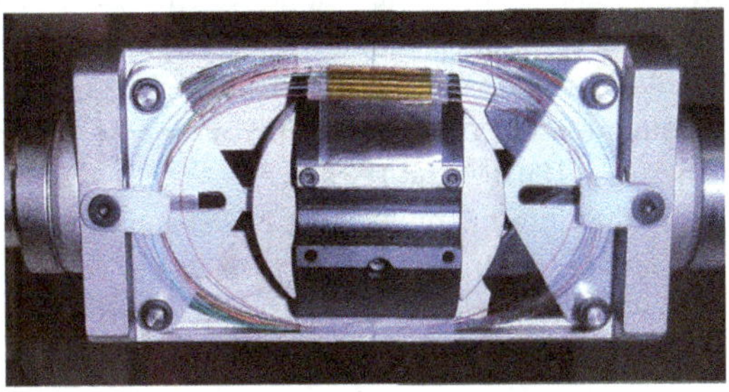

Figure 6.10 A subsea joint with fibers spliced and ready for assembly. (Photograph courtesy of Keith Ford-Ramsden)

A rigid outer protection shell and bend restrictors are secured to the joint to ensure the minimum bend radius of the cable is not compromised. In the case of an armored cable repair the armored wires are keyed into the outer protection shell to ensure the joint has a similar tensile strength to the parent cable prior to the damage.

VI. Law and Policy Challenges for the Repair of Cables

The importance of cable repair ships being able to repair a cable fault as expeditiously as possible cannot be underestimated. Time is of the essence, not only in the repair of a damaged cable, but also because each international cable functions as the back-up restoration path for other cables should they suffer a fault, whether through human activities or as a result of natural disasters such as tsunamis or earthquakes.[2] Cable faults can have a significant impact on the

[2] An excellent discussion of the historical data involving disruption of international telecommunication cables by earthquakes and typhoons is found in L. Carter *et al.*, "Submarine Cables and the Oceans: Connecting the World" (2009) Report of the United

connectivity of a State or States served by that cable system and can consequently affect a wide range of activities such as internet, phone and banking services. For example, during the 2006 Hengchun Earthquake, nine out of eleven cables in the area were severed and a total of 21 faults were discovered in the nine damaged cable systems.[3] It took eleven cableships a total of forty-nine days to complete the cable repair work.[4] According to one report, Taiwan's international calling capacity went down to 40 per cent and 98 per cent of Taiwan's communications with Malaysia, Singapore, Thailand and Hong Kong were also disrupted.[5] Similarly, in 2008, cables located in the Middle East were damaged in two locations, reportedly by two vessels dragging their anchors. This resulted in disruptions to five different cable systems, including cables that connected Southeast Asia, the Middle East and Western Europe, a cable system that connected Europe and Asia, a cable system that connected Dubai, a cable system that served Alexandria and a cable system that served Bandar Abbas in Iran.[6] The cuts affected "at least 60 million users in India, 12 million in Pakistan, 6 million in Egypt and 4.7 million in Saudi Arabia".[7] Another reason for speedy repairs is the cost—the average cost of a single repair is between USD1M and USD3M.[8]

During the 163 years that international cables have been utilized, States have developed practices that are in harmony with provisions of the United Nations Convention on the Law of the Sea (UNCLOS). Specifically, the vast majority of coastal States do not require permits for emergency repair of cables either within their territorial waters, their exclusive economic zones (EEZ), or on their continental shelves. This remains the norm in most parts of the world.

There are solid reasons that support this practice. The norm in most countries, including Australia, Canada, Japan, the United States, and throughout Europe, is for repair ships to sail unimpeded by the coastal State in order to carry out repairs

---

Nations Environment Program and the International Cable Protection Committee (UNEP-WCMC-ICPC) at 38–42. Available at http://www.unep-wcmc.org/medialibrary/2010/09/10/352bd1d8/ICPC_UNEP_Cables.pdf (last accessed 7 June 2013). The Great Tohoku Earthquake struck Honshu, Japan on 11 March 2011 and knocked out six international cables; repairs were delayed by concerns of radioactive contamination of cableships.

3 ICPC Press Release dated 21 March 2007 available at http://www.iscpc.org/information/ICPC_Press_Release_Hengchun_Earthquake.pdf (last accessed 7 June 2013).

4 *Ibid.*

5 Proceedings of the Reliability of Global Undersea Cable Communications Infrastructure: Study and Global Summit (ROGUCCI) Report, (2010) Issue 1, at 173–174.

6 *Ibid.* at 175.

7 A.A. Zain, "Cable Damage Hits 1.7 m Internet Users in UAE" *Khaleej Times*, 5 February 2008.

8 D. Burnett, "Recovery of Cable Repair Ship Cost Damages from Third Parties that Injure Submarine Cables" (2010) 35 *Tulane Maritime Law Journal* 103, 108.

on an expedited basis without the need to obtain permits or escort vessels, and without payment of any fees.

Repair operations are transparent, with the local landing party receiving all reports and documentation for a repair, which it may share with the coastal State upon receipt of a request from the latter. The repair ships themselves are well known, as there are only approximately 40 repair vessels in operation around the world. They sail from known base ports in specified regions, are conspicuous in their appearance, and display the day shapes or night lights required for ships engaged in cable operations.

Standard repair procedures for cableships normally involve notification being given to the coastal State so that the State may issue appropriate notices to mariners advising them of impending repair operations. In short, repair operations are carefully planned deliberate operations that are never secret.

Notwithstanding the widespread acceptance by coastal States of the need for expeditious repair of international cables, some coastal States have, since approximately the late 1990s, required that permits be granted, both inside and outside of territorial seas, and include onerous conditions which cableships must fulfill prior to commencing repairs. These permits and conditions can add weeks or months to the time needed to repair a damaged cable and increases the cost of the repair by millions of dollars. For example, in 2011 Indonesia initiated a requirement that only Indonesian flag ships owned and crewed by Indonesians can carry out cable repairs in its vast archipelagic waters. The problem was that there were no such ships in operation when the requirement was issued and efforts to obtain ships, at a cost many times higher than the market price, encountered major delays. Even if the ships could be obtained, there is a major problem with sourcing a trained crew in Indonesia. The backlog of repairs far exceeds the limited capacity of a single repair ship. This, combined with a high fault rate, including theft,[9] and unreasonable delay periods for obtaining at least five separate permits has made Indonesia a choke point in international communications.[10] Importantly, as noted above, these delays and costs impact not only the coastal State that impose them, but every other coastal State connected to the damaged cable in need of repair. The challenges to cable repair operations within each maritime zone are set out below.

[9] Kompas.com, 27 March 2013, "Indonesian Internet Cable is Frequently Being Stolen Scrap Metal"; Bangka Tribune News, 5 March 2013, "16 Tons of Fibre Cable Suspected as Stolen Now Secured".

[10] A. Palmer-Felgate *et al.*, "Marine Maintenance in the Zones—A Global Comparison of Repair Commencement Times" SubOptic Conference Presentation May 2013; H.B. Nugroho, "The Legal Regime of Submarine Cables", dissertation on file with the University of Virginia School of Law, May 2013 at 171–186.

Within Territorial Waters

The majority of coastal States do not require permits for repairs undertaken in territorial waters (i.e. an area over which they have full territorial sovereignty) as they recognize the importance of repairing critical international infrastructure in the shortest time possible. However, some coastal States, although perfectly entitled to do so, require permits for repairs which are as onerous as the permits for laying described in Chapter 5. One such example is Indonesia's cabotage regulations, which have been extensively detailed in Chapter 5. For present purposes, it suffices to say that Indonesia recently changed its cabotage laws to require that, from 2014, as a condition for issuing a repair permit, only Indonesian flagged vessels will be able to carry out repairs in its archipelagic waters and territorial seas.[11] The goal of this law is to boost Indonesia's flagging maritime industry. While under international law, Indonesia has the right to condition any repairs in its territorial and archipelagic waters, exercising this right creates harm to Indonesia and other nations. Such a precedent does not bode well for the promotion of cost efficient and rapid repairs of international cables. The expected increased costs associated with this regulatory change will impact all States whose nationals participate in cables that land or transit Indonesian archipelagic waters. Restrictions and delays for repairs will also substantially increase the risk of major internet disruptions, the impact of which will be felt in the many countries served by these cables.

As described in this Chapter, cableships are hired by cable owners through Zone and private agreements. Key considerations are reliability and quality of the cable repair ship, its base port and supporting depots, and its competitive cost. The flag of the vessel is normally unimportant as long as the flag State complies with UNCLOS and requirements of the International Maritime Organization. If every State where an international cable lands were, like Indonesia, to require that only repair vessels flying its flag could repair the cable, the current successful and cost effective arrangements would no longer be possible. Instead scores of new repair vessels would be required, at greatly increased cost to all cable owners and their customers worldwide. Such a precedent would also hamper responses to natural disasters where multiple systems require multiple emergency repairs and prompt repair requires the use of repair ships that are immediately available, irrespective of which State they are flagged in.

---

11 Minister of Transportation Regulations No 48 of 2011 on Procedure of Permit to Operate Foreign Flag Vessel in Activities other than Domestic Transport of People and Goods, 18 April 2011, Article 2(1) ("Foreign flagged vessels are allowed to perform activities other than transport of people and goods in Indonesia waters only if there is no Indonesian flagged ship able or sufficient to do so") and Article 2(2) ("Pursuant to Article 2(1), to operate in Indonesian waters, the foreign flagged ship is required to obtain a permit from the Minister [of Transportation]"). Other permits issued by the Ministry of Foreign Affairs and the Ministry of Defense are also required.

Within the EEZ/Continental Shelf

As explained in Chapter 3, the repair of submarine cables is an "internationally lawful use of the sea" related to the operation of submarine cables in the EEZ.[12] With regard to the continental shelf, Article 79(1) affirms that all States are entitled to lay submarine cables in accordance with the provisions in Article 79 but does not refer specifically to the right to repair and maintain cables. However, the remaining provisions of Article 79 assume that the right to lay submarine cables includes the right to repair and maintain them.[13] The challenge of coastal State encroachment on the freedom to repair and maintain cables, as provided for in UNCLOS, is illustrated in the following prominent examples of this recent deviation from the norm of permit-free repairs of cables outside of territorial seas.

China

China imposes permitting requirements that delay repairs of international cables outside of Chinese territorial seas and add significant costs by way of permit fees and delays in allowing cableships to sail to carry out repairs in areas beyond China's sovereignty. For example, between January 2005 and April 2009, there were 19 cable faults caused by fishing vessels in China's EEZ in the East China Sea between China, Japan, Korea and Taiwan.[14] These repairs were delayed by between one to two weeks as a result of permit requirements of the Chinese government. The delays also resulted in significant costs being incurred, by way of the permit fees and the costs incurred from the cable repair ships remaining idle awaiting the permits.

These injuries are compounded by the fact that repairs in the eastern Pacific region of China's EEZ are far more numerous than in almost any other area of the world because of destructive stow net fishing techniques employed by Chinese fishing vessels. Stow net fishing involves large steel structures and anchors being dropped into the seabed to hold nets to a depth of up to 3 m, snapping any cable they hit (Figure 6.8). These techniques and natural disasters (such as earthquakes and typhoons), particularly off of Taiwan, are responsible for the high number of faults.[15]

---

[12] See UNCLOS Art 58(1).

[13] UNCLOS Art 79(2) refers to the "laying or maintenance" of submarine cables and Art 79(5) refers to "repairing" existing cables.

[14] For additional information, *see* CIMA/COLP/ICPC Regional Workshop on Submarine Cables, Beijing, PR China, 7–8 May 2009, Workshop Report available at http://cil.nus.edu.sg/wp/wp-content/uploads/2009/10/Workshop_Report_on_Submarine_Cables.pdf (last accessed 7 June 2013).

[15] Carter *et al., supra* note 2, at 38–48. In the Hengchun earthquake in December 2006 eleven cables were severed, and during Typhoon Morokot in August 2009 nine cables were severed; in most cases the damage occurred as a result of the effects of landslides

The Chinese policy of requiring permits for repairs undertaken to international cables outside of its territorial seas violates the freedom to maintain and repair cables clearly set out in UNCLOS.[16] Following successful regional workshops, the cable industry attempted to establish a workshop with the State Oceanic Agency, which is the government agency of China largely responsible for regulating submarine cable repair and survey permits. The cable industry proposed establishing a protocol or practical arrangement with the State Oceanic Agency so as to accommodate the undefined 'security issues' which are purportedly behind the permitting requirements. The protocol or arrangement would include crew pre-clearance and certification, provision for government observers to be present on cableships during the repair operations on a no-delay basis, and other measures. The protocol would then be employed on a trial basis and subsequently reviewed. The workshop proposal was not accepted and remains on the table for China's authorities to consider. In the meantime all States using fiber optic cables to China face increased risks from cables that are out of service awaiting permission to carry out the emergency cable repairs.

India

As discussed above, purpose-built cable repair ships that service many cable systems in a region are stationed at strategic base ports around the world, ready to sail and carry out repairs on 24 hour notice. Cable repair ships contracted to cover cable systems in the Indian Ocean are based in Singapore and the United Arab Emirates.

As with the laying of cables (discussed in Chapter 5), India has established extremely onerous permit requirements that require cableships to obtain seven different permits before they can begin repairs, not only in the territorial seas but also in the EEZ.[17] A particularly onerous requirement is that any cableship, prior to carrying out a repair in the Indian EEZ must, in effect, first enter an Indian port for a physical security clearance inspection. This requirement was communicated to cableship operators in a directive of 2006, "Apprehension of Vessels Violating

---

that generated turbidity currents triggered with breaking impact over hundreds of kilometers of seabed.

16 UNCLOS Arts 58(1), 58(2), 78, 79(2), 79(5) and 112.

17 R. Rapp *et al.*, "India's Critical Role in the Resilience of the Global Undersea Communications Cable Infrastructure" (2012) 36(3) *Strategic Analysis* at 375–383. The permits include: Research, Survey, Exploration and Exploitation Permit issued by the Ministry of Defence and Integrated Headquarters; Ministry of Home Affairs Clearance for all personnel/crew; Specified Period License issued by the Director General of Shipping; Naval Clearance from the Indian National Shipowners' Association; Crew Visa Clearance from the Immigration Office; Vessel Importation, as assessed by Indian Customs Officials. *See* Rapp *et al.*, at 381.

Provisions of MZI Act 1976 and MoD Guidelines", issued by India's Principal Director of Naval Operations, which states:

> 1. In the recent past, there have been a marked increase in offshore exploration and production activities, resulting in a number of Indian and foreign (flagged/manned) chartered vessels operating in our EEZ. This has led to an increase in number of violations of laid down conditionalities as specified in Defense Clearance letter issued by Integrated Headquarters of MoD (Navy) from time to time, MoD Guidelines 1996 and MZI Act of 1976 and regulations in force. *It has also come to light that some vessels operate without valid security clearance.*
> 2. In order to sift the violator(s) from the rule-abiding ones, a system of periodic checks of vessels involved in "Exploration and Production" activities in the Indian Offshore region is being brought into force with effect from 15 June 06. *Under this system, vessels that are found to be operating without the necessary clearance will be escorted to harbor and handed over to the Coast Guard/Police for contravening the provision of the MZI Act of India, 1976.*
> 3. The above is for information and *compliance.*[18]

Failure to comply with the above missive can result in the repair ship being forced into port by Indian naval vessels. As this is not a risk that cableship operators are prepared to take they opt to send the cable repair ship into an Indian port first instead of going directly to the fault location to begin repairs.

By entering an Indian port on, in essence, a *de facto* involuntary basis, India's custom duties regulations are triggered. The cableship and the spare cable aboard it are considered to have been 'imported' into India. In order to leave the Indian port, the ship must first post security based on a percentage of the value of the vessel and spare cable. Months after the repair is completed, the security is returned to the ship owner with unexplained deductions. These deductions may amount to USD350,000 to USD500,000. The time taken to secure the seven permits, including the naval inspection permits, may be 90–94 days.[19] Adding the additional costs of permit fees and daily hire costs for the repair ship for these extended stays (which total between USD45,000 to USD70,000 per day),[20] the unnecessary repair costs and delays amount to millions of dollars.

In 2013, India promulgated a new requirement that foreign flag cableships spending 30 days or more in India's territorial seas or EEZ (a *de facto* requirement by Indian law) must have Indian national cadets comprising 1/3 of their total crew, and that the Indian national crew be divided evenly between deck and engineering.[21] This requirement usurps the exclusive flag State authority

---

18 OP/OMD/5106/MoD/Guidelines dated 9 June 2006 (emphasis added).

19 2011, ICPC Survey Results for India Government Permitting from 11 Repairs to International Submarine Cables from 7/2005 to 6/2011. A repair for a fault on 1 May 2011 required 94 days. The repair for the fault on 27 June 2011 took 90 days.

20 *See* Burnett *supra* note 8 at 109–110.

21 Shipping Development Circular No 1 of 2013 dated 18 January 2013 issued by the Ministry of Shipping of India.

under Article 94(3)(b) to specify manning and crew training of cableships flying its flag.

The resulting delays and costs also have an effect on India's worldwide connectivity, which is a direct threat to the off-shore call-centers and global outsourcing business that have featured so prominently in India's recent economic growth. Equally important is the fact that the delays mean that all States connected by the cable incur the same heightened vulnerability risk during the time it is out of service.

Under UNCLOS, coastal States have no right to require cable vessels to enter their ports prior to carrying out repairs to international cables in their EEZ or to impose customs duties or permit requirements that impede repairs.[22] Over the past two years the cable industry, represented by the ICPC, has been working with the EastWest Institute[23] to meet with Indian officials and encourage India to comply with UNCLOS. The ICPC and the EastWest Institute have sought to provide practical options to India, such as obtaining vessel pre-clearance and notification, to avoid the need for cable repair ships to enter ports prior to commencing emergency repairs in India's EEZ. To date, the efforts have not produced any meaningful changes to India's excessive regulations. A noteworthy consequence of India's onerous permitting environment was the unprecedented 2012 decision of the new Sea-Me-We 5 cable system to exclude India as a landing party.[24] The Sea-Me-We cable system is the world's largest consortium owned cable system. The exclusion of India as a landing party is likely to have been a consequence of the Indian government's policies and the imposition of excessive costs that are hostile to international cables and the international law upon which their success is based.

VII. The Way Forward—Recommendations

Many of the Recommendations suggested in Chapter 5 apply to the repair of cables. In particular, given the importance of speedy repairs coupled with the fact

---

22 UNCLOS Arts 58(1), 58(2), 78, 79(2), 79(5) and 112.

23 EastWest Institute is a global think tank, *see* http://www.ewi.info/ (last accessed 7 June 2013).

24 S. Tagare, (from Sunil Tagare's Personal Views on the Telecom Industry, 17 February 2012) "The recent decision by the Indian Government to impose a $40 million fine to be deposited as a bond by any new submarine cable wanting to land in India has infuriated a lot of foreign carriers. (Officially they call it Advance bond for taxes, surcharges, custom duties, etc. to be imposed at the discretion of the government). Even though technically it is not a fine that is how it is perceived in the carrier community. The first casualty of this law is that Sea-Me-We-5 has decided not to land in India. Sea-Me-We-5 was going to be the main artery that would connect Southeast Asia, Middle East and the Gulf region to Europe with the latest 100Gbps technology".

that the need for repairs is often unforeseen, an important step forward would be for both the industry and governments to cooperate to develop a set of 'best practices' which 'prioritizes' timely cable repairs[25] both within and outside territorial waters, while at the same time taking into consideration the legitimate rights of coastal States within these maritime zones. First, coastal States should consider minimizing repair permit requirements in the territorial sea/archipelagic waters and establishing infrastructure which allows the streamlining of all permit processes in these zones by one lead agency. In the EEZ and continental shelf, coastal States should consider removing all permit requirements that are inconsistent with UNCLOS. A reasonable notification requirement should be sufficient. Alternatively, for the small number of States that will not accept the usual notification scheme common in most of the world for emergency repairs, other reasonable steps could be implemented in partnership with the industry. Coastal States working with industry should establish best practices with regard to the exchange of information on cable repairs. This could include a protocol or practical arrangement (discussed above) whereby there would be crew pre-clearance and certification, optional provision for government observers to be present on cableships during the repair operations on a no-delay basis, or other measures. It could also include cable repair ships informing a pre-designated lead government agency of the details of the ship, its location and details of repair. Efficiency and simplicity must be the paramount considerations, which could be fulfilled by a simple form submitted by the cable repair ship or landing party with the requisite information as soon as the repair operation is commenced.

To conclude, historically State practice has allowed cable repairs to take place both within and outside of territorial waters without permits or preconditions that delay emergency repairs and increase costs. This norm has stood the test of time and served the international community well. Recently, a small number of States have deviated from this norm and required permits and imposed conditions prior to allowing the cableships to commence emergency repairs. This is a detrimental practice which has repercussions for the coastal State itself as well as for other States that have cable systems landing or transiting the coastal State. Hopefully, States adopting these deleterious practices will reconsider their national laws and policies and bring them into line with UNCLOS and traditional State practice which embraces rapid, efficient, and reliable repair of international cables.

---

25 ROGUCCI Report, *supra* note 5 at 105.

CHAPTER SEVEN

THE RELATIONSHIP BETWEEN SUBMARINE CABLES AND THE MARINE ENVIRONMENT

Lionel Carter, Douglas Burnett and Tara Davenport

Introduction

There has been a perception that submarine cables and cable operations have a negative impact on the marine environment.[1] While there are inevitably interactions between the environment and cables, they are not necessarily detrimental. This Chapter examines those cable/environment interactions. It then discusses whether the trend of increasing coastal State environmental regulations on cable operations are consistent with international law and the 1982 UN Convention on the Law of the Sea (UNCLOS) and, equally as important, whether these regulations are necessary to protect the marine environment.

I. Interactions Between Submarine Cables and the Environment

The interactions of submarine power and telecommunications cables with the marine environment can be viewed in the context of water depth and cable size. In depths > ~2000 m, i.e. a nominal limit for bottom trawl fishing,[2] the diameter of a telecommunications fiber optic cable is between 17–22 mm, which is about the size of a garden hose (see Figure 1.4). These cables are laid directly on the seabed. Hence they have a small physical footprint, especially when viewed in a global context as depths > ~2000 m constitute 84 per cent of the world's ocean. Telecommunications cables in waters < ~2000 m depth can be up to 50 mm diameter due to the addition of protective wire armor. Submarine power cables,

---

[1] See for example, A. Freiwald *et al.*, "Cold-water Coral Reefs: Out of Sight—No Longer out of Mind" (2004) UNEP–WCMC, United Nations Environment Programme (UNEP) Biodiversity Series, at 84, available at http://www.unep-wcmc.org/medialibrary/2010/09/10/29fefd54/CWC.pdf (last accessed 7 June 2013).

[2] For example, P. Mole *et al.*, "Cable Protection—Solutions Through New Installation and Burial Approaches" Conference Proceedings of SubOptic, 11–16 May 1997, San Francisco at 750–757.

which at present are laid no deeper than ~1600 m, are larger than telecommunications cables, with diameters ranging between 70–150 mm although they can be up to 300 mm. The following sections will examine the different operations relating to these small, deep and larger, shallow cables and their interactions with benthic settings.

Cable Route Surveys

The main tools used for cable route surveys are acoustic instruments such as echo-sounders, multibeam or seabed mapping systems, commercial side-scan sonars and, in areas where cables are to be buried, acoustic sub-bottom profilers. These survey tools are guided by accurate navigation from satellite-based Global Positioning System (GPS) and Differential GPS. High frequency and low energy survey systems are generally used during the surveys. This is because surveys focus on the depth, topography and composition of the seabed surface and also gather information on sediments immediately below the seabed. Our knowledge of the effects of surveys and other human-made acoustics on marine animals is incomplete,[3] but available data suggest that the risk associated with cable route survey instruments is minor.[4] This contrasts with some high energy naval sonar systems that produce prolonged acoustic pulses in mid-range frequencies. Such systems have been implicated with the stranding of certain whale species[5] and are the focus of considerable and ongoing research.[6]

To verify the acoustic data and imagery, photographic or video records of the seabed may be collected by survey vessel or divers (where depths permit). These tools are non-invasive. Sediment samples from the seabed may also be required and these are collected by small grabs that typically recover samples of up to a few kilograms. To verify seismic records, physical testing of the substrate is becoming the norm. The prime instrument is the cone penetrometer, which measures the sediment strength by the resistance encountered as a rod is pushed

---

3 Refer National Research Council, "Ocean Noise and Marine Mammals" (2003) Committee on Potential Impacts of Ambient Noise in the Ocean on Marine Mammals, Ocean Studies Board, National Academies Press, Washington DC, www.nap.edu (last accessed 7 June 2013).

4 "Impacts of Marine Acoustic Technology on the Antarctic Environment" Version 1.2 July 2002, SCAR Ad Hoc Group of Marine Acoustic Technology and the Environment, available at http://www.geoscience.scar.org/geophysics/acoustics_1_2.pdf (last accessed 7 June 2013).

5 A. Fernández *et al.*,"'Gas and Fat Embolic Syndrome' Involving a Mass Stranding of Beaked Whales (Family Ziphiidae) Exposed to Anthropogenic Sonar Signals" (2005) 42 *Veterinary Pathology* at 446–457.

6 D.E. Claridge, "Providing Field Support for the Behavior Response Study (BRS-07)" (2007) available at http://www.dtic.mil/cgi-bin/GetTRDoc?AD=ADA505862; see also Physorg.com "Sperm Whales in Gulf Seemingly Unaffected by Distant Seismic Sounds" (2008) Oregon State University, http://www.physorg.com/news138545651.html (last accessed 7 June 2013).

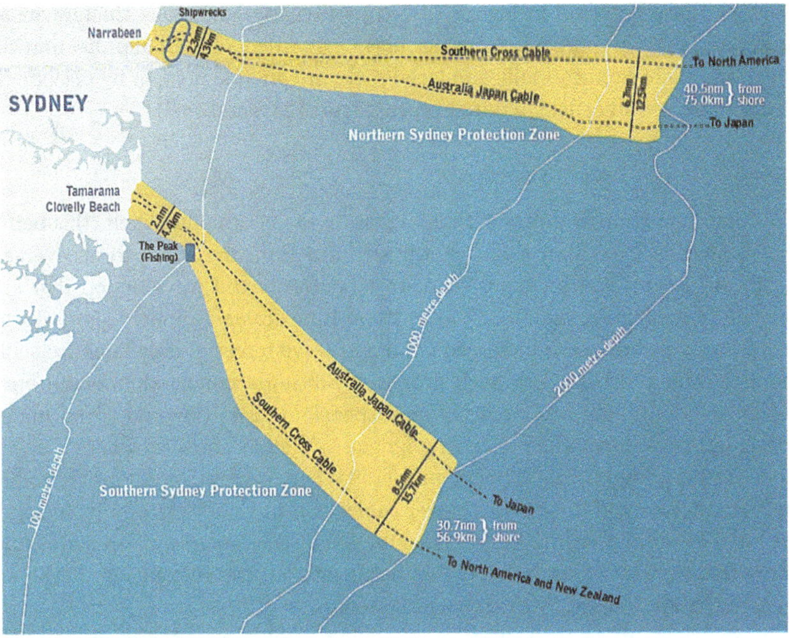

Figure 7.1 Established telecommunications cable protection zones (yellow) extending from shore to the 2000 m depth contour off New South Wales, Australia. Locations of zones are freely available on charts and brochures for all seabed users. (Image courtesy of the Australian Communications and Media Authority (ACMA))

~5 m into the seabed. If required, sub-seabed samples may be collected by coring, but this is usually kept to a minimum reflecting the high quality of modern acoustic and penetrometer information. Accordingly, the physical impact to the seabed and subsoil by cable route surveys is minimal.

Surface Laid Cables

Telecommunication and power cables are routinely laid on the seabed in water depths > ~2000 m, which is beyond the main zone of human activities. However, surface-laid telecommunications and power cables may also occur in depths shallower than ~2000 m where the seabed is unsuitable for burial, such as in areas of submarine rock outcrops and high ecological sensitivity. They may also be located in an effectively policed and legally designated *cable protection zone*[7]

[7] Australian Communications and Media Authority (ACMA), New South Wales Protection Zones, available at http://www.acma.gov.au/Industry/Telco/Infrastructure/Submarine-cabling-and-protection-zones/nsw-protection-zones-submarine-cable-zones-i-acma (last accessed 7 June 2013).

(such as those depicted in Figure 7.1). In particularly hazardous shallow areas, power and telecommunications cables may be afforded additional protection via coverings made from carefully emplaced rocks or concrete mats or by placement of the cable within articulated iron pipes (Figures 13.5 and 13.6).[8]

Interactions with Water and Sediment

Telecommunication and power cables crossing the continental shelf (this being the submarine plain that slopes gently seaward from the shore to an average depth of ~130 m) can be exposed to strong currents and wave action that are capable of instigating sediment transport on time scales of hours to months to years. During storms, the increased wind and wave forces greatly enhance sediment erosion and transport. Such processes can undermine, exhume and bury cables.[9] Undermining can create cable suspensions that sway and strum under strong currents thus inducing serving fatigue as well as abrasion where suspensions are supported by rocky promontories.[10] Where cable suspensions are stable and long-lived, as in the case of power cables laid in Cook Strait, New Zealand, they can become cemented to the rock by encrusting organisms—a stabilizing effect that may be offset by increased water drag on the suspension due to an enlarged profile caused by the biological growth.

In zones of moderate wave/current action, cables may self-bury into soft sediment under turbulence induced by passing currents. Burial also occurs as a migrating sand wave passes across a cable; a process that may be followed by exhumation under the next sand wave trough. The temporary nature of burial is apparent in water depths of < ~30 m where storm-forced waves and currents can temporarily remove the sand blanket to expose a cable; a process that may be followed by fair-weather burial as sediment naturally accumulates.[11] Finally, high discharge rivers produce zones of high sediment accumulation that enhances

---

8 Transpower and Ministry of Transport, Cook Strait Submarine Cable Protection booklet (2011) at 16 available at https://www.transpower.co.nz/resources/cook-strait-cable-booklet; see also International Cable Protection Committee (ICPC), 'About Submarine Power Cables', accessed through the "Information" button on the ICPC webpage at http://www.iscpc.org/ (last accessed 7 June 2013); and CEE, "Basslink Marine Biological Monitoring, McGauran's Beach" (2007) Report to Enesar Consulting at 43. Available on request from Senior author.

9 P. Allan, "Cable Security in Sandwaves" Paper presented at the International Cable Protection Committee Plenary, May 2000, Copenhagen; L. Carter and K. Lewis, "Variability of the Modern Sand Cover on a Tide and Storm Driven Inner Shelf, South Wellington, New Zealand" (1995) 38 *New Zealand Journal of Geology and Geophysics* 451–470; see also L. Carter *et al.*, "Seafloor Stability along the Cook Strait Power Cable Corridor" (1991) Proceedings of the 10th Australasian Conference on Coastal and Ocean Engineering 565–570.

10 I. Kogan *et al.*, "ATOC/Pioneer Seamount Cable After 8 Years on the Seafloor: Observations, Environmental Impact" (2006) 26 *Continental Shelf Research* 771–787.

11 L. Carter and K. Lewis *supra* note 9.

natural cable burial. High sediment accumulation is frequently associated with mountainous regions near actively colliding tectonic plates.[12] Unfortunately, such areas tend to be earthquake prone thus raising the spectre of hazardous submarine landslides and turbidity currents[13] as well as cable-damaging floods.[14]

Interactions with Marine Biota

Any effect of cables on marine biota can be assessed (i) by biological census taken before and after a cable installation,[15] or in the case of an existing cable, (ii) by comparative analyses of the biota near to and distant from a cable.[16] These quantitative studies show that surface laid cables have little or no impact on the resident fauna and flora. On the basis of extensive video coverage and 138 sediment cores Kogan *et al.* in a 2006 study found no statistical difference in the distribution and abundance of animals dwelling within 1 m and 100 m of a coaxial telecommunications cable.[17] Grannis also recorded no significant change in fauna living in the soft sediment near to and away from a fiber optic cable.[18] Likewise, Andrulewicz *et al.* revealed no change in the composition, biomass and abundance of benthic animals preceding and following deployment of a power cable.[19]

By providing a firm substrate, cables can become sites of marine encrustation (see Figure 7.2), as observed (i) off California where anenomes, typical of rocky substrates, were found confined to a coaxial cable traversing a soft muddy substrate that is unsuitable for such animals, (ii) in Cook Strait, (iii) in Bass Strait where articulated pipe cable protection has a biological mantle similar to that of surrounding rocks,[20] and (iv) in many other areas.

---

12 J.D. Milliman and J.P.M. Syvitski, "Geomorphic/Tectonic Control of Sediment Discharge to the Ocean: the Importance of Small, Mountainous Rivers" (1992) 100 *Journal of Geology* 525–544.

13 S.-K. Hsu *et al.*, "Turbidity Currents, Submarine Landslides and the 2006 Pingtung Earthquake off SW Taiwan" (2006) 19(6) *Terrestrial, Atmospheric & Oceanic Science* 767–772.

14 L. Carter *et al.*, "Near-synchronous and Delayed Initiation of Long Run-out Submarine Sediment Flows from a Record-breaking River Flood, Offshore Taiwan" (2012) 39 *Geophysical Research Letters* L12603.

15 E. Andrulewicz *et al.*, "The Environmental Effects of the Installation and Functioning of the Submarine *SwePol Link* HVDC Transmission Line: A Case Study of the Polish Marine Area of the Baltic Sea" (2003) 49(4) *Journal of Sea Research* 337–345.

16 Kogan *et al.*, *supra* note 10; see also B.M. Grannis "Impacts of Mobile Fishing Gear and a Buried Fiber-optic Cable on Soft-sediment Benthic Community Structure" (2001) M. Sc. Thesis, University of Maine, at 100.

17 See Kogan *et al.*, *supra* note 10.

18 See Grannis *supra* note 16.

19 See Andrulewicz *et al.*, *supra* note 15.

20 See ICPC *supra* note 8 'About Submarine Cables'.

Figure 7.2 Delicate encrustations of coral and coralline algae on a fiber optic telecommunications cable. (Photograph courtesy of G. Rivera and S. Drew)

Historically, organisms encrusted on recovered communications cables have contributed to our knowledge of the marine biota, especially that of the deep ocean (> ~2000 m depth), which occupies almost 60 per cent of the planet's surface and even today, is still largely unexplored.[21]

Surface laid cables are exposed to fish and marine mammals; a situation that came to the fore during the telegraphic cable era when 16 cable faults were attributed to whale entanglements recorded between 1877 and 1955. Thirteen faults resulted from sperm whales, which were identified by their entangled remains; the remaining faults were attributed to a humpback whale, killer whale and an unknown species.[22] Most faults occurred near the edge of the continental shelf and adjacent continental slope where telegraphic cables had been repaired. This led to speculation that the repairs produced coils or loops that subsequently ensnared the whales. However, with the replacement of submarine telegraphic cables by coaxial cables in the 1950s, whale entanglements ceased. This continued to be the case throughout the fiber optic cable era, which began in the mid

21 P.M. Ralph and D.F. Squires, "The Extant Scleractinian Corals of New Zealand" (1962) 29 Zoology Publications from Victoria University of Wellington 1–19; see also C.D. Levings and N.G. McDaniel, "A Unique Collection of Baseline Biological Data: Benthic Invertebrates from an Underwater Cable Across the Strait of Georgia" (1974) Fisheries Research Board of Canada, Technical Report 441 at 19.

22 B.C. Heezen, "Whales Entangled in Deep Sea Cables" (1957) 4 *Deep-Sea Research* 105–115; see also B.C. Heezen and G.L. Johnson, "Alaskan Submarine Cables: A Struggle with a Harsh Environment" (1969) 22(4) *Arctic* 413–424.

1980s.[23] The marked change reflects technological advances in cable design, surveying and laying: (i) cables are now torsionally balanced—an improvement that reduces the tendency to self coil on the seabed; (ii) cables are laid under tension over accurately charted seabed topography; (iii) repaired cables are relaid without slack and, in shallow water, the repaired sections are usually buried and (iv) cables on the continental shelf and upper continental slope are often buried below the seabed.

Exposed telecommunications cables can be damaged by sharks, barracuda and other fish as identified from teeth embedded in the cable serving.[24] Bites cut the serving and insulation allowing seawater to ground the cable's power conductor. The first deep-ocean fiber optic cable sustained a series of shark attacks in 1985–1987. The culprit was the deep dwelling crocodile shark, which caused cable faults in depths of 1060–1900 m. It was speculated that the sharks were attracted by electromagnetic fields or cable vibrations, but later experiments were inconclusive. Nevertheless, the episode instigated design improvements that have greatly reduced the bite problem.

Chemical Stability

The basic fiber optic telecommunications cable consists of: (i) one or more pairs of glass fibers; (ii) a sheath of steel strands for strength; (iii) a copper conductor for power transmission and (iv) an insulating sheath of high density polyethylene. In shallow water, one or more layers of galvanized steel wire may be added for protection. Anti-fouling agents are not used.[25] The behavior of some of these cable components in seawater has been investigated in the laboratory and coastal sea by Collins.[26] Sections of various cable types, some with their cut ends exposed and others sealed, were immersed in 5 liters of seawater and any leaching from the copper conductor and iron/zinc galvanized armor was analyzed at set time intervals. Only zinc was detected in the seawater where it registered

---

23 M.P. Wood and L. Carter, "Whale Entanglements with Submarine Telecommunications Cables" (2008) 33 *IEEE Journal of Oceanic Engineering* at 445–450.

24 International Cable Protection Committee, "Fish and Shark Bite Database" Report of the International Cable Protection Committee (October 1988) at 5; see also L.J. Marra, "Sharkbite on the S.L. Submarine Lightwave Cable System: History, Causes and Resolution" (1989) 14(3) *IEEE Journal of Oceanic Engineering* 230–237.

25 See L. Carter *et al.*, "Submarine Cables and the Oceans—Connecting the World" Report of the United Nations Environment Program and the International Cable Protection Committee (2009) 'UNEP/ICPC Report' at 33. Available online at http://www.unep-wcmc.org/medialibrary/2010/09/10/352bd1d8/ICPC_UNEP_Cables.pdf (last accessed 7 June 2013); see also Emu Ltd, "Subsea Cable Decommissioning: a Limited Environmental Appraisal" Report No 04/J/01/06/0648/0415. Open file report is available at email@ukcpc.org.uk.

26 K. Collins, "Isle of Man Cable Study—Preliminary Material Environmental Impact Studies" (2007) Preliminary Report, University of Southampton.

Figure 7.3 A power cable in the tide-swept Cook Strait, New Zealand, where currents move gravel (fragments > 2 mm diameter) on a daily basis. Cables are protected by steel armoring with an outermost serving of polypropylene yarn, which is intact and retains its yellow-black markings after more than a decade of exposure to frequent sediment abrasion. (Photograph courtesy of Transpower New Zealand)

concentrations < 6 parts per million (ppm) for cables with sealed ends and < 11 ppm for those with exposed ends. Leaching reduced after ~10 days. As the tests were carried out in closed containers, the concentrations of cable-sourced zinc in the open ocean can be expected to be much lower because of dilution by the oceanic circulation. Furthermore, zinc occurs naturally in the ocean where it is essential for biological processes such as the production of plant plankton.[27]

In contrast to everyday plastic debris such as polystyrene and polycarbonate fragments, which are known to degrade in the ocean,[28] cable grade, high density polyethylene sheathing is basically non-reactive with seawater. It would take centuries to fully convert this material to carbon dioxide and water via oxidation, hydrolysis and mineralization.[29] Plastic debris degrades in the presence of ultra-violet light, but cable polyethylene is light stabilized with further protection provided by steel armoring and burial on the continental shelf where light can penetrate to the seabed surface. The depth of light penetration depends on the presence of sediment and plankton in the seawater, but the so-called photic

---

27 F.M.M. Morel and N.M. Price, "The Biogeochemical Cycles of Trace Metals in the Oceans" (2003) 300(5621) *Science* 944–947.

28 K. Saido *et al.*, "New Contamination Derived from Marine Debris Plastics" 238th ACS National Meeting, 22–26 August 2009, Washington, DC.

29 A.L. Andrady, "Plastics and their Impacts in the Marine Environment" Proceedings of the International Marine Debris Conference on Derelict Fishing Gear and the Ocean Environment, 6–11 August 2000, Hawaii.

zone is typically < 150 m below the ocean surface. The abrasive effect of mobile sand and cable movement in strong waves and currents may abrade and release particles that may affect marine organisms.[30] However, physical breakdown is minimized by cable burial, steel armor and advanced polypropylene servings that are abrasion resistant (Figure 7.3). The limited information regarding power cables also suggest that modern coatings such as polypropylene yarn are abrasion resistant as demonstrated by observations of cables in the current swept Cook Strait (Figure 7.3).

Buried Cables

Because of hazards posed by shipping, bottom trawl fishing, dredging and other activities on the continental margin, power and telecommunications cables may be buried below the seabed for extra protection.[31] Such burial measures can disturb the bottom or benthic environment of the continental shelf and uppermost continental slope. However, compared to repetitive fishing, ships' anchoring, and dredging, cable burial is usually a one-off operation for the 20–25 year design life of the cable. Further disturbance can occur, however, (i) when a cable fails and requires repair (during which time disturbance will be localized to the fault location) and (ii) when a decommissioned cable is removed (which in the context of a cable's life time, is still an infrequent event).[32] Another consideration is the limited extent of burial, which is confined to a designated cable route. This contrasts with, for example, bottom trawl fishing, which is so widespread and repetitive that it has been described recently as the submarine equivalent of industrial-scale agricultural plowing on land.[33] The following section examines seabed disturbance under various aspects of burial followed by a brief review of seabed recovery.

Cable Route Clearance

Debris that may impede burial operations is removed by a grapnel towed by a ship along the proposed cable route.[34] Depending on the grapnel size, penetration is typically 0.5 to 1.0 m in soft muddy sediment. Accurate positioning of grapnel

---

30 M. Allsop *et al.*, "Plastic Debris in the World's Oceans" (2006) Greenpeace Publication at 10, available online at http://www.unep.org/regionalseas/marinelitter/publications/docs/plastic_ocean_report.pdf (last accessed 7 June 2013).

31 L. Carter *et al.*, *supra* note 25 at 30.

32 For further information, refer to Chapter 8 on Out-of-Service Submarine Cables.

33 P. Puig *et al.*, "Ploughing the Deep Sea Floor" (September 2012) 489 *Nature* 286–289.

34 National Oceanographic and Atmospheric Administration (NOAA), "Final Environmental Analysis of Remediation Alternatives for the Pacific Crossing-1 North and East Submarine Fiber-optic Cables in the Olympic Coast National Marine Sanctuary" (2005) at 77 and Appendix, see http://sanctuaries.noaa.gov/library/alldocs.html (last accessed 7 June 2013).

tows is essential to define the burial route and hence minimize unnecessary seabed disturbance. In addition, debris can also be identified by high resolution, side-scan sonar that may detect (i) objects as small as cables and wires with diameters of a few centimeters, depending upon equipment settings and (ii) seabed composition. In that case, debris removal is localized.

Cable Burial

Mechanical plowing is the more common burial process. During this process a plow is towed across the seabed, the plowshare opens a furrow into which the cable is placed, and the cable is covered as the furrow sides close.[35] Further covering comes from the natural rain of sediment, which, in regions of high river input, can locally exceed 1 cm/year.[36] This generalized picture of plowing varies in accordance with the nature of the substrate and the type of plow being used.[37] For ecologically sensitive coastal zones such as marshlands, tidal flats and eelgrass/seagrass meadows, specialist plows are available that have a minimal footprint.[38] Directional drilling beneath sensitive coastal areas will also reduce disturbance.[39] On the continental shelf, cables are buried to depths according to the seabed type and the nature of the hazard.[40] For soft to firm sediments, ships' anchors bite 2 to 3 m into the seabed and bottom trawl fishing gear penetrates ~0.5 m. Thus to address the latter hazard, cables are buried to 1 m, in which case the appropriate plowshare would leave a strip ~0.3 m wide (plowing to larger sub-seabed depths may leave a wider strip). For soft to firm substrates, the furrow will self-heal, but for harder materials only partial closure may result.[41] In addition to the plow furrow, the seabed is likely to be compacted and biota disturbed by the passage of the skids or wheels that support the plow. Again the nature of disturbance depends upon seabed type, associated organisms and plow size. Overall, the plowshare plus skid/wheel disturbance can range from ~2 to

35 P.G. Allan, "Geotechnical Aspects of Submarine Cables", IBC Conference on Subsea Geotechnics, November 1998, Aberdeen.

36 C.A. Huh *et al.*, "Modern Accumulation Rates and a Budget of Sediment off the Gaoping (Kaoping) River, SW Taiwan: A Tidal and Flood Dominated Depositional Environment Around a Submarine Canyon" (2009) 76(4) *Journal of Marine Systems* 405–416.

37 R. Hoshina and J. Featherstone, "Improvements in Submarine Cable System Protection" Conference Paper presented at SubOptic, Kyoto 2001, Paper P6.7 at 4; R. Rapp *et al.*, "Marine Installation Operations: Expectations, Specifications, Value and Performance" Conference Poster presented at SubOptic, Monaco 2004, Poster We 12.5 at 3.

38 Ecoplan, Monitoring der salz-wiesen vegetation an der Bautrasse im Ostheller von Norderney 1997–2002, (2003) Ecoplan Report for Deutsche Telekom.

39 S. Austin *et al.*, "A Comparative Analysis of Submarine Cable Installation Methods in Northern Puget Sound, Washington" (2004) 7 *Journal of Marine Environmental Engineering* 173–183.

40 Hoshina and Featherstone *supra* note 37.

41 Refer to NOAA *supra* note 34.

Figure 7.4 SMD remotely operated vehicle with manipulating arms and other equipment for cable repairs and burial (right) and positioning propellers (left). (Photograph courtesy of TE SubCom)

8 m width. Following plowing, coastal State permit conditions may require that a post-burial survey be conducted, which may involve single or repeated surveys to monitor seabed recovery.

High pressure water injection or *jetting* is commonly used to bury cables that are already laid, although water jets may also be incorporated into plows to facilitate burial during the cable laying process. Jets are commonly incorporated onto remotely operated vehicles (ROVs) (see Figure 7.4) that can operate in water depths over 1000 m, in steep topography and in very soft sediments—essentially conditions that are unfavorable for plowing.[42] ROV-equipped jets liquefy sediment along a cable causing it to sink to a pre-determined depth below the seabed. This technique is ideal for burying repaired sections of cable as well as sectors where a cable is only partially buried by plowing. The width of jetting disturbance tends to be wider than plowing due to the formation of turbid plumes that may occur in sands but are more noticeable in soft muds. Jet-induced liquefaction may displace or damage the marine biota inhabiting the path for the trench, whereas turbid plumes may affect more distant sites.[43] Any impact of plumes is best assessed on a case-by-case basis because plume dispersal and settling depends upon local currents and waves, the nature and concentration of the sediment in plumes, the composition and resilience of the resident biota, seabed topography and the frequency of natural perturbations such as storms.

---

42 Hoshina and Featherstone *supra* note 37, see also M. Jonkergrouw, "Industry Developments in Burial Assessment Surveying" Paper Presented at SubOptic, Kyoto 2001, Paper P6.3.1 at 4.

43 Refer to NOAA *supra* note 34.

Even though burial has markedly reduced the numbers of fiber optic cable faults in water depths < 200 m,[44] failures still occur. Where possible, a damaged cable is recovered by grapnel (Figure 6.4) or ROV operated from a cable repair ship (Figure 7.4). The damaged section is removed and replaced by new cable (*splice*) onboard the repair ship. The repaired section is then relaid perpendicular to the original cable and where appropriate is buried by jetting using a ROV (Figures 6.5 and 6.6).[45]

Seabed disturbance may result from the recovery of a buried decommissioned cable. According to Emu Ltd,[46] recovery will dislodge the overlying sediment and benthic biota. Local seabed conditions dictate the nature of that disturbance. If soft, muddy sediment prevails, any disruption will be small, as these sediments are physically weak and cannot maintain any relief. In contrast, consolidated materials may form cohesive fragments that form a blocky surface whose longevity depends on the strength and frequency of wave/current action, rates of natural sedimentation and any biological activity.

Environmental Recovery

From the previous discussion it is apparent that benthic disturbance associated with cable burial, repair and recovery occurs primarily on the continental shelf and upper continental slope. This is confirmed by the occurrence of 50 to 70 per cent of all cable faults (and related repairs) in water depths < 200 m.[47] This depth range covers several environmental settings with distinct oceanographic, geological and biological features that dictate the rate of seabed recovery. Human activities, in particular bottom trawl fishing, also play a prominent role by influencing patterns of erosion and siltation.[48]

Substrate recovery in sheltered coastal areas may be facilitated by their accessibility. This permits the use of laying techniques that minimize disturbance and also allows better access for remedial measures. In the case of seagrass beds in Botany Bay, Australia, remedial actions have been proposed that involve the removal of plants from the cable route and their replanting after cable emplacement.[49]

---

44 M.E. Kordahi and S. Shapiro, "Worldwide Trends in Submarine Cable System Faults" Conference Paper presented at SubOptic, Monaco 2004, Paper We A2.5 at 3.

45 Hoshina and Featherstone *supra* note 37.

46 Emu Ltd, "Subsea Cable Decommissioning: A Limited Environmental Appraisal" (2004) Report No 04/J/01/06/0648/0415. Open file report available from email@ukcpc.org.uk.

47 J. Featherstone *et al.*, "Recent Trends in Submarine Cable System Faults", Conference Proceedings SubOptic, Kyoto 2001 at 5; see also M.E. Kordahi *et al.*, "Trends in Submarine Cable System Faults" Conference Proceedings SubOptic, Baltimore 2007 at 4.

48 Puig *et al.*, *supra* note 33.

49 Molino-Stewart Pty, Botany Bay Cable Environmental Impact Assessment (2007) available at http://www.molinostewart.com.au/index.php?option=com_content&view=article&id=95:botany-bay-cable-environmental-impact-&catid=41:environmental-impact-assessment-and-approvals&Itemid=82 (last accessed 7 June 2013).

For other seagrass/eelgrass restorations, the sowing of grass seed can be effective. Cable burial in salt marshes along Germany's north coast was facilitated by a low impact, custom built vibrating plow.[50] Post-deployment monitoring of the salt marsh showed that vegetation in disturbed areas re-established within one to two years and fully recovered within five years. In soft sediment settings such as mangrove swamps, recovery from human-caused disruptions can range from two to seven months.[51]

Seaward of the coast, the continental shelf descends gradually to the shelf edge at an average depth of ~130 m. Accompanying this deepening is a decline in wave energy and its potential to move sediments. That, together with ocean currents and tides, weather events, seabed biology and geology, collectively influence the rate and nature of seabed restoration.[52] Restoration can be assessed in the context of three depth zones, each with their own hydrodynamic character. However, assessments are generalized and actual seabed recovery will be affected by local conditions. This is exemplified by tide-dominated continental shelves where tidal currents can force sediment movement at most shelf depths, for example, the Channel between the United Kingdom and France, Straits of Messina (Italy) and Cook Strait (New Zealand). Returning to the depth zone approach, the *inner shelf* (0 to ~30 m) is exposed to frequent wave and current action, especially during storms. As a result the seabed is typically mobile sand except in the vicinity of high discharge rivers where muddy deposits may prevail as off the Mississippi River delta in the United States. Both physical and biological recovery from cable burial in inner shelf sands commonly occurs within weeks to months.[53] Substrates of the *middle shelf* (~30 to 70 m) are less frequently disturbed by waves and swell with the main bouts of sediment erosion and transport associated with storms.

---

50 Ecoplan Report *supra* note 38.

51 K.M. Dernie *et al.*, "Recovery of Soft Sediment Communities and Habitats Following Physical Disturbance" (2003) 285–286 *Journal of Experimental Marine Biology and Ecology* at 415–434.

52 For example, C.A. Nittrouer *et al.*, *Continental Margin Sedimentation—From Sediment Transport to Sequence Stratigraphy* (International Association of Sedimentologists, Special Publication 37 Blackwell Publishing, 2007) at 549 and references therein.

53 CEE, Basslink Project Marine Biology Monitoring, McGauran's Beach, (2006) Report to Enesar Consulting; see also J. DeAlteris *et al.*, "The Significance of Seabed Disturbance by Mobile Fishing Gear Relative to the Natural Processes: A Case Study in Narragansett Bay, Rhode Island" in L. Benaka, ed, *Fish Habitat: Essential Fish Habitat and Rehabilitation* (American Fisheries Society, 1999) at 400; see also S. Bolam and H. Rees, "Minimizing Impacts of Maintenance Dredged Material Disposal in the Coastal Environment: A Habitat Approach" (2003) 32(2) *Environmental Management* 171–188; see also National Oceanic and Atmospheric Administration (NOAA), Stellwagen Bank National Marine Sanctuary Report, 2007 at 41, available at http://sanctuaries.noaa.gov/science/condition/sbnms/welcome.html (last accessed 7 June 2013).

The nature of the substrate also reflects the amount of sediment supply with sediment-nourished shelves having muddy substrates whereas sediment-starved systems have middle shelves mantled by sand and gravel that are essentially current-modified deposits from the last ice age when sea level was 120 m lower than present.[54] In the case of muddy substrates off Massachusetts, United States, a cable trench had not completely in-filled one year after laying due to slow natural deposition, but the benthic fauna appeared to have recovered within that year.[55] In contrast, physical evidence of power cable burial in sandy deposits of the Baltic Sea was erased within one year of burial—a consequence of local waves and currents.[56] In addition, there were no significant changes in the benthic animal communities. Those examples and others[57] indicate that the rate of burial trench recovery reflects: (i) the amount of sediment supplied to the middle shelf; (ii) the physical nature of the sediment cover whereby loose mobile sand is unable to retain physical evidence of burial in contrast to well consolidated mud substrates; (iii) the frequency of natural disturbances such as storms, and (iv) the depth of the trench incision. On the *outer shelf* (~70 to ~130 m) and *upper continental slope* (> ~130 m), reduced sediment supply and infrequent wave/current action suggest that any trench scar may last longer than on the middle shelf. However, local conditions will ultimately dictate the recovery rate. If, for example, the seabed is composed of unconsolidated sand/gravel then recovery can be rapid, especially in the presence of tidal or ocean currents that are commonly intensified along the continental shelf edge. Alternatively, a consolidated substrate, low current action and weak sediment supply may bring about a slower recovery.[58]

Cable Recycling and Life Cycle

The robust nature of submarine cables and the commercial value of their materials make them attractive for recycling. However, it has only been in recent years that effective recycling schemes have been developed.[59] In the case of Mertech

---

54 Nittrouer *et al.*, *supra* note 52.

55 Grannis *supra* note 16 at 100.

56 Andrulewicz *et al.*, *supra* note 15.

57 For example, California Coastal Commission, Coastal Development Permit Application and Consistency Certification, (2005), E-05-007 at 50, available at http://www.coastal.ca.gov/energy/Th6b-9-2005.pdf (last accessed 7 June 2013); see also California Coastal Commission, Coastal Development Permit Amendment and Modified Consistency Certification E-98-029-A2 and E-00-0004-A1, http://documents.coastal.ca.gov/reports/2007/11/Th8a-s-11-2007.pdf (last accessed 7 June 2013).

58 See NOAA *supra* note 34.

59 For example, see MTB, Cable Recycling, http://mtb-recycling.fr/en/index-2cables_EN.html (last accessed 8 April 2013) and Mertech Marine, SAT-1 "Proof of Out-of-Service Deep Sea Cable Recovery and Dismantling as a Viable Business Case" Presentation at the International Cable Protection Committee Plenary, 2011 Lisbon, Portugal.

Marine,[60] decommissioned coaxial and fiber optic telecommunications cables are recovered by cableships or other specially adapted vessels at a quoted rate of ~40 km/day. Onshore, mechanical processors break the cables down and separate out their main components; copper, polyethylene plastic and steel. Clearly, the metal components are valuable but so too is the high quality polyethylene, which is recycled through plastic conversion plants. Interestingly, the plastic retains its quality even though recovered coaxial cables have been on the seabed for over 30 years. Recycling offers several benefits that include clearing the seabed of non-operational infrastructure, recovery of valuable materials plus remuneration and employment for recyclers. As discussed in Chapter 8, management of out-of-service cables, including recycling and salvage, has been addressed by the industry in the form of an International Cable Protection Committee recommendation and adherence to UNCLOS.

Recycling is part of an analysis that identifies the net amounts of carbon produced during the cradle-to-grave life cycle of a submarine fiber optic cable. Such analyses help gauge a cable's overall environmental footprint. A study by Donovan[61] showed that the main potential environmental effects related to (i) electrical power used at land-based terminal stations—127 gigawatt hours, and (ii) fuel consumed during all ship operations including cable laying and maintenance—1515 tonnes of fuel. Fuel and power consumption were calculated for the lifetime of a cable. Donovan estimated that 7 grams of carbon dioxide equivalents would be released into the atmosphere for every 10,000 gigabit kilometers (note, (i) 10,000 Gb/km is a designated unit of reference of a cable's functionality and (ii) *carbon dioxide equivalent* is the warming potential of all greenhouse gases if expressed as CO_2). The significance of 7 grams is brought into perspective by comparing a cable-based teleconference between New York and Stockholm with the equivalent face-to-face meeting. For a two day teleconference of 8 hours/day, 5.7 kg of CO_{2eq} would be released compared to 1920 kg emitted for the face-to-face meeting, which involved 16,000 km of air travel. While such life-cycle analyses rely on some assumptions that may be debatable, Donovan's study nonetheless highlights the small carbon footprint of submarine telecommunications and their positive contribution to reducing greenhouse emissions by reducing the need for transoceanic air travel and other high carbon-use activities.

60 Refer http://www.mertechmarine.co.za/ (last accessed 7 June 2013).

61 C. Donovan, "Twenty Thousand Leagues Under the Sea: A Life Cycle Assessment of Fibre Optic Submarine Cable Systems" (2009) Degree Project, SoM EX2009-40 KTH Department of Urban Planning and Environment, Stockholm. See http://cesc.kth.se/submarine-cable-systems/ Thesis at 97.

Electromagnetic Fields (EMF) from Power Cables

Apart from the interactions described above, electromagnetic fields (EMF) generated by power cables may also have an impact on the environment.

The offshore expansion of submarine power grids associated with wind-turbine farms has raised the possibility that associated EMF fields may affect marine animals.[62] By way of background, it is well established that DC and AC submarine power cables produce a magnetic field, the intensity of which is directly related to the applied voltage. Model studies conducted by Normandeau *et al.*[63] in 2011 show that a DC field is more intense than its AC counterpart (see Chapter 13). Intensity is strongest directly above a cable but reduces rapidly in horizontal and vertical directions, for instance at 10 m horizontal distance, the field is about two per cent of the peak intensity above the cable. When several cables are present, the strength of their collective magnetic field appears to be influenced by (i) the separation distances of cables; (ii) the direction of current flow with opposing currents having a cancelling effect on the field, and (iii) cable voltage. As a water current or a swimming animal passes through a magnetic field, it induces an electric field whose strength depends upon (i) the direction of the water current/swimming organism (the maximum electrical field being induced when the path is perpendicular to the cable's magnetic field); (ii) the speed of the water current/organism, and (iii) the strength of the magnetic field.

The biological literature records a range of marine organisms that are sensitive or potentially sensitive to magnetic, electric or combined fields. The list includes some sharks and rays i.e. *Elasmobranchs*, other types of fish (e.g. mackerel, cod, salmon), sea turtles, some marine invertebrates (e.g. sea urchins, snails, lobsters) and possibly whales.[64] EMF may influence an animal's navigation, feeding, orientation, and/or detection of other animals. Such potential responses are based on studies of an animal's behavior, anatomy or functioning, as well as theoretical analyses. Experiments with sandbar sharks, whose shallow coastal habitat is also favored by offshore wind turbine farms, show that they respond to low intensity electric fields, suggesting they could also be affected by power cables. However, field trials with cables have not been undertaken. In the case of sockeye salmon, the young partly rely on Earth's geomagnetic field for navigation. However, because the fish are *pelagic* or free swimming and the EMF of cables

---

62 OSPAR Commission "Assessment of the Environmental Impacts of Cables" (2009) available online at http://qsr2010.ospar.org/media/assessments/p00437_Cables.pdf (last accessed 7 June 2013). The document merely raises the possibility, but contains no research to support this possibility.

63 Normandeau Associates Inc *et al.*, "Effects of EMFs from Undersea Power Cables on Elasmobranchs and other Marine Species" US Department of the Interior, Bureau of Ocean Energy Management, Regulation and Enforcement, Pacific OCS Region, Camarillo, CA. OSC Study BOEMRE 2011-09, May 2011.

64 *Ibid.*

is limited in extent, juvenile salmon are unlikely to be adversely affected. Even if cables influence the local geomagnetic field, salmon can compensate by relying more on sight and smell, which are also used to aid their navigation.

Normandeau *et al.*[65] note substantial gaps in our knowledge of animal behavior in the presence of EMF, especially relating to actual field studies involving power cables. Thus the report notes that any conclusions about the effects of EMF on organisms must be regarded as speculative. Nevertheless, if mitigation is required on the basis of evidenced-based science, options are available to lower EMF. These include:

- Use of AC cables, which have magnetic fields lower than DC cables for the same voltage;
- Use of higher voltage cables that generate magnetic fields less intense than lower voltage systems for the same power output;
- Change the conductivity and permeability of the cable sheathing or serving;
- Place cables closer together so that opposing current flows cancel one another's EMF (however, this option could produce unacceptable constraints on cable maintenance);
- Cable burial;
- Align a cable to minimize the collective effect of the cable EMF and local geomagnetic field.

II. UNCLOS, Coastal State Environmental Regulations and Submarine Cables

The past forty years have seen an exponential growth in concern for the marine environment, coupled with a growing body of law known as international environmental law.[66] This concern for the marine environment is reflected in the increasing number of coastal State regulations aimed at protecting the marine environment from harm arising from activities in the oceans. In line with this development, there has been a growing trend by coastal States to subject cable operations to environmental regulations. However, such environmental regulations may be inconsistent with UNCLOS, can delay or impede laying/repair operations and most importantly, may be unnecessary. In this regard, the following sections will first examine the extent to which UNCLOS allows coastal States to impose environmental regulations on cable operations and then discuss specific examples of environmental regulations that have posed challenges to the industry.

---

65 *Ibid.*

66 For a general overview of the growth of international environmental law, see P. Birnie *et al.*, *International Law and the Environment* (3rd ed, New York, 2009) at 1–43.

UNCLOS and the Protection of the Marine Environment

The protection of the marine environment received a significant boost with the adoption of UNCLOS. As noted by two scholars, UNCLOS establishes:

> a unifying framework for marine environmental protection that seeks to address all sources of marine pollution, incorporates by reference the latest international rules and standards, strengthens the enforcement capacity of port and flag States, and gives coastal States extensive jurisdiction with regard to the protection and preservation of the marine environment within their territorial seas and EEZs.[67]

While there are provisions on environmental protection scattered throughout UNCLOS, there is also a whole part devoted to the protection of the marine environment. Part XII of UNCLOS contains general obligations on States to protect and preserve the marine environment,[68] which are supplemented by specific articles addressing different sources of marine pollution from land-based sources, from seabed activities subject to national jurisdiction, pollution from activities in the deep seabed, ship-source pollution and pollution from the atmosphere.[69] Generally, these specific provisions place an obligation on States to establish global and regional rules, standards and practices to prevent, reduce and control pollution from a particular source, either through competent international organizations or through diplomatic conferences. These specific provisions then enjoin States to adopt national laws and regulations to prevent marine pollution from these particular sources. The laws and regulations should "take into account"[70] or "be no less effective"[71] or "at least have the same effect"[72] as global and regional rules, standards and practices. In this way, UNCLOS incorporates by reference the latest internationally agreed rules, standards and recommended practices and procedures. Part XII also extends coastal States' specific enforcement powers in respect of pollution from the various sources.[73]

The critical question for present purposes is to what extent does UNCLOS allow coastal States to impose environmental regulations on cable operations? The following sections will attempt to answer this.

---

67 D. Rothwell and T. Stephens, *The International Law of the Sea* (Oregon, 2010) at 338.
68 See UNCLOS Sections 1–4, Part XII.
69 UNCLOS Section 5, Part XII.
70 UNCLOS Art 207(1) and Art 212(1).
71 UNCLOS Arts 208(3), 209(2) and 210(6).
72 UNCLOS Art 211(2).
73 UNCLOS Section 6, Part XII.

Cable Deployment and Marine Pollution

The first point to note is that the laying or repair of cables does *not* fall within the definition of marine pollution provided in UNCLOS. Article 1(4) defines "pollution of the marine environment" as:

> the introduction by man, directly or indirectly, of substances or energy into the marine environment, including estuaries, which results or is likely to result in such deleterious effects as harm to living resources and marine life, hazards to human health, hindrance to marine activities, including fishing and other legitimate uses of the sea, impairment of quality for use of sea water and reduction of amenities.[74]

While the definition of marine pollution is deliberately wide so as to accommodate any type of pollution where it results in harmful effects, it is highly unlikely that the laying or repair of cables *per se* can be deemed to be "marine pollution". While arguably the deployment of cables is an "introduction by man, directly or indirectly, of substances or energy into the marine environment", as explained in Part I, cables cause minimal disturbance to the seabed and the surrounding marine environment. While power cables do generate EMF, there has been no conclusive evidence to show that they actually cause harm to "living resources and marine life". Fiber optic cables certainly do not cause any "harm to living resources and marine life". Neither power nor telecommunications cables result in "hazards to human health, hindrance to marine activities, including fishing and other legitimate uses of the sea, impairment of quality for use of sea water and reduction of amenities".

Coastal State Authority to Impose Environmental Regulations on Cables/Cable Operations

Within territorial seas and archipelagic waters, coastal States have extensive authority to take measures to protect the marine environment pursuant to their sovereignty over these zones,[75] including measures *vis-à-vis* cable operations. In the EEZ and continental shelf, however, the situation is less clear. In the EEZ, the coastal State is given jurisdiction with regard to the protection and preservation of the marine environment "as provided for in the relevant provisions of this Convention".[76] There does not appear to be any express provision giving coastal States the jurisdiction to impose regulations on cable operations or cables in the EEZ/continental shelf in order to protect the marine environment, apart from coastal State measures to prevent ship-source pollution. The latter would apply

---

74 UNCLOS Art 1(4).
75 UNCLOS Arts 2 and 49.
76 UNCLOS Art 56(b)(iii).

to cableships like any other vessel.[77] The only other arguably relevant provision is Article 208 on pollution from seabed activities which states the following:

> Coastal States shall adopt laws and regulations to prevent, reduce and control **pollution** of the marine environment arising from or in connection with **seabed activities subject to their jurisdiction** and from artificial islands, installations and structures under their jurisdiction, pursuant to articles 60 and 80.[78]

It is unlikely that coastal States could use this provision to impose environmental regulations on cables and/or cable operations. As argued above, submarine cables and cable operations do not constitute "pollution" under UNCLOS. Further, cables and cable operations in the EEZ/continental shelf are not "seabed activities" under the "jurisdiction" of the coastal State, but rather are one of the freedoms that other States are permitted in the EEZ and continental shelf of the coastal State.

Similarly, the provision on the freedom to lay cables on the continental shelf also seems to imply that coastal States may not subject cable operations to environmental regulations. Article 79(2) of UNCLOS states that:

> Subject to its right to take reasonable measures for the exploration of the continental shelf, the exploitation of its natural resources and the prevention, reduction and control of pollution from pipelines, the coastal State may not impede the laying or maintenance of such cables or pipelines.

This suggests that a coastal State may subject the laying and repair of cables on its continental shelf only to reasonable measures related to the exploration of the continental shelf and exploitation of its natural resources, and that only *pipelines* may be subject to measures for preventing, reducing and controlling pollution.

States or companies exercising the right to lay and repair cables must, of course, have due regard to the rights and duties of the coastal States, including those rights and duties related to the marine environment.[79] However, this is an obligation on other States undertaking activities in the EEZ to take into consideration the coastal State's rights and duties with regard to its jurisdiction over the protection and preservation of the marine environment (something which is arguably done during the Survey Phase of cable operations when efforts are made to avoid environmentally sensitive areas). This does not translate into coastal State authority to impose environmental measures on cable operations in the EEZ and continental shelf. This is reinforced by Article 194(4) which states that:

> In taking measures to prevent, reduce or control pollution of the marine environment, States shall refrain from unjustifiable interference with activities carried out by other States in the exercise of their rights and in pursuance of their duties in conformity with this Convention.

---

77 UNCLOS Art 211.

78 UNCLOS Art 208(1). Emphasis added.

79 UNCLOS Art 58(3).

The next three sections will discuss three specific examples of environmental regulations imposed by coastal States on cable operations, namely environmental impact assessments (EIAs), regulations imposed pursuant to Marine Protected Areas (MPAs) or Marine Spatial Planning (MSP) policies and certain environmental practices in cable laying and operation adopted in the Northeast Atlantic Ocean.

Environmental Impact Assessments

Environmental Impact Assessments under International Law

As States seek to protect their coasts and oceans through various legislative measures there are increasing requirements to assess the actual and potential effects of offshore activities on the marine environment. Such requirements are well established in Australasia, Europe, North America and parts of Asia.[80]

Evaluation of an activity's effect on the marine environment is typically covered by an Environmental Impact Assessment (EIA) report. EIAs are important tools for coastal States in the protection of their marine environment. Indeed, under general international law, it has been held that an EIA has gained so much acceptance among States that it is now considered obligatory to carry out an EIA where there is a risk that a proposed activity may cause considerable trans-boundary environmental effects (i.e. effects extending across national boundaries).[81] For activities that do not cause trans-boundary effects (i.e. environmental effects confined within national boundaries), there is wide consensus that there is also an obligation to carry out an EIA if the activity is likely to have an impact on the environment and the impact is significant.[82] Article 206 of UNCLOS itself states that:

> When States have reasonable grounds for believing that planned activities under their jurisdiction or control may cause substantial pollution of or significant and harmful changes to the marine environment, they shall, as far as practicable, assess the potential effects of such activities on the marine environment and shall communicate reports of the results of such assessments in the manner provided in article 205.

---

80 For example, the Hong Kong Environmental Protection Department, "Environmental Impact Assessment Ordinance" (2002). See also C2C Cable Network Hong Kong Section at http://www.epd.gov.hk/eia/register/profile/latest/e_dir46.pdf (last accessed 7 June 2013).

81 See *Pulp Mills on the River Uruguay* (Argentina v Uruguay), International Court of Justice (ICJ) Judgment of 20 April 2010, available at http://www.icj-cij.org/docket/files/135/15877.pdf at para 203.

82 See generally, A. Epiney, "Environmental Impact Assessment" in R. Wolfrum (ed), *The Max Planck Encyclopedia of Public International Law* Volume III (Oxford University Press, 2012) at 587–589.

However, neither general international law nor UNCLOS specifies the precise scope and content of an EIA and there is no formulation under international law on the manner and procedure to be applied when requesting an EIA.[83] Further, both international law and UNCLOS lack clarity on who has the responsibility for conducting an EIA. Article 206 of UNCLOS, for example, appears to place the responsibility for undertaking an EIA on the State, but in practice the State usually leaves it to the company or enterprise that is undertaking the activity.

Environmental Impact Assessments for Cables/Cable Operations

EIAs, in the case of submarine cables, assess any environmental impacts of cable route surveys, cable laying and maintenance. This information is required before permission is granted to deploy a cable. Typically, an EIA covers the following aspects of a cable project: (i) the nature of the proposed project incorporating basic information on the route, type and length of cable, cable laying information including burial, timing and duration of the operation plus other factors; (ii) documentation on the environment that commonly encompasses relevant information on the oceanography, seafloor geology, biology, natural hazards, human activities and social aspects such as avoidance of sites of historical or cultural significance; (iii) potential effects of the project encompassing cable route survey and laying operations; (iv) measures required to reduce any negative impacts to an acceptable level (which may involve restrictions on timing and location of operations, requirements to restore any disturbed setting, installation of observers to prevent collisions with marine mammals, etc) and (v) monitoring to ensure that remedial measures are effective. Major EIAs are likely to be substantial technical documents. Accordingly, they often contain a non-technical summary that is accessible by the public for consultation.

However, the nature of the required information can vary between States and even within a single State. Requirements may involve (i) a brief review of environmental conditions and potential impacts; or (ii) an appropriate technical assessment accompanied by a statement of compliance with environmental accreditation or (iii) a comprehensive analysis that requires additional research, e.g. field measurements, computer modeling and formal consultation with government, other seabed users and the public.[84] In that context, completion of EIAs may take weeks or even years in extreme cases.[85] For example, California has some of the most onerous environmental permitting requirements of any state in the United States and obtaining an environmental impact report can take

83 *Pulp Mills on the River Uruguay* (Argentina v Uruguay) *supra* note 81 at para 205.

84 Carter *et al.*, *supra* note 25.

85 See for example, Monterey Bay National Marine Sanctuary, EIR/EIS for MBARI MARS Cabled Observatory, refer http://www.mbari.org/staff/linda/MARS%20Cable%20Environmental%20Impact%20Report%20through%202010.pdf (last accessed 7 June 2013).

several years, causing considerable delay to the deployment and landing of a cable system.[86]

Permission to deploy a submarine cable is often conditional on cable operators submitting an EIA to the coastal State. But is this condition consistent with UNCLOS? Within territorial waters, coastal States certainly have the authority to request an EIA before cable laying operations commence. Whether it is an *obligation* is not as clear-cut. Both international law and UNCLOS require that there must be, at the very least, reasonable grounds for believing that activities may cause a substantial adverse impact to the marine environment.[87] Given the relatively benign nature of submarine cables and cable operations, including the fact that they do not cause significant harm to the marine environment, there are grounds for arguing that there is no obligation to require an EIA. In any event, regardless of whether it is a right or obligation under international law and/or UNCLOS, coastal States may wish to reconsider extensive EIA requirements that unduly interfere with cable operations or result in an impractical cost benefit analysis.

Within the EEZ/continental shelf, the coastal State does not appear to have authority to request an EIA for cable deployment in these zones. First, as noted above, the right of coastal States to impose environmental regulations on cable operations appears to be limited. Second, Article 206 provides that States shall conduct an EIA when they have "reasonable grounds for believing that planned activities **under their jurisdiction or control** may cause substantial pollution of or significant and harmful changes to the marine environment" (emphasis added). Cable operations in the EEZ/continental shelf are not "under the jurisdiction or control" of the coastal State but rather are one of the freedoms allocated to other States.[88]

Nonetheless, even if an EIA for cable operations in the EEZ is not recognized under UNCLOS, this does not mean to say that cable companies should not be cognizant of the potential environmental impacts of their operations. Indeed,

---

86 See A. Lipman and Nguyen T. Vu, "Building a Submarine Cable: Navigating the Regulatory Waters of Licensing and Permitting" Submarine Telecoms Forum, Finance and Legal Edition (March 2011), available at http://www.bingham.com/Publications/Files/2011/04/Building-a-Submarine-Cable-Navigating-the-Regulatory-Waters-of-Licensing-and-Permitting (last accessed 7 June 2013).

87 See UNCLOS Art 206 and A. Epiney, "Environmental Impact Assessment" in R. Wolfrum (ed), *The Max Planck Encyclopedia of Public International Law* Volume III (Oxford University Press, 2012) at 587–589.

88 This raises an interesting question as to whether the State or company of the State which is conducting cable operations are obliged under UNCLOS Art 206 to conduct an EIA in the EEZ/continental shelf on the basis that these activities are under their jurisdiction or control. In any event, such States/companies can always argue that there are no reasonable grounds for believing that cable activities may cause substantial pollution or harm to the marine environment.

cable companies may themselves wish to avoid environmentally sensitive areas to avoid any potential damage and controversy. In practice cable companies normally take steps to avert this issue during the initial Survey Phase during which they ensure that fragile ecosystems are identified and bypassed as possible locations for cable laying.[89]

Marine Protected Areas and Marine Spatial Planning

Another recent trend, which has had an impact on the freedom to lay, repair and maintain submarine cables, is the tendency of coastal States to designate areas outside of territorial waters as marine protected areas (MPAs) or conservation areas. This section will give an overview of MPAs, and a related tool for the protection of the marine environment, Marine Spatial Planning (MSP). It will also discuss the basis under international law for these protective regimes and examine the ways in which they have impacted cable operations.

Overview of Marine Protected Areas and Marine Spatial Planning

The marine environment has come under pressure from an increased human presence offshore. Area-based management is recognized as one of the ways in which the marine environment can be protected from some uses. Area-based management encompasses a range of tools, which can have a wide variety of objectives, such as the conservation and management of species or protection of fragile habitats and are "designed to achieve these objectives by managing human activities within a spatially defined area".[90]

One of the most widely recognized area-based management tools is the concept of the MPA.[91] An MPA has been defined as "any area of intertidal or subtidal terrain together with their overlying waters and associated flora, fauna, historical and cultural features, which has been reserved by law or other effective means to protect all or part of the enclosed environment".[92] In practice, there are many types of MPAs, which can either address protection of a single species or a fragile habitat. For instance, one of the more dramatic examples which prompted the need for an MPA was the indiscriminate fishing of deep-water corals. These corals form fragile communities that reside in water depths ranging from 40 m

---

89 See Chapter 4 on the Planning and Surveying of Submarine Cable Routes.

90 J. Roberts *et al.*, "Area-based Management on the High Seas: Possible Application of the IMO's Particularly Sensitive Sea Area Concept" *International Journal of Marine and Coastal Law* 25 (2010) 483–522 at 484.

91 *Ibid.*

92 World Conservation Union, Resolution 17.38 of the 17th General Assembly of the International Union for Conservation of Nature and Natural Resources (IUCN), Proceedings of the 17th Session of the General Assembly of IUCN and 17th Technical Meeting, 1–10 February 1988, San Jose, Costa Rica 104–106.

to more than 1000 m and provide habitats for fish and other organisms.[93] One suite of deep-water coral communities—the Darwin Mounds off northeast Scotland—were found to be extensively damaged by bottom trawling.[94] This led the European Commission to close off the area to bottom trawl fishing and assign it protective status.

MPAs have grown in size and number. The majority of MPAs are designated within the 200 nm EEZ.[95] As of 2012, the total area assigned to MPAs was 11,254,389 km^2 or about 3.2 per cent of the world's ocean.[96]

Concomitant with the expansion of MPAs is the development of policies to regulate offshore activities in ocean spaces around the world, including Europe, the United Kingdom, Australia, New Zealand, Canada, the United States and other regions. Commonly referred to as *Marine Spatial Planning* (MSP), it is designed to provide frameworks to address various marine and coastal issues relating to environmental conservation and sustainability, commercial and recreational activities, and conflicts between offshore stakeholders. Such a need is perceived from an increased recreational, industrial, scientific and security presence offshore. In the case of the United States, a National Ocean Council is now in place to implement policy concerning stewardship of coasts and oceans (plus the Great Lakes).[97] Its aims include:

- Improving the resiliency of ecosystems, communities and economies;
- Protecting, maintaining and restoring the health and biological diversity of oceans;
- Advancing scientific knowledge and understanding to improve decisions relating to a changing global environment and other issues;
- Supporting sustainable, safe, secure and productive access to, and uses of the ocean;
- Exercising rights and jurisdiction in accordance with applicable international law that involve respect for and preservation of navigational rights and freedoms;

---

93 Freiwald *et al.*, *supra* note 1; see also UNEP, Ecosystems and Biodiversity in Deep Waters and High Seas (2006) UNEP Regional Seas Report and Studies, No 178 at 58, available at http://www.unep.org/pdf/EcosystemBiodiversity_DeepWaters_20060616.pdf (last accessed 7 June 2013).

94 A.J. Wheeler *et al.*, "The Impact of Demersal Trawling on NE Atlantic Deepwater Coral Habitats: the Case of Darwin Mounds, United Kingdom" in P.W. Barnes and J.P. Thomas, eds, *Benthic Habitats and the Effects of Fishing* (American Fisheries Society, 2004) 807–817.

95 Roberts *et al.*, *supra* note 90 at 485.

96 Marine Reserves Coalition (2012), see 'Marine Protected Areas' http:///www.marinereservescoalition.org/ (last accessed 7 June 2013).

97 National Ocean Council (2013) available at http://www.whitehouse.gov/administration/eop/oceans (last accessed 7 June 2013).

- Increasing scientific understanding of ocean ecosystems including their relationships to humans and their activities; and
- Fostering public understanding of the oceans.

While the wording may differ between different national MSP policies, most strive to maintain a balance between sustainable offshore development and the maintenance of a healthy and sustainable marine environment.[98]

Basis of Marine Protected Areas under International Law

Within territorial waters, MPAs can be established by the coastal State pursuant to their sovereignty over this zone. Within the EEZ, UNCLOS has several specific provisions, which arguably provide the basis for the establishment of different types of MPAs. First, UNCLOS gives coastal States sovereign rights for the purpose of conserving and managing the natural resources, whether living or non-living[99] and also places an obligation on coastal States to "ensure through proper conservation and management measures that the maintenance of the living resources in the exclusive economic zone is not endangered by over-exploitation".[100]

Second, UNCLOS also provides some basis for coastal States to take measures to "protect and preserve rare or fragile ecosystems as well as the habitat of depleted, threatened or endangered species and other forms of marine life", and such measures would include MPAs.[101] Whilst this is a "separate and independent legal obligation" distinguishable from the obligation to prevent pollution, it has been said that it is not a jurisdictional rule which creates jurisdiction of the coastal State with regard to its EEZ.[102]

Third, there are three categories of MPAs, which specifically target ship-source pollution.[103] The first category are MPAs designated pursuant to Article 211(6) of UNCLOS where the adoption of special mandatory measures for the prevention of pollution is required in a certain area of the EEZ for "recognized technical reasons in relation to its oceanographical and ecological conditions, as well as its utilization or the protection of its resources and the particular character of its traffic". The second category is the designation of Particularly Sensitive Sea Areas

98 Department of Environment, Food and Rural Affairs (DEFRA) UK Marine Policy Statement (2011) available at https://www.gov.uk/government/publications/uk-marine-policy-statement (last accessed 7 June 2013).
99 UNCLOS Art 56(1)(a).
100 UNCLOS Art 61(2).
101 UNCLOS Art 194(5).
102 R. Lagoni, "Marine Protected Areas in the Exclusive Economic Zone" in A. Kirchner (ed), *International Marine Environmental Law: Institutions, Implementation and Innovations* (Kluwer Law, 2003) at 160.
103 *Ibid.*, at 160–161.

(PSSAs) under the auspices of the International Maritime Organization (IMO).[104] The third category is Special Areas adopted under the International Convention for the Prevention of Pollution from Ships and its protocol (MARPOL 73/78).[105]

It is clear that coastal States do have some legal basis for adopting MPAs in their EEZs, however, the question is to what extent can coastal States restrict the rights of other States within these MPAs including the right to lay and repair submarine cables. This will be dealt with in the next section.

Cable Operations within Marine Protected Areas

It seems reasonably clear that within a MPA, coastal States can restrict activities over which they are given sovereign rights and jurisdiction under UNCLOS,[106] such as fishing and resource exploration and exploitation. However, it is also widely agreed that the ability of coastal States to restrict recognized freedoms such as navigation within MPAs is limited.[107] The IMO, as the competent international organization responsible for shipping, is the only body that can control navigation through MPAs.[108] While there is no equivalent body for submarine cables, by the same reasoning it can be argued that as with navigation, coastal States do not have the authority to impose blanket prohibitions on the laying or repairing of cables within MPAs.[109]

Despite this, there have been several instances of MPAs being adopted in the EEZ which restrict cable operations in these areas. For example, the United Kingdom sought to control the routing of a new cable system[110] initially planned to enter designated Special Areas of Conservation outside its territorial waters.[111] Another example is the designation by US authorities of additional areas in the

---

104 Information on the designation of PSSAs is available on the IMO website at http://www.imo.org/OurWork/Environment/pollutionprevention/pssas/Pages/Default.aspx (last accessed 7 June 2013).

105 1978 Protocol Relating to the 1973 International Convention for the Prevention of Pollution from Ships (including Annexes, Final Act and 1973 International Convention), adopted 17 February 1978, 1340 UNTS 61 (entered into force 2 October 1983) (MARPOL 73/78).

106 See generally UNCLOS Art 56.

107 E.J. Goodwin, *International Environmental Law and the Conservation of Coral Reefs* (Routledge, 2011) at 52.

108 *Ibid.*, at 53.

109 Also refer to the discussion above on the ability of coastal States to impose environmental regulations on cable/cable operations.

110 D. Toombs and R. Carryer, "Jurisdictional Creep and the Retreat of UNCLOS" Paper Presented at the 2010 SubOptic Conference, Yokohama, Japan 11–14 May 2010 (personal copy with authors).

111 See United Kingdom Marine Conservation (Natural Habitats) Regulation 2007 available at the UK Legislation Web site, available at http://www.legislation.gov.uk/uksi/2007/1842/contents/made (last accessed 7 June 2013).

US EEZ as a "critical habitat" for the protection of the leatherback sea turtle.[112] Given the already wide-ranging environmental permitting requirements imposed by the US for cable operations, the industry, represented in this case by the North American Submarine Cable Association (NASCA)[113] was concerned that the designation would add an additional requirement for cable operations in these designated areas.[114] In particular, NASCA was apprehensive that some agencies may interpret this regulation as imposing a requirement on Federal Agencies to initiate a consultation before taking any action (including permit requirements) that may affect an endangered species or its critical habitat.[115] In their view, this would "impose substantial additional permitting costs and delays on undersea cable operators without any corresponding increase in the protection of leatherback sea turtles"[116] and they requested that the US authorities confirm that such a consultation would not be required.[117] While the US authorities considered the comments by NASCA, it ultimately did not clarify this requirement as they stated that it was the responsibility of the relevant agency to determine if such a consultation was required.[118]

There have also been examples of MPAs, which protect the marine environment and permit cable deployment and related operations. For example, Australia has been particularly active in the protection of the marine environment, with the federal government announcing a proposal to increase the number of MPAs from 27 to 60 thus creating the world's largest network of MPAs, known locally as *reserves*.[119] The network covers 3.1 million km^2 and extends out to the

---

112 See generally Comments of the North American Submarine Cable Association (NASCA) Before the National Oceanic and Atmospheric Administration, US Department of State, In the Matter of *Endangered and Threatened Species: Proposed Rule to revise the Critical Habitat Designation for the Endangered Leatherback Sea Turtle,* Docket No 0808061067-91396-01 RIN 0648-AX06, dated 23 April 2010, available at http://www.n-a-s-c-a.org/app/download/2942651913/NASCA+Comments+re+Leatherback+Sea+Turtle+Habitat.pdf?t=1272296427.

113 The North American Submarine Cable Association (NASCA) is a regional cable protection committee which consists of an "organization of companies that own, install or maintain submarine telecommunications cables in the waters of North America". For more information, see the NASCA Website available at http://www.n-a-s-c-a.org/ (last accessed 7 June 2013).

114 *Ibid.*, at page 3.

115 This is a requirement under Section 7 of the Endangered Species Act.

116 See NASCA Comments, supra note 112 at 1.

117 *Ibid.*

118 See Response to Comment 48 in Federal Register, Volume 77 Issue 17 dated 26 January 2012 available on line at http://www.nmfs.noaa.gov/pr/pdfs/fr/fr77-4170.pdf (last accessed 7 June 2013).

119 Australian Government, Announcement of the Final Commonwealth Marine Reserves Network Proposal (2012), available at http://environment.gov.au/coasts/mbp/reserves/index.html (last accessed 7 June 2013).

200 nm limit of the EEZ. Three levels of protection are designated: (i) *Marine National Parks*, which have the highest level of protection whereby commercial activities are prohibited apart from vessel passage and some aspects of tourism; (ii) *Multiple Purpose Zones* that allow activities such as recreational fishing, some types of commercial fishing and resource exploration while maintaining designated conservation values, and (iii) *Special Purpose Zones* that permit additional commercial activities, but still exclude those deemed damaging to an ecosystem.

Generally, the laying and maintenance of submarine telecommunications and power cables are generally permitted or allowable activities in multi-purpose protected areas such as those adopted by Australia, especially in light of their designation as *critical infrastructure*, their special status under UNCLOS and their low environmental impact.[120] Similarly, commercial and science cables reside in National Marine Sanctuaries in the United States.[121] It is clear that MPAs and cable operations are not mutually exclusive.

Cable Protection Zones as Marine Protected Areas?

For those States with designated cable protection zones,[122] there has been the suggestion that such areas may act as MPAs as they prohibit potentially hazardous and environmentally damaging activities such as bottom trawl fishing, ships' anchoring and seabed mining.[123] To assess the validity of that suggestion, a study was undertaken of a cable protection zone off Auckland, New Zealand.[124] No statistically valid difference was found in fish species inside or outside the zone—an observation attributed to the short four-year existence of the protection zone and illegal fishing within the protected zone. However, where the zone included reefs, there was a preferred concentration of fish within the zone, suggesting some protective effect. While the results are inconclusive, they demonstrate that for a cable protection zone to act as an MPA, it must have suitable fish habitats and effective policing to prevent poaching.

---

120 See Carter *et al., supra* note 25; see also OSPAR *supra* note 62.

121 See Grannis, *supra* note 16; see also NOAA (2005) *supra* note 34 and NOAA (2007) *supra* note 53; see also Kogan *et al., supra* note 10; see also Monterey Bay National Marine Sanctuary *supra* note 85.

122 See Chapter 11 on Protecting Submarine Cables from Competing Uses.

123 See for example, ACMA, *supra* note 7; V.A. Froude and R. Smith, "Area-based Restrictions in the New Zealand Marine Environment" (2004) Department of Conservation MCU Report, available at http://www.marinenz.org.nz/nml/files/documents/7_minfish/minfish_froude_04.pdf (last accessed 7 June 2013); see also Cook Strait Submarine Cable Protection booklet *supra* note 8.

124 N.T. Shears and N.R. Usmar, "The Role of the Hauraki Gulf Cable Protection Zone in Protecting Exploited Fish Species: De Facto Marine Reserve?" (2005) Department of Conservation Research and Development Series 253 at 27.

OSPAR Commission Guidelines on Best Environmental Practice in Cable Laying and Operation

Perhaps one of the more striking examples of environmental regulations which encroach upon the freedom to lay and repair cables is the *Guidelines on Best Environmental Practice in Cable Laying and Operation* (BEP Guidelines) issued by the OSPAR Commission in 2012.[125] The OSPAR Commission was established to administer and implement the Convention for the Protection of the Marine Environment of the Northeast Atlantic (the OSPAR Convention).[126] The OSPAR Convention is a "mechanism by which fifteen Governments of the western coasts and catchments of Europe, together with the European Community, cooperate to protect the marine environment of the North-East Atlantic".[127] It covers a wide area that is divided into five regions comprising Arctic waters, the Greater North Sea, Celtic Sea, the Bay of Biscay and the Iberian Coast, and the Wider Atlantic.[128]

The BEP Guidelines are based on two reports issued by the OSPAR Commission on the environmental impact of submarine cables.[129] Its purpose is, *inter alia*, to set out possible measures to avoid and mitigate any ecological impacts of construction, operation and removal of underwater cables, differentiate between possible measures regarding various types of sea cables and identify remaining gaps in knowledge and the resulting specific research needs.[130] However, there are several issues with both the assumptions underlying the BEP Guidelines and the recommendations contained in it, some of which are highlighted in the following paragraphs.

---

125 Agreement 2012-2, OSPAR 12/22/1, Annex 14. Available for download from the OSPAR Commission website http://www.ospar.org/v_measures/browse.asp?menu=00750302260125_000002_000000 (last accessed 7 June 2013).

126 Convention for the Protection of the Marine Environment of the North-East Atlantic, adopted 22 September 1992, 2354 UNTS 67 (entered into force 25 March 1998) (OSPAR Convention).

127 See the OSPAR Commission Website available at http://www.ospar.org/content/content.asp?menu=00010100000000_000000_000000 (last accessed 7 June 2013).

128 For more information on the Regions that are covered by the OSPAR Convention, see http://www.ospar.org/content/content.asp?menu=00420211000000_000000_000000 (last accessed 7 June 2013).

129 OSPAR Commission, "Background Document on Potential Problems Associated with Power Cables other than those for Oil and Gas Activities" (2008) Publication Number 370/2008 available online at http://www.ospar.org/documents/dbase/publications/p00370_cables%20background%20doc.pdf (last accessed 7 June 2013); OSPAR Commission "Assessment of the Environmental Impacts of Cables" (2009) Publication Number 437/2009 available online at http://qsr2010.ospar.org/media/assessments/p00437_Cables.pdf (last accessed 7 June 2013).

130 BEP Guidelines, *supra* note 125 at para 1.

- The BEP Guidelines cite oil leaks from power cables as the basis for regulation.[131] The use of oil in ocean cables became obsolete in the 1990s.[132] Modern ocean power cables use mass impregnated paper or XPLE (cross-linked polyethylene) for insulation, and no oil at all is used.[133] The modern plastics used in HVDC cables for insulation are environmentally benign.
- The BEP Guidelines state that modern cable installation techniques like burial have a "lethal effect on some [unnamed] species".[134] However, no citations or studies are provided to support this claim.
- The BEP Guidelines require that existing cables and pipelines be bundled.[135] This requirement is problematic on several levels. First, submarine cables when repaired in the ocean are usually located by grapnel runs.[136] Cable owners will be very unwilling to drag grapnels over oil and gas pipelines where contact could result in an oil spill or pipe rupture. Second, cables are critical international infrastructure upon which the internet and international communications depend. Co-locating cables with pipelines increases the risk of damage to both the cables and pipelines, makes any repair more complicated and dangerous, and increases the risk of marine pollution from a damaged pipeline. In addition, the effects of a single event, such as a terrorist attack or natural weather phenomenon, may have wider repercussions if cables and pipelines are located in close proximity. If a number of telecommunication cables are damaged in a single event then there will be reduced re-routing options, and repairs will be hampered because multiple repair vessels will need to be sourced and will be required to operate simultaneously in the same area.
- The BEP Guidelines require that cables laid in any area be buried to a depth of 1–3 m to reduce speculative heat impacts.[137] Modern fiber optic cables, especially those laid beyond the continental shelf and upper continental slope (< ~1500 m water depth), are simply draped on the seabed with minimal impact. Cables are only buried on the continental shelf/upper in order to avoid bottom trawlers and ship anchors. Burial of cables in great ocean depths (nominally taken as > 2000 m) is not technically possible at the present time.
- The BEP Guidelines mandate that explosives not be used for burial.[138] Cable owners never use explosives to lay or maintain cables for the simple reason that to do so would damage or destroy the cable. The cable industry itself, represented by the International Cable Protection Committee (ICPC), recommends

---

131 *Ibid.*, BEP at 2.0 and 3.6 (Insulation of power cables).
132 See Figure 13.1 in Chapter on Power Cables.
133 See "About Power Cables" www.iscpc.org, under 'Publications'.
134 BEP Guidelines, *supra* note 125 at para 3.2.
135 BEP Guidelines, *ibid.*, at para 5.2.1.
136 See Figure 6.4 in Chapter on Submarine Cable Repair and Maintenance.
137 BEP Guidelines, *supra* note 125, at para 5.3.1.
138 BEP Guidelines, *ibid.*, at para 5.2.2.

that air guns and other seismic techniques not be used near cables as their use will damage sensitive optical amplifiers.[139]

- The BEP Guidelines require removal of out-of-service cables.[140] However, any removal should be made with due regard to the impact of removal on the benthic environment. Removal may produce a disturbance that has a more negative impact than leaving the cable on the seabed, bearing in mind that the environmental impact of cables is so benign that they have been used by governments for artificial reefs for years.[141] (Figures 7.2 and 8.1.) However, where removal has little environmental impact, especially for surface laid cables in water depths > ~1500 m, cable recovery can be undertaken (see above, Cable Recycling and Life Cycles and Chapter 8).[142]
- The BEP Guidelines also require formal Environmental Impact Statements (EIS) be prepared for the high seas.[143] This is not consistent with UNCLOS, which provides that waters beyond of national jurisdiction, especially the high seas, are not subject to regulation by any single State or group of States such as OSPAR.
- The BEP Guidelines require cable companies to pay for mitigation for not complying with the above requirements,[144] presumably with the payments being made to OSPAR itself, but the specifics of who will receive the 'ecological compensation measures' funds and how amounts will be determined and divided by OSPAR are not provided.

One of the major reasons for the above problems with the BEP Guidelines is the fact that they were conceived and issued without input from stakeholders in the telecommunication and power cable industries. A survey of all of the major telecommunications companies and power cable companies by the ICPC in August 2012 revealed that none of these companies had been contacted nor were

---

139 ICPC Recommendation No. 7A, Offshore Seismic Survey Work in the Vicinity of Active Submarine Cable Systems.

140 BEP Guidelines, *supra* note 125 at para 5.3.3.

141 See http://www.state.nj.us/dep/fgw/artreef.htm (last accessed 7 June 2013). The Division of Fisheries and Wildlife, New Jersey Department of Environmental Protection, New Jersey has an extensive artificial reef programme on the continental shelf. Fourteen sites are defined, each site containing clusters of artificial reefs. Since 1984, over 2100 artificial reefs have been built, covering about 2 per cent of the 65 square kilometers enclosed by the designated sites. The website also leads to the *New Jersey Reef News*, containing useful summaries of scientific investigations of artificial reefs, e.g. *NJ Reef News 2000.*

142 ICPC Recommendation No. 1 Recovery of Out-of-Service Submarine Cables; see also D. Burnett, "The Legal Status of Out-of-Service Submarine Cables" (July/August 2004) 137 *Journal of Maritime Studies* 22.

143 BEP Guidelines, *supra* note 125 at para 5.2.1.

144 BEP Guidelines, *ibid.*, at para 4.0 [Implementation of ecological compensation measures].

they even aware of BEP. These included major companies in the OSPAR region such as British Telecom, France Telecom, TDC (Denmark) and Deutsche Telecom. Similarly, there was no consultation with industry trade groups such as the ICPC, CIGRE, Subsea Cables UK, and the Danish Cable Protection Committee. As a result, the BEP recommendations inevitably contain misconceptions about cables and cable operations. Moreover, the OSPAR Commission itself consists of environmental ministries, but not ministries associated with telecommunications or power cable infrastructure, which meant that the latter's input was also omitted from the drafting of the BEP Guidelines. In an effort to open dialogue with OSPAR, Subsea Cables UK applied for observer status with OSPAR on 25 March 2013. OSPAR informed Subsea Cables UK on 4 July 2013 that the application had been vetoed by Germany.

Another possible explanation for the various issues in the BEP Guidelines is that it heavily relies on the precautionary principle, which has acquired special significance in international environmental law.[145] The precautionary principle is based on the concept that "where there are threats of serious or irreversible damage, lack of full scientific certainty shall not be used as a reason for postponing cost-effective measures to prevent environmental degradation".[146] While the legal status and content of the precautionary principle is still evolving and its status as a principle of customary international law has not been confirmed with any degree of certainty,[147] the OSPAR Convention obliges States Parties to apply the precautionary principle.[148] Accordingly, even though the BEP Guidelines acknowledge that there are gaps in the knowledge of noise impacts on fauna, heat impacts on benthic species and electromagnetic impacts on the orientation of fish and marine mammals,[149] it concludes "there is sufficient evidence that placement and operation of submarine cables *may* affect the marine environment" and hence, the precautionary principle should be applied.[150] The BEP Guidelines include no

145 Birnie *et al.*, *supra* note 66 at 154.

146 Principle 15 of the Rio Declaration on Environment and Development available online at http://www.unep.org/documents.multilingual/default.asp?documentid=78&articleid=1163 (last accessed 7 June 2013).

147 For a description of the debate on the status of the precautionary principle, please refer to M. Schroder, "Precautionary Approach/Principle" in R. Wolfrum (ed), *The Max Planck Encyclopedia of Public International Law* (Oxford University Press, 2012) at 400–405.

148 See the OSPAR Convention, Art 2(2)(a) which states that the Contracting Parties shall apply "the precautionary principle, by virtue of which preventive measures are to be taken when there are reasonable grounds for concern that substances or energy introduced, directly or indirectly, into the marine environment may bring about hazards to human health, harm living resources and marine ecosystems, damage amenities or interfere with other legitimate uses of the sea, even when there is no conclusive evidence of a causal relationship between the inputs and the effects".

149 BEP Guidelines, *supra* note 125 at 3.2–3.5 and 8.0.

150 BEP Guidelines *ibid.*, at 8.0.

cost versus benefit analysis of its controversial measures. For the cable industry and the international community, the application of the precautionary principle and its accompanying consequences for cable operations in such a wide expanse of ocean area has potentially far-reaching consequences.

Conclusions

Submarine cables are the arteries through which global commerce and human telecommunications pass. Cables also underpin the rapidly developing offshore renewable energy sector as well as scientific ocean observatories. In these critical roles, cables have nonetheless been recognized as having low to negligible impact on the marine environment. As has been made clear in this Chapter, the marine environment can be sufficiently protected without the need to unduly restrict the freedom to lay and repair submarine cables. The few environmental issues surrounding cables should be resolved by meaningful scientific research documented in quality peer-reviewed journals. When imposing environmental regulations on cables and cable operations, coastal States should carefully consider whether the regulations are consistent with international law and whether they achieve the purpose of protecting the marine environment. The coastal State should also consider the impact of such regulations on cable deployment. Perhaps most importantly, there should be a dialogue between industry, governments and environmental organizations. After all, companies and industry organizations such as the ICPC have shown themselves to be open to working with environmental organizations and independent scientists to better determine the interactions of submarine cables and the marine environment. Such fruitful cooperation between science and industry extends back over 150 years and will undoubtedly continue in the future.

CHAPTER EIGHT

OUT-OF-SERVICE SUBMARINE CABLES

Douglas Burnett

Introduction

This Chapter answers the question of what happens to cables that are redundant or are out-of-service, and what are the legal requirements applicable to them under international law. Before examining the legal requirements, it is helpful to understand how out-of-service cables are reused.

I. Understanding Out-of-Service Submarine Cables

Factors that Result in Out-of-Service Status

Commercial considerations govern when a cable's status is changed from 'active' to 'out-of service'. These include the following factors:

- The cable system may have reached the end of its design life, which is typically 20–25 years;
- The increased cost of operating and maintaining the cable may have become such that the owners of the cable system agree to decommission it;
- The need to remove the risk of liability for sacrificed gear and anchor claims and coastal State legal requirements in territorial seas;
- Improved cable technology may cause a cable system to become non-competitive with newer systems. For example, overbuilding may have resulted in a glut of capacity on the cable route making the operation of a cable commercially non-viable, notwithstanding the fact that it may only have been used for as little as 40 per cent of its design life.[1]

[1] D. Burnett, "The Legal Status of Out-of-Service Submarine Cables" (2004) 137 *Maritime Studies* 22–27, reporting on 13 submarine cables taken out-of-service in 2003–2004

Frequently it is a combination of these factors that results in the cable owner deciding to decommission the cable system and change its status to out-of-service.

Reuse of Out-of-Service Cables Commercially, for Scientific Purposes or as Artificial Reefs

The fact that a cable is out-of-service for one purpose, such as telecommunications, does not mean that its life is over or that it has no value to its owners. In some cases, an out-of-service cable is reused for other commercial services. For example in 1988 sections of two analogue cables were reused to form a 327 km cable system between the United States and Cuba.[2] In 2007, a section of the 1287 km Gemini-Bermuda fiber optic cable system was created by cutting an out-of-service cable outside the United States continental shelf boundary and re-laying it so as to land in Bermuda instead of the United Kingdom. By doing this the new owners were able to avoid obtaining new landing permits for a cable landing in the United States. The possibility of reusing cables in this manner has led some in the cable industry to take the view that out-of-service cables are in 'deep storage'.

Out-of-service cables have also been reused for environmental monitoring purposes. For example, the H20 program cable is an out-of-service telecommunications cable that was sold for a token price in 1984 to Incorporated Research Institutions for Seismology (IRIS) and subsequently laid between San Luis Obispo and Oahu Hawaii in order to measure water movements and pressure.[3] Similarly in 1988 the Hawaii-2 Observatory, used for seismology research, was installed on an out-of-service telephone cable halfway between California and Hawaii.[4] Another section of an out-of service cable was reused in 2007 to connect a 10 km radius sensor ring, known as the Aloha Cabled Observatory, which is used to measure temperature, pressure, current, salinity and provides video images.[5] As described in Chapter 15, military acoustic sensor cables have also been reused to increase knowledge of marine mammals.

---

that had an average lifespan of 12 years 4 months of age, well short of their 25 year design life.

2 B. Glover, "History of the Atlantic Cables & Undersea Communications: from the First Submarine Cable of 1850 to the Worldwide Fiber Optic Network" available online at http://atlantic-cable.com/Cables/CableTimeLine/index1951.htm (last accessed 6 June 2013).

3 D. Burnett, "International Law Considerations for Owners and Operators of Cabled and Buoy Observatories" (2006) 31(1) *IEEE Journal of Oceanic Engineering* 230–235 at footnote 17; see also Global Seismic Network http://www.iris.washington.edu/about/GSN/index.htm (last accessed 6 June 2013).

4 Information regarding the Hawaii-2 Observatory is available at http://www.whoi.edu/science/AOPE/DSO/H2O/ (last accessed 6 June 2013).

5 Information about the Aloha Cabled Observatory is available at http://aco-ssds.soest.hawaii.edu/index.html (last accessed 6 June 2013). The observatory lies 100 kilometers north of the island of Oahu, Hawaii (22 45'N, 158W).

Figure 8.1 Recycled out-of-service submarine telecommunication cables provide an artificial reef habitat for fish and mussels in the Ocean City Reef Foundation project off the Maryland coast. The donated cable was fashioned into bundles and laid on the seabed in about 110 feet of water some nine years ago.
(Photograph courtesy of Rick Younger, Ocean City Reef Foundation (OCRF))

Recovered cables have also been successfully reused as artificial reefs by the state governments of Maryland[6] and New Jersey[7] in the United States, confirming the environmentally benign impact of cables in the environment.[8]

Submarine Cables Salvage and Abandoned Cables

Out-of-service analogue cables may have value for commercial salvage because of the value of the copper or other metal components of the cable. In the past, treasure hunters claiming to be cable recovery experts or simply stealing cables have created underwater hazards through unlawful cable recovery activities. Recovery

---

6 Artificial reefs on the Maryland continental shelf are managed by the Ocean City Reef Foundation, a non-profit organization which, like its New Jersey counterpart, aims to restore marine life to once productive waters through emplacement of artificial reefs. And like New Jersey, it has reefs composed of submarine telecommunications cable. See http://www.ocreeffoundation.com (last accessed 6 June 2013).

7 The Division of Fisheries and Wildlife, New Jersey Department of Environmental Protection, New Jersey, has an extensive artificial reef program on the continental shelf. Fourteen sites are defined, each site containing clusters of artificial reefs. Since 1984, over 2100 artificial reefs have been built, covering about two per cent of the 65 square kilometers enclosed by the designated sites. For more information see http://www.state.nj.us/dep/fgw/artreef.htm (last accessed 6 June 2013).

8 Please refer to Chapter 7 on the Relationship between Submarine Cables and the Marine Environment.

of a cable by a third party without the owner's permission is conversion (theft) and any proceeds belong to the cable owners, together with a claim for damages. Even when submarine cables are no longer in use, they continue to be the property of the cable owner and the proceeds of any recovery or reuse are retained by the owner.

Although the practice is not common, there are instances of cable owners entering into commercial salvage contracts in order to recover segments of out-of-service cables. (See Chapter 7, Cable Recycling and Life Cycle at footnote 59.) Experience has shown that great care must be undertaken in the selection of the salver. It is important to ensure that the salver has the requisite level of skill, access to specialized vessels, appropriate salver's liability extensions of the vessel's hull insurance and Protection & Indemnity (P&I) policies, undertakes appropriate record keeping and has an adequate performance history to demonstrate their professional credentials and reliability. The salvage plan should be carefully negotiated and provide for disposal of non-metallic cable components in an environmentally sound manner ashore that avoids ocean dumping. Because salvage is governed by maritime law, experienced admiralty counsel should be consulted by cable owners before entering into a cable salvage agreement. These norms are considered standard cable industry recommendations.[9]

Damages for negligent salvage include increased risk by cable owners by unauthorized or incompetent salvers who selectively retrieve easy to recover sections of cable while leaving other sections with masses of twisted armor wires called 'bird cages' and cable displacements that expose cable owners to increased risk for indemnity claims for sacrificed fishing gear and damage to the seabed environment.

It should be borne in mind that salvage operations in maritime zones under the sovereignty of coastal States (i.e. internal waters, territorial seas and archipelagic waters) are subject to the laws and regulations of the coastal State. Therefore, the coastal State can require consent before salvage operations are carried out in these zones. Salvage operations which take place outside territorial waters (i.e. the exclusive economic zone (EEZ), continental shelf and high seas) are not subject to coastal State consent.

There are instances in which cables have been abandoned and their former owners no longer claim an interest in them. This is more common with regard to abandoned *telegraph* cables. Third parties may obtain title to such cables through admiralty proceedings in a court of competent jurisdiction. The third party must demonstrate to the admiralty court that despite reasonable efforts to locate the owners of the cable, they are no longer in existence, not traceable, or have explicitly renounced their interests in the cable. Due to the continuing international

9 International Cable Protection Committee (ICPC), Recommendation No 1, 11 May 2011, Management of Redundant and Out-of-Service Cables, available upon request from the ICPC at www.iscpc.org (last accessed 6 June 2013).

legal obligations discussed below, salver arguments of implied ownership waivers by cable owners should be treated with caution and in most cases are rejected. In many cases, historical review of even old telegraph cables will determine that the interests of the original owners have been assumed by successor companies who remain in operation. (See Appendix 2) If an admiralty court does determine that a cable has been abandoned, title of the abandoned cable is vested in the third party who can then proceed to recover the cable in accordance with standard industry recommendations.

II. International Law Requirements

Is there a Legal Requirement for the Cable Owners to Remove Out-of-Service Cables?

In maritime zones under the sovereignty of the coastal State—internal waters, territorial seas and archipelagic waters—the laying of cables is subject to the consent of the coastal State. Therefore, the coastal State can require that cables in these maritime zones be removed when they are no longer in service. The more interesting issues arise with regard to cables laid outside the outer limit of the territorial sea, i.e. in the EEZ or on the continental shelf.

International telecommunication cables enjoy unique status under international law and treaties. The freedom to lay, maintain and repair international cables is well established.[10] The United Nations Convention on the Law of the Sea (UNCLOS) addresses the decommissioning of structures and installations, but essentially such installations and structures are constructed to enable the coastal State to exploit the natural resources on the seabed. Even the provisions on the removal of installations and structures do not include pipelines,[11] let alone cables.

---

10 Convention for the Protection of Submarine Telegraph Cables, adopted 14 March 1884, TS 380 (entered into force 1 May 1888) (1884 Cable Convention); the Geneva Convention on the High Seas, adopted 29 April 1958, 13 UST 2312, 450 UNTS (entered into force 30 September 1962) (Geneva Convention); and the United Nations Law of the Sea Convention, adopted 10 December 1982, 1833 UNTS 3 (entered into force 16 November 1994) (UNCLOS), discussed in Chapter 3.

11 UNCLOS Art 60(3) provides for decommissioning of installations or structures on the seabed:

> Any installations or structures which are abandoned or disused shall be removed to ensure safety of navigation, taking into account any generally accepted international standards established in this regard by the competent international organization. Such removal shall also have due regard to fishing, the protection of the marine environment and the rights and duties of other States. Appropriate publicity shall be given to the depth, position and dimensions of any installations or structures not entirely removed.

According to Anderson, "It is implicit within Article 60(3) that the complete removal of all structures is not always required and that partial removal may be acceptable. It

Therefore, the prevailing view is that there is no legal obligation under UNCLOS to remove cables that have been placed on the seabed in the EEZ or on the continental shelf when they are no longer in service.

This widely accepted view is supported by examination of the various UNCLOS provisions dealing with cables and pipelines in the context of the treaty.[12] For example:

- Article 21(1) refers to 'facilities or installations' under (b) and separately to 'cables and pipelines' under (c).
- In Part V on the EEZ a distinction is made between 'artificial islands, installations and structures' (compare with Articles 56(1)(b)(i) and Article 60 throughout) in respect of which the coastal State has *exclusive* rights (Article 60(1) and (2)); and cables and pipelines (compare with Article 58(1)), in respect of which *all* States have rights, so that the rights of the coastal State cannot be exclusive. Moreover, the wording in Article 60(4) and (5) regarding a [500 meter] safety zone *around* installations etc. could hardly apply to pipelines [or cables].
- The same dichotomy can be found in Part VI on the continental shelf where separate provisions are given in respect of artificial islands, installations, and structures (Articles 79(4) and (80)) and in respect of cables and pipelines (Articles 79(1), (2) and (4)) or pipelines only (Article 79(3)).
- Again in Part VII on the High Seas, a similar distinction is made: Articles 87(1)(c) and 112–115 deal with cables and pipelines, while Article 87(1)(d) deals with 'artificial islands and other installations' [in a context that excludes pipelines].... [I]t may be noted that the special regime laid down for high seas cables and pipelines in Articles 112–115 also applies to cables and pipelines in the EEZ (compare with Article 58(2)) and on the continental shelf (compare with Article 78(2)); this supports the conclusion that [cables and] pipelines are not covered by the provisions of Article 60 (and Article 80).
- In Part IX on the Area, Article 145(a) refers to 'installations, pipelines and other devices' from which it follows that a pipeline is not an installation. Article 147(2) refers only to installations which, from the context, do not include pipelines (compare with the provision of safety zones *around* the installations). Article 147(2)(a) contains a specific provision on the *removal* of the installations, which is to be solely in accordance with the provisions of Part XI (and therefore not subject to the mandatory rule of Article 60(3)).[13]

---

is also generally accepted that 'installations and structures' do not include subsea pipelines". See J.M. Anderson, "Decommissioning Pipelines and Subsea Equipment: Legislative Issues and Decommissioning Processes" (2002) 25(2) *Underwater Technology: The International Journal of the Society for Underwater* 105–111.

12 P. Peters *et al.*, "Removal of Installations in the Exclusive Economic Zone", (1984) 15 *Netherlands Yearbook of International Law* 167–207 at 189–190.

13 *Ibid.*

In light of the above, and the fact that the right to lay, repair and maintain cables is a recognized freedom in the EEZ and continental shelf, it appears clear that if any State was to unilaterally require the removal of cables outside its territorial sea, it would be exceeding is jurisdiction under UNCLOS. No State has been given the power to instruct other States to remove their cables outside of territorial seas. Accordingly, under international law, there is no requirement to remove out-of-service cables in maritime zones beyond the territorial sovereignty of any State.

Is the Abandonment of Cables 'Dumping'?

An issue does arise as to whether the abandonment of cables on the seabed would be pollution of the marine environment by 'dumping'. The definition of dumping in Article 1 of UNCLOS is the same as that contained in the 1972 Convention on the Prevention of Marine Pollution by Dumping of Wastes and Other Matter (London Convention). The definition of dumping includes any deliberate disposal into the sea of vessels, aircraft, platforms or "other man-made structures at sea".[14] The definition of dumping states that it does not include "placement of matter for a purpose other than the mere disposal thereof, provided that such placement is not contrary to the aims of this Convention".[15] However, there was some question as to whether the deliberate abandonment of cables or pipelines could constitute dumping because cables and pipelines are structures, and the decision to abandon a cable no longer in use constitutes the disposal of a man-made structure at sea.

The 1996 Protocol to the 1972 London Convention clarified this issue. It provides that dumping includes any abandonment of man-made structures at sea for the sole purpose of deliberate disposal. However, it also states that dumping does not include the abandonment in the sea of matter such as cables which were placed in the sea for a purpose other than the mere disposal.[16]

On this issue, the definition of dumping in UNCLOS should be interpreted as clarified in the 1996 Protocol to the 1972 London Convention.[17] This makes it

---

14 UNCLOS Art 1(5)(a)(ii).

15 UNCLOS Art 1(5)(b)(ii).

16 See Art 1(2)(2), 1996 Protocol to the 1972 Convention on the Prevention of Marine Pollution by Dumping of Wastes and Other Matter, adopted 7 November 1996, 2006 ATS 11 (entered into force 24 March 2006) (London Convention). This Convention was adopted under the auspices of the International Maritime Organization (IMO).

17 It should be noted that under Art 210, UNCLOS provides authority for 'competent international organizations' to adopt rules and regulations that are legally binding on States. The International Maritime Organization is the United Nations specialized agency charged with responsibility for the safety and security of shipping and the prevention of marine pollution by ships. It is therefore has the competence to adopt rules and regulations relating to dumping, as it has done in the London Convention *supra* note 16, that should be taken into account by States Parties to UNCLOS.

absolutely clear that the abandonment of communications cables on the seabed does not constitute dumping.

Are Other Treaties Applicable?

The Convention of the Protection of Underwater Cultural Heritage explicitly excludes submarine cables from the definition of underwater cultural heritage and classifies cable laying and repair as activities only 'incidentally' affecting underwater cultural heritage.[18]

Is There Liability to Third Parties for Out-of-Service Cables beyond Territorial Seas?

The fact that there are no legal requirements for removal of an out-of-service cable does not mean that the cable owners do not continue to have other legal responsibilities for their property. By analogy, the owners of a vessel that sinks on the high seas can still be liable for claims by coastal States for pollution from leaking bunkers decades after the vessel is lost. This liability is tied to the fact that the vessel owner (or its hull underwriter) continues to own the sunken vessel. Since cables are benign in the ocean environment, pollution liability is not a concern. But other liabilities and responsibilities applicable to cables continue. The principal third party liability for cable owners of out-of-service cables relates to third party claims for sacrificed fishing gear or anchors, as discussed in Chapter 3. A related concern is whether cable operators are under an obligation to ensure that out-of-operation cables are marked on navigational charts (see Chapter 11, Figure 11.2).

How Does the Cable Industry Address Out-of-Service Cables?

UNCLOS sets out the rights and obligations of States with respect to cables. However, as discussed above, removal of out-of-service cables is primarily a decision made by the cable owners. The cable industry has published a recommendation on the various factors to be evaluated in deciding whether or not to remove all or part of an international cable system.[19] The best industry practice with respect to out-of-service cables is to consider the following factors:[20]

i. Any potential effect on the safety of surface navigation or other uses of the sea, including a comparison of whether removal is reasonable or realistic given the

---

[18] Convention on the Protection of Underwater Cultural Heritage, adopted 2 November 2001, (2002) 41 ILM 37, UNESCO Doc 31(C)/Res 24 (entered into force 2 January 2009) Arts 1 and 7.

[19] ICPC, Recommendation No 1, 11 May 2011, *supra* note 9.

[20] *Ibid.*, at 3.1.

presence of other man-made objects on the seabed such as shipwrecks, debris, and oil and gas structures and installations.

ii. Present and possible future effects on the marine environment. If the cable is composed of material that is inert or environmentally benign, consideration should be given to leaving the cable in place.
iii. The risk that the cable will significantly shift position at some future time.
iv. The costs and technical feasibility associated with removal of cables.
v. The determination of a new use or other reasonable justification for allowing the cable or parts thereof to remain on the seabed.
vi. The environmental impact of leaving the cable in place compared to the disruption caused by attempting to remove the cable.
vii. The management of out-of-service cables as part of the cable protection program.
viii. The potential socio-economic and economic benefits of recovering the cable.

If a decision is made to retain the out-of-service cable for future use or leave it in place, cable owners should then carry out the following actions:[21]

i. Notify international and national charting authorities that the cable is out-of-service.
ii. Notify local fishermen and other seabed users of the change in status and confirm that future claims for sacrificed gear shall be considered on their merits.
iii. Confirm that the cable owner remains responsible to any party by insurance cover or otherwise.
iv. Consider alternative uses for the cable such as transfer to a scientific research body.

Fair evaluation of the above factors and undertaking the recommended actions is consistent with UNCLOS requirements for international cables that are out-of-service.

Conclusion

Cables in maritime zones subject to the territorial sovereignty of the coastal State are subject to the laws and regulations of that State.

The salvage of cables in maritime zones under the sovereignty of coastal States is governed by the law of the coastal State, but the salvage of cables outside of the territorial sea/archipelagic waters of any State is governed by the law of salvage.

[21] *Ibid.*, at 3.2.

The international regime governing out-of-service cables beyond the limits of the territorial sea/archipelagic waters of any State is more complicated. States have the right to lay, repair and maintain cables in these areas, and they are not subject to regulations by coastal States. There was a question of whether the decision to abandon an out-of-serve cable constitutes dumping, but the 1996 Protocol to the 1972 London Convention clarified this issue to make it clear that it would not constitute dumping.

Although there is no obligation for cable owners to remove out-of-service cables outside the territorial sea of any State, this does not mean that the rights and obligations of cable owners cease when the cable is out-of-service. The potential liabilities of the cable owner for interference with the rights of other users of the oceans continues. Finally, it is recommended that cable owners follow best international practice with respect to out-of-service cables beyond the limits of the territorial sea. Although there is no obligation under international law to remove them, cable companies are advised to follow the practices recommended by the ICPC for out-of-service cables.

PART IV

PROTECTING CABLESHIPS AND SUBMARINE CABLES

CHAPTER NINE

PROTECTING CABLESHIPS ENGAGED IN CABLE OPERATIONS

Mick Green and Douglas Burnett

Introduction

There are a number of fundamental differences between the operation of cableships and the operation of the vast majority of merchant vessels that ply the world's oceans. The main difference being that cableships conduct their work at sea, usually in a stationary position or moving very slowly, whilst most mercantile vessels conduct their work in port and use the oceans as a means of transporting their cargo to different destinations. This difference is recognised in two international legal instruments that seek to provide protection for cableships when engaged in their lawful operations at sea, these being:

- the 1884 Convention for the Protection of Submarine Telegraph Cables (the 1884 Convention);[1] and
- the Convention on the International Regulations for Preventing Collisions at Sea 1972 (COLREGS).[2]

However, as will be demonstrated below, the current legal regime for the protection of cableships engaged in cable operations is not as effective as it should be given the critical function of cableships in ensuring the resilience of the world's

---

1 Convention for the Protection of Submarine Telegraph Cables, adopted 14 March 1884, TS 380 (entered into force 1 May 1888) (1884 Cable Convention). The United States generally considers the provisions of the 1884 Cable Convention as customary international law, *see* American Law Institute, *Restatement (Third) of Foreign Relations Law of the United States*, Vol. 1 (1987), section 521, comment f. As at 31 January 2013 there are 40 State Parties to the 1884 Cable Convention.

2 Convention on the International Regulations for Preventing Collisions at Sea, 1972, as amended, adopted 20 October 1972, 1050 UNTS 16 (entered into force 15 July 1977) (COLREGS). As at 31 January 2013 there are 155 State Parties.

telecommunications systems. This Chapter will examine the current international legal regime on the protection of cableships and explore the gaps or weaknesses in the legal regime. It will also discuss the steps that governments, industry and other relevant stakeholders can take to enhance the protection of cableships engaged in cable operations.

I. International Law on Protection of CableShips Engaged in Cable Operations

As noted above, the 1884 Cable Convention and COLREGS contain rules which seek to protect cableships engaged in cable operations. These instruments address two aspects of protection, namely safe working distances and cable marker buoys.

Safe Working Distances

The 1884 Cable Convention was the first international treaty governing submarine telegraph cables and deals solely with submarine cables. It not only recognized that submarine cables were a vital means of communication that needed to be protected, but that cableships engaged in the laying and repair of these cables also needed to be protected.

Accordingly, Article V of the 1884 Cable Convention states that:

> Vessels engaged in laying or repairing submarine cables shall conform to the regulations as to signals which have been, or may be, adopted by mutual agreement among the High Contracting Parties, with the view to preventing collisions at seas.
>
> When a ship engaged in repairing a cable exhibits the said signals, other vessels which see them, or are able to see them, shall withdraw to or keep beyond a distance of one nautical mile at least from the ship in question, so as not to interfere with her operations.
>
> Fishing gear and nets shall be kept at the same distance.

Therefore when a cableship is engaged in the repair of a submarine cable and is displaying the appropriate day or night signals, all other vessels should maintain a distance of at least 1 nm from the cableship. This provision further extends to vessels engaged in fishing and limits the minimum distance that any fishing gear or nets can approach the cableship to 1 nm. The 1884 Cable Convention only applies in areas outside of territorial waters.[3] COLREGS also has provisions on cableships. A cableship engaged in the laying or repair of a submarine cable is defined under Rule 3(g)(i) General Definitions of COLREGS, as a vessel that is

---

[3] At the time the 1884 Cable Convention was adopted the breadth of the territorial sea was generally accepted as being 3 nm.

Figure 9.1 (left) Day shapes on a cableship involved in repairs.
Figure 9.2 (right) Night lights on a cableship involved in repairs.
(Photographs courtesy of BT)

restricted in her ability to manoeuvre and is therefore unable to keep out of the way of another vessel. Under Rule 27(b) vessels that are restricted in their ability to manoeuvre are required to exhibit the appropriate day shapes or night signals, these being:

> *Day Shapes*
> [T]hree shapes in a vertical line where they can best be seen. The highest and lowest of these shapes shall be balls and the middle one a diamond;[4]
>
> *Lights*
> [T]hree all-round lights in a vertical line where they can best be seen. The highest and lowest of these lights shall be red and the middle light shall be white[.][5]

Whilst the 1884 Cable Convention specifies the minimum distance that other vessels should maintain from a cableship engaged in a repair, Rule 18 of COLREGS simply requires that vessels "keep out of the way of" the cableship. There is, therefore, no objective measure of distance that must be maintained between vessels, which presents a problem for cable repair ships engaged in repairs, as other vessels can come as close as they want to the cableship so long as there is no physical contact. COLREGS applies both within and outside territorial waters.

---

[4] COLREGS Rule 27(b)(ii).
[5] COLREGS Rule 27(b)(i).

Cable Marker Buoys

Cableships use cable marker buoys during repair operations. After the cableship has located the vicinity of the fault where the cable has been cut, one end of the cut cable is brought to the surface using a grapnel and the end is marked by attaching a cable marker buoy (Figure 6.1). The cableship then uses the grapnel to recover the other end of the cable, removes the fault and completes a joint to spare cable (Figure 6.4). The cableship returns to the marker buoy, paying out spare cable as it proceeds. The buoy is then recovered, the cable end is brought to the surface and a second joint is completed to join to the spare cable (Figure 6.6). Each of the joints can take between 12–24 hours and the cableship can remain at the repair site for 3–5 days, depending upon weather and other conditions.

Article VI of the 1884 Cable Convention provides that:

> Vessels which see, or are able to see, the buoys showing the position of a cable when the latter is being laid, is out of order, or is broken, shall keep beyond a distance of one-quarter of a nautical mile at least from the said buoys.
>
> Fishing nets and gear shall be kept at the same distance.

In contrast to Article VI of the 1884 Cable Convention, COLREGS is silent on submarine cable marker buoys, and makes no provision for safe distance that vessels should maintain from these buoys.

Provisions on the Protection of Cableships in UNCLOS?

It should be borne in mind that while some of the provisions of the 1884 Cable Convention were incorporated into UNCLOS, the two key provisions relating to the protection of cableships (Article V) and interference with a cable marker buoy (Article VI) were omitted. As discussed in Chapter 3, there was some debate during the drafting of the 1956 Draft Articles on the Law of the Sea by the International Law Commission (which formed the basis for both UNCLOS and its predecessor, the 1958 Geneva Conventions) as to whether the provisions of the 1884 Cable Convention should be included in their entirety.[6] It was ultimately decided that only three articles from the 1884 Cable Convention would be included and that the remaining provisions (including Articles V and VI) would not be incorporated.[7] This was on this basis that the three articles being included in UNCLOS were essential principles on the law of the sea and were consequently necessary to include in any codification efforts.[8] Accordingly, and somewhat unfortunately,

---

[6] *Yearbook of the International Law Commission*, Volume I, UN Doc.A/CN.4/Ser.A/1951 (1951) at 363.

[7] Articles II, IV and V of the 1884 Cable Convention were incorporated in Arts 27, 28 and 29 of the 1958 High Seas Convention. Copies of these articles are contained in Appendix 3.

[8] There were initial misgivings that the provisions on the protection of submarine cables proposed for adoption were too detailed and that the International Law Commission (ILC) should only state general principles. However, the ILC ultimately adopted three

there are no express provisions on the protection of cableships engaged in cable operations in UNCLOS.

II. Disregard for Safe Working Distances from Cableships and Cable Buoys

This section will examine one of the major challenges to the protection of cableships, namely the fact that other vessels fail to keep a safe distance from both cableships and cable buoys, thereby endangering cable operations. It will then examine the steps that the cable industry, governments and international organizations can take to address this challenge.

Disregard for Safe Working Distance

The biggest risks presented to cableships and their personnel during repair operations is from collision with other ships, particularly in confined areas such as in traffic separation schemes.

There are many instances of fishing vessels sailing far too close to cableships or placing their fishing gear in the vicinity of cableships. These actions place both vessels at risk and in extreme cases have prevented cableships from completing repairs. For example, this was the case during a repair off the coast of France where fishing vessels intentionally deployed nets in very close proximity to the cableship.[9]

---

provisions from the 1884 Cable Convention based on the rationale that the articles chosen contained essential principles, *see Yearbook of the International Law Commission*, Volume I, UN Doc A/CN.4/Ser.A/1955 (1955) at 20–21.

[9] *Ninety-Four Consortium Cable Owners vs Eleven Named French Fishermen*, Tribunal de Grande Instance de Boulogne Sur Mer (1st Chamber) 28 August 2009, [File No 06/00229 DG/LM]. (Judgment in favor of cable consortium cable owners against 11 French fishermen for damages caused by interference in cable repair ship operations by French fishing vessels.) The French court found that the actions of the French fishermen to extract financial payments to allow the cable repair ships to repair a cable fault violated Articles R46 and R47 of the French Civil Code (CDFE), which require fishing vessels to keep all of their equipment and nets at least one nautical mile from the vessel repairing an underwater cable. (Note, this French domestic statute implements Article V of the 1884 Cable Convention.) The argument by the French fishermen that the cable was laid in French territorial seas without legal authorization was rejected. Instead, the court found that "the measures taken against [the two cableships] were part of a concerted attempt to obstruct the operation of underwater cables in fishing areas in return for financial compensation." The court further found "that each [fishing vessel] in question individually contributed to the incorrect interception of the two cable ships and sailors, as part of this action, played a part in an act of personal, even concerted misconduct that gave rise to damages." Damages were awarded against the 11 named fishermen with interest for the delay of several days in carrying out the repairs. But damages do not remove the disruption caused by leaving vital international cables broken for weeks.

Removal of Cable Buoys

In the past, the problem of missing cable buoys presented an issue. Polyethylene buoys used by the cableship to mark the cut end of a cable during a repair, went missing. These events generally occurred when the cableship moved away from the repair site due to weather or for other reasons. It is unlikely that the buoys were not attached correctly as buoying off a cable end is a standard procedure. The number of reported incidences of missing buoys has reduced in recent years, which could be due to some cableships using modular buoys which are more difficult to recover using small boats. However there remain instances in which fishing vessels have not only disregarded the provision to maintain one quarter nm distance from a cable buoy, as required by the 1884 Cable Convention, but have also stolen the buoy.

Steps by Governments to Protect Cableships: Adoption of International Regulations

The issue of disregard for safe working distances may be attributable in part to the fact that the international law on the protection of cableships is limited. While the 1884 Cable Convention provides that the distance for maintaining a safe space between cable vessels and other vessels is 1 nm, this requirement is only applicable to States Parties to the Convention, which currently stands at 40. In contrast, Rule 18 of COLREGS, which currently has 155 State Parties, simply requires that vessels 'keep out of the way of' the cableship. The COLREGS provision includes no objective distance for ensuring the safety of cableships undertaking repairs. This allows other vessels a wide measure of discretion as to the distance they should keep from cableships and, in principle, there will be no violation of COLREGS so long as there no physical contact between a vessel and a cableship.

One possible solution to address the inadequacy of the current situation would be to amend Rule 18 of COLREGS to require all vessels (including fishing vessels and their gear) to maintain a distance of 1 nm from vessels restricted in their ability to manoeuvre, i.e. cableships. This would essentially incorporate the minimum distance provided for in Article V of the 1884 Cable Convention. Similarly, COLREGS Rule 18 should also be amended to require that fishing vessels and their gear should keep at least one quarter nm away from cable buoys, as provided for in Article VI of the 1884 Cable Convention.

The proposed amendments to COLREGS outlined above could carry the objective standards of safety distances contained in Articles V and VI of the 1884 Cable Convention into modern practice. This would provide a means of addressing the problem of disregard for safe working distance and cable marker buoys. It would also reduce the risk of collision between cableships engaged in laying and repair of submarine cables and help safeguard cable marker buoys. But more importantly, it would speed up the repair of critical submarine cable infrastructure, as

Figure 9.3 Interior structure of a purpose built cableship of the Reliance class. (Image reproduced with permission from Tyco Electronics Subsea Communications LLC (TE Subcom), © TE SubCom 2013 all rights reserved)

it would minimize the chances of other vessels interfering with time-sensitive cable repair operations.

As COLREGS is an international convention any amendments must be government-driven. Governments need to reach out to other like-minded governments to promote and support the amendment of COLREGS.[10] Political will is required to initiate and negotiate any amendment to a multilateral treaty. This can be difficult, particularly where there are a large number of States Parties, but governments must be able to generate and sustain the political momentum required to ensure the amendments are adopted.

Steps by International Organizations

The International Maritime Organization (IMO) is the specialized agency of the United Nations responsible for ensuring the safety and security of shipping. It also acts as the secretariat and depositary for COLREGS. The IMO therefore has an important role to play in ensuring that cable laying and repairs are carried out efficiently. Other international organizations, such as the International Cable

10 The United States formally submitted a proposal to the IMO Sub-Committee on Safety of Navigation to amend COLREGS to provide safety distances from cableships and cable repair buoys during repair operations. (NAV 59/INF.5, 26 June 2013).

Protection Committee (ICPC)[11] also play a role in protecting cableships, their crews and submarine cables by developing best practices for the industry. ICPC seeks to raise awareness of the critical nature of submarine cables and the challenges faced by the industry in the protection of these assets.

Steps for Industry to Protect Cableships

There are also several measures that the cable industry can take to prevent against other vessels sailing too close to cableships engaged in cable operations. These will be elaborated upon below.

Use of Automatic Identification Systems to Protect Cableships

The use of the Automatic Identification System (AIS) was introduced by the IMO with the prime purpose of improving safety of vessels. The regulation adopted by the IMO[12] requires that AIS be fitted aboard all ships over 300 gross tonnage engaged on international voyages and all passenger vessels. The system transmits information about the vessel on VHF[13] which includes the unique identification of the vessel, its course and speed. COLREGS caution against the use of AIS as a direct and sole tool in collision avoidance due to the limitations of VHF radio communications and because not all vessels are equipped with AIS. However it is a powerful tool as it has the ability to instantly identify any vessel that is on a collision course or may cause a near miss with a cableship engaged in a repair operation. It is used sparingly but to good effect during repair operations when the repair vessels are operating in confined waters or in traffic separation schemes.

Use of Automatic Radar Plotting Aids to Protect Cableships

In combination with AIS, the use of radar, specifically Automatic Radar Plotting Aids (ARPA), is a valuable tool in the determination of cableship security from passing marine traffic. ARPA provides the cableship with the ability to plot

---

11 The International Cable Protection Committee (ICPC) plays an essential role in providing leadership and guidance on issues related to submarine cable security and reliability. Since its formation in 1958 by BT and Cable & Wireless (C&W) membership has grown to over 136 members from more than 63 countries. Members include owners and operators of submarine cable, submarine cable system suppliers, submarine cable suppliers, survey companies, cableship operators and governments, as well as several HVDC cable system owners. Ninety eight per cent of installed fiber optic cables are owned and operated by ICPC members. Further, virtually all the marine service suppliers that own and operate cableships used to install and repair these systems are members of the ICPC.

12 The requirements were adopted by the IMO in 2000, as part of a revised new Chapter V of the International Convention for Safety of Life at Sea (1974/1988) (SOLAS). See http://www.imo.org/ourwork/safety/navigation/pages/ais.aspx (last accessed 6 June 2013).

13 Very High Frequency (VHF) is the range of radio frequency electromagnetic waves from 30 MHz to 300 MHz designated by the International Telecommunications Union (ITU). This radio frequency is used for maritime communication and is installed on all large ships but is limited to line of sight.

the course and speed of approaching vessels. In conjunction with AIS this then enables the cableship to make direct contact with any vessels that are not listening to the securité messages issued by the cableship.

Pyrotechnic Maroons

As a last measure cableships may try to avert a collision with another vessel through the use of pyrotechnics. The use of *maroons* is generally undertaken when all other means have been exhausted and the two vessels are in close proximity. A maroon is a pyrotechnic device that is fired in the direction of the offending ship in order to draw the crew's attention to the fact that their vessel is straying into danger.

Notices to Mariners (Submarine Cable Works Notice)

In addition to the usual Notices to Mariners, which are issued to identify cable installations, similar Notices are occasionally issued for extensive or large scale repairs. Submarine Cable Works Notices are also issued for each repair operation. The ship operator will also routinely contact national and local fishery federations and the coast guard to advise that repair operations are being undertaken in certain locations.

VHF Radio Warning Messages—Securité

Securité messages are broadcast by cableships on VHF to alert mariners of navigational warnings that may be the location of a cable buoy deployed by the repair ship or the location of the repair operation. A general broadcast for a navigational warning is issued on VHF channel 16 with an instruction to switch to a VHF working channel to receive the full message. These messages advise mariners to keep a safe distance from the cableship, being a vessel restricted in her ability to manoeuvre and can include a request to maintain a distance of at least 1 nm from the repair ship. These messages are typically broadcast by the cableship in areas of high traffic density or, more specifically, in response to ARPA radar information notifying the cableship of a potential near miss.

III. Threat to Cableships and Personnel from Piracy and Armed Robbery

Cableships engaged in laying and repair operations are more at risk of piracy or armed robbery attacks than other types of vessels. This is because during cable operations, cableships are usually travelling at a slow speed or remain stationary which makes them easier to board. In addition, cableships with their high value and large crews make a lucrative target for pirates and armed robbers.

Over recent years many submarine cables have been installed and repaired in areas that are designated as presenting a risk of piracy. The Horn of Africa is the most notable example, but the east coast of Africa, the Malacca Straits, and other areas may lead to situations where cables are laid or repaired in waters in which there is a higher risk of piracy and armed robbery attacks. There have been instances in East Africa where cable routes have been altered to avoid high risk areas and the cableships have been accompanied by special security forces, resulting in a substantial increase in the costs of operations.[14]

Steps by Industry

To mitigate this risk of piracy and other illegal acts, and to comply with the requirements of the International Ship and Port Facility Security (ISPS) Code[15] every company/ship must develop, implement and maintain a Ship Security Plan (SSP). The purpose of the SSP is to contribute to the prevention of illegal acts against ships and their crew. In essence, the ISPS Code takes the approach that ensuring the security of ships and port facilities is a risk management activity and that in order to determine what security measures are appropriate, an assessment of the risks must be made in each particular case.

Cable companies also take other precautions before carrying out operations in waters where there is a piracy risk. These include 'hardening' of the ship with armor, razor wire, Long Range Acoustic Devices, extra lookouts, a citadel space (a secure room for the crew to lock themselves into to avoid being taken hostage) and intense crew training. In addition to these measures, in high threat areas, armed guards or armed crew and armed escort vessels are evaluated and included as necessary to allow the cable operations to proceed with reduced risks. These

---

14 It was reported that the East African Marine Cable (TEAM) between Kenya and Fujairah in the United Arab Emirates had to be shifted an additional 200 km from Kenya's coastline due to concern about pirate attacks: See "Navies to Guard Undersea Cable from Somali Pirates" Reuters, 16 April 2009 available online at http://www.reuters.com/article/2009/04/16/idUSLG73912 (last accessed 14 May 2013). Similarly, the East African Submarine Cable System (EASSy) also had to be re-routed 400 km to avoid Somali pirates and the two cableships conducting the laying operations had to be accompanied by teams of 24 highly trained French security forces to deter pirate attacks. This increased the cost of operations by US$6 million dollars: See Kui Kinyanjui, "EASSy Fibre Cable Finally Set for Landing" Mars Group Kenya, 23 February 2010 available online at http://www.marsgroupkenya.org/multimedia/?StoryID=281927&page=2 (last accessed 6 June 2013).

15 The International Ship and Port Facility Security (ISPS) Code is an amendment to the Safety of Life at Sea Convention (1974/1988) (SOLAS) on minimum security arrangements for ships, ports and government agencies. Having come into force in 2004, it prescribes responsibilities for governments, shipping companies, shipboard personnel, and port/facility personnel to "detect security threats and take preventative measures against security incidents affecting ships or port facilities used in international trade".

costs can amount to millions of dollars in the case of cable installation and significant costs and delays added to repair operations.

In addition, many ships that transit (or operate) in high risk areas are obliged to take out additional insurance, including kidnap and ransom insurance, as well as cover for the ship and crew and are also required to make security arrangements for the ship during transit. The latter measures usually take the form of a citadel, a security team on board with appropriate methods for dealing with the potential issue and other measures described above. The level of security measures should allow the cableship operator to obtain a corresponding reduction in premiums from its underwriters, however, these are normally worked out on a voyage by voyage basis. At the end of the day, increased insurance costs and security measures are costs that are passed on to the companies and maintenance agreements that employ these specialized vessels.

Steps by Governments

UNCLOS affords certain rights and obligations to States in the suppression of piracy, which includes a duty to cooperate to the fullest extent in the repression of piracy,[16] the right to board vessels where there is reasonable ground for suspecting that the ship is engaged in piracy[17] and the right to arrest a pirate ship, the pirates and seize the property on board.[18] These rights only apply in the exclusive economic zone and high seas[19] and cannot be exercised by States in the maritime zones under the territorial sovereignty of another State (such as territorial seas, archipelagic waters and straits used for international navigation).[20]

States should utilize the rights that they have in relation to piracy to protect cableships from such attacks. Further, in view of the critical impact of disruptions to submarine cables, States might consider the use of naval vessels as escorts for cableship engaged in repairs in high risk waters. A naval warship escort will undoubtedly deter pirates and armed robbers and speed repairs and restoration of vital communications.

Conclusions

As can be seen from the above, the protection of cableships is an essential component of ensuring the security and availability of the world's telecommunications infrastructure. In particular interference from other competing activities

16 UNCLOS Art 101.
17 UNCLOS Art 110.
18 UNCLOS Art 105.
19 UNCLOS Arts 101 and 58(2).
20 For additional information on jurisdiction in maritime zones refer Chapter 3.

and from threats such as piracy and armed robbery cause delays to repair operations and result in disruptions to critical telecommunications services. Moreover, additional costs caused by such delays are passed on to the ultimate users of submarine cables. It is therefore in the interest of governments, the cable industry and international organizations to work together to ensure the adequate protection of cableships.

CHAPTER TEN

SUBMARINE CABLES AND NATURAL HAZARDS

Lionel Carter

Introduction

Damage to submarine cables and cable networks are sustained mainly through three causes; negligent human activities, intentional human activities and natural hazards. Faults caused by human activities are discussed in Chapters 11 and 12. The present Chapter examines the damage that may be caused to cables as a result of natural phenomena occurring in the marine environment where the cables operate. A basic knowledge of relevant natural phenomena is essential for planning cable routes and for considering governmental and industry issues about infrastructure protection, reliability and route diversity. For current purposes the term *natural hazards* is defined as "*Naturally-occurring physical phenomena caused by either rapid or slow onset events having atmospheric, geologic and hydrologic origins at the global, regional, national or local scale*".[1] The different phenomena that comprise a natural hazard are discussed in this Chapter.

I. Cable Faults Caused by Natural Hazards

As the majority of submarine cables are owned by private consortia, there is no single repository for collecting and analyzing information regarding cable faults. Furthermore, where data are privately collected it may not be publically available. However, within the cable industry there are several organizations that maintain databases on cable faults and publish data and interpretative findings.[2] Cable

---

1 UNESCO, 2006 Disaster preparedness and mitigation, see http://www.unesco.org/new/en/natural-sciences/special-themes/disaster-preparedness-and-mitigation/natural-hazards/ (last accessed 1 June 2013).

2 For example, M. Kordahi *et al.*, "Trends in Submarine Cable System Faults" Study on behalf of the Submarine Cable Improvement Group (2007) at 1, available online at http://www.suboptic.org/Uploads/Files/WeA1.2.pdf (last accessed 1 June 2013).

owners distinguish between requests for data from their competitors and those from scientists or scientific organizations. In the latter case, data have been made available upon request to assist in research such as whale interactions with cables, the carbon footprint of cables in the marine environment, and the velocity and range of sediment-laden flows or *turbidity currents*.

Analyses of cable databases reveal a division between faults caused by human activities and those by natural hazards. Of the 150 to 200 faults affecting fiber optic cables annually, over 65 per cent occur on the continental shelf in water depths of less than 200 m and are the result of damage caused by ships' anchors, fishing practices and dredging/mining activities. Faults caused by natural phenomena comprise less than 10 per cent of the total faults.[3] This is not surprising given that human activity on the continental shelf has intensified over the past decades and is expanding into deeper waters as fish stocks are depleted, advances in technology enable greater exploitation of hydrocarbons and other minerals, and large areas of continental shelf are being occupied to generate renewable energy.

Bottom fish trawling now extends to depths of approximately 1500 m, but in the period between 2000–2006 faults caused by fishing in depths >1000 m were only one to three per year.[4] In waters deeper than 1000 m, natural hazards become more prominent, accounting for >30 per cent of cable faults. This percentage excludes damage attributed to 'unknown' causes that are likely to include some natural effects.[5] In addition to their prominence in deeper water, another important characteristic of some natural hazards is their potential to damage multiple cables, especially in cases of submarine landslides and turbidity currents, which can break or damage ten or more cables over hundreds of kilometers.[6] Multiple cable breaks may also occur by a ship dragging its anchor across the seabed, but damage tends to be less extensive compared to that sustained as a result of landslides and turbidity currents.

---

3 Kordahi *et al., ibid.*; see also M.P. Wood and L. Carter, "Whale Entanglements with Submarine Telecommunications Cables" (2008) 3(4) *IEEE Journal of Oceanic Engineering* at 445–450.

4 Kordahi *et al., supra* note 2.

5 L. Carter *et al.*, "Submarine Cables and the Oceans—Connecting the World" (2009) UNEP-WCMC Biodiversity Series No 31, ICPC/UNEP/UNEP-WCMC at 38. Available online at http://www.unep-wcmc.org/medialibrary/2010/09/10/352bdld8/ICPC_UNEP_Cables.pdf (last accessed 6 June 2013) 43–48.

6 B.C. Heezen *et al.*, "Further Evidence for a Turbidity Current Following the 1929 Grand Banks Earthquake" (1954) 1 *Deep-Sea Research* 193–202 at 193; and S.-K. Hsu *et al.*, "Turbidity Currents, Submarine Landslides and the 2006 Pingtung Earthquake off SW Taiwan" (2008) 19 *Terrestrial, Atmospheric and Oceanic Science* 767–772, available online at http://tao.cgu.org.tw/pdf/v196p767.pdf (last accessed 6 June 2013).

II. What are Natural Hazards and Where do they Occur?

The definition of natural hazards, as noted earlier,[7] encompasses a wide range phenomena that include earthquakes, tsunamis, volcanic eruptions, weather-related disturbances, and climate change.[8] The definition embraces their multiple and often inter-related causes that can produce an abrupt or long-term impact, for example atmospheric (wind) forcing of the ocean produces currents and waves with the power to move sand, which can cause long-term abrasion of a cable's protective sheath resulting in its general degradation and ultimate failure.[9]

Hazard occurrence reflects the various environmental forces working alone or in combination with one another. The coast and adjacent continental shelf and uppermost continental slope down to approximately 200 m water depth are most frequently exposed to hazards that are generated by weather and shelf/slope current systems.[10] In calm weather, waves and currents alone are unlikely to be hazardous, except in regions of exceptional flows that are capable of eroding the seabed, as in the case of tide-dominated seaways.[11] More pervasive is the impact of storms that cause waves and surges to erode coasts as well as reinforcing existing tides and ocean currents on the shelf to enhance their erosional power. Such an erosional regime is in contrast to coasts that receive an abundant supply of sediment from rivers, thus favoring deposition.[12] Again, this situation is affected by storms that cause rivers to flood and dump large volumes of sediment on the continental shelf and beyond.[13]

Less frequent, but nonetheless a major threat to coasts and shelves, are earthquakes and tsunamis. These are capable of causing widespread devastation, as

---

7 UNESCO 2006, *supra* note 1.

8 L. Carter *et al.*, *supra* note 5 at 38.

9 L. Carter *et al.*, "Seafloor Stability along the Cook Strait Power Cable Corridor" Proceedings of the 10th Australian Conference on Coastal and Ocean Engineering, 1991, 565–570, reprint available from the Senior chapter author.

10 C.A. Nittrouer *et al.*, eds, *Continental Margin Sedimentation: from Sediment Transport to Sequence Stratigraphy* (Blackwell Publishing, 2007) at 549.

11 For example, the inner shelf of Wellington, New Zealand, see L. Carter and K. Lewis, "Variability of the Modern Sand Cover on a Tide and Storm Driven Inner Shelf, South Wellington, New Zealand" (1995) 38 *New Zealand Journal of Geology and Geophysics* 451–470.

12 For example a terrestrial erosional regime is evident in Taiwan, where large amounts of sediment are deposited by rivers into the ocean, see S.J. Kao and J.D. Milliman, "Water and Sediment Discharge from Small Mountainous Rivers, Taiwan: The Roles of Lithology, Episodic Events, and Human Activities" (2008) 116 *Journal of Geology* 431–448.

13 L. Carter *et al.*, "From Mountain Source to Ocean Sink—the Passage of Sediment Across an Active Margin, Waipaoa Sedimentary System, New Zealand" (2010) 270 *Marine Geology* 1–10.

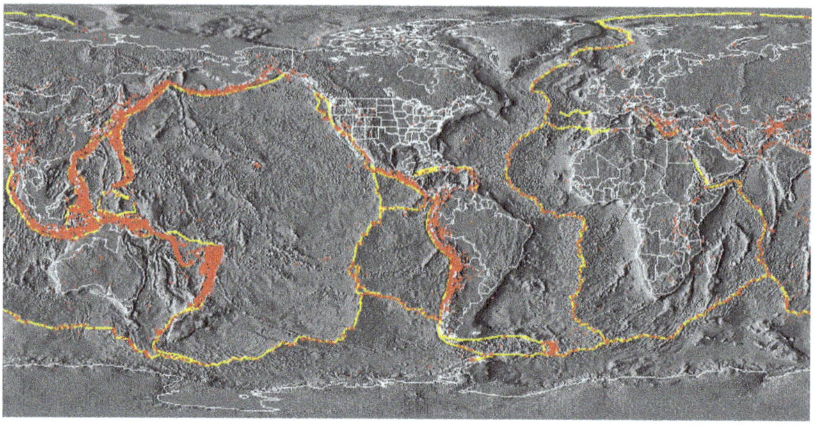

Figure 10.1 Crustal plate boundaries. Tectonic plates outlined in yellow with epicenters of earthquakes of magnitudes > 5 (red) concentrated where plates collide especially around the Pacific Ocean, North Indian Ocean and southern Europe for the decade 1980–1990. Mid-ocean earthquakes, located along plate boundaries, result from plates moving apart. (Image courtesy of Western Washington University http://www.smate.wwu.edu/teched/geology/technology.html)

evinced by the earthquake and tsunami in Indonesia, 2004, and in Japan, 2011.[14] Earthquakes can produce major displacements of the shelf seabed through liquefaction, faulting and landsliding; the latter occurring on slopes as gentle as 1°. Tsunamis can also disturb the seabed as well as erode the coast when the waves extend onshore and when waters later drain back to sea.

The continental shelf steepens at about 120–200 m depth onto the continental slope, which descends to 1500–3500 m depth at an average inclination of 4° but which can locally steepen to 30° or more. The presence of a slope enhances gravitational effects so that sediment disturbed by earthquakes and other perturbations is prone to landsliding and the formation of turbidity currents.[15] The hazard posed by sediments passing down-slope into the abyssal ocean is exemplified around the circum-Pacific Rim and other regions where tectonic plates are actively converging (see Figure 10.1).

Converging plates provide ideal conditions for submarine landslides and turbidity currents. Firstly, the convergence increases mountain building, which generates large amounts of detritus through earthquake-triggered landslides and river erosion. Secondly, the mountains interact with prevailing moisture-laden

14 H. Tanaka *et al.*, "Coastal and Estuarine Morphology Changes Induced by the 2011 Great East Japan Earthquake Tsunami" (2012) 54(1) *Coastal Engineering Journal* 1250010-1-1250010-25.

15 M.A. Hampton *et al.*, "Submarine Landslides" (1996) 34(1) *Reviews of Geophysics* 33–59.

winds to accentuate rainfall and bolster the capacity of rivers to carry detritus to the coast. Thus small, seismically active, mountainous islands such as Taiwan, Papua-New Guinea, New Zealand and others collectively account for over half the land-derived sediment entering the world ocean.[16] Thirdly, earthquakes destabilize the offshore deposits, formed from the high influx of sediment, to produce landslides and turbidity currents that pass down slopes. Once triggered, a turbidity current can flow for hundreds of kilometers with sufficient power to damage submarine cables.[17] Occasionally, earthquake generated turbidity currents may form in regions of infrequent seismic activity, as was the case with the 1929 Grand Banks event off the eastern seaboard of US-Canada. However, such events are rare compared to the highly seismic circum-Pacific rim where turbidity currents form frequently, for example, more than one per annum off Taiwan.[18]

Continental margins also accommodate deposits of methane that are commonly bound with ice to form methane hydrates whose stability is controlled by water pressure and temperature.[19] The transformation from solid methane hydrate to gas appears to be rapid, even explosive, as suggested by the presence of craters in the seabed, some with diameters exceeding 10 km.[20] Such rapid escape of subsea gas has the ability to set off submarine landslides. In view of their potential as a hydrocarbon source of energy, methane hydrates may also be the focus of seabed mining thus further increasing risk to any proximal cables.

Despite its depth, nominally taken here as being greater than 3000 m, the abyssal ocean is still prone to hazards that may disrupt cables. In addition to periodic incursions by landslides and turbidity currents, the abyssal ocean is also subject to locally fast flowing and turbulent currents as well as volcanic eruptions.

While over 40 per cent of the ocean is abyssal plains and hills, at least another 35 per cent is comprised of significant elevations notably plateaus, rises, submarine volcanoes or *seamounts* and mountain chains, as exemplified by the vast Mid-Atlantic Ridge that challenged the laying of the first submarine cable.[21] Such topography interacts with deep ocean currents, especially along the western

---

16 J.D. Milliman and J.P.M. Syvitski, "Geomorphic/Tectonic Control of Sediment Discharge to the Ocean: The Importance of Small Mountainous Rivers" (1992) 100 *Journal of Geology* 525–544.

17 S.-K. Hsu *et al.*, *supra* note 6; and P.J. Talling *et al.*, "Onset of Submarine Debris Flow Deposition Far from Original Giant Landslide" (2007) 450(22) *Nature* 541–544.

18 R. Gavey, "An Evaluation of Modern Hyperpycnal Processes and their Relevance to the Geological Records" (2012) Unpublished Ph.D. thesis lodged at Southampton University, United Kingdom.

19 J.P. Kennett *et al.*, *Methane Hydrates in Quaternary Climate Change: The Clathrate Gun Hypothesis* (AGU Books Board, 2003).

20 B. Davy *et al.*, "Gas Escape Features off New Zealand: Evidence of Massive Release of Methane from Hydrates" (2010) 37(21) *Geophysical Research Letters*.

21 J.S. Gordon, *A Thread Across the Ocean: the Heroic Story of the Transatlantic Cable* (Simon and Schuster, 2002) at 239.

margins of the major oceans where forces related to the Earth's rotation intensify currents against an elevation. There, in water depths down to 5000 m and deeper, strong turbulent currents exist that are capable of eroding the seabed and moving suspended cables.[22] Superimposed on these deep flows are large eddies, generated where major currents collide with the subsea topography. Detectable by satellites, the eddies can extend to the seabed and stir up mud and sand.[23]

Volcanic activity is common throughout the deep ocean where it is associated with (i) converging tectonic plates, which in the case of the Pacific is highlighted by the *Pacific Ring of Fire*; (ii) *hot spots* in the Earth's oceanic crust, above which magma has erupted to form island chains such as Hawaii; (iii) tectonic plate spreading zones where new magma comes to the surface to form the great mid-ocean ridges from where the new volcanic crust spreads laterally to drive the tectonic plates, and (iv) isolated seamounts formed at some local weakness in the crust. Hazards to cables can result from lava flows, hot debris avalanches, landslides, rough volcanic topography and hydrothermal vents.[24]

This mainly coast-to-abyss depiction of natural hazards is accompanied by marked variations associated with geography. In polar latitudes, for example, icebergs and sea ice pose threats to cable systems as do the tropical storm centers where hurricanes, typhoons and cyclones cause widespread disruption of coast and shelf environments. Tropical to mid latitudes also witness intense rainfalls and floods whose frequency and intensity are moderated by climate modes such as *El Niño-La Niña*. And, as noted previously, there is the powerful influence of tectonic plates especially in the circum-Pacific region. These geographic and depth-related factors demonstrate the variability of natural hazards and emphasize the need for hazard assessment that is site specific for a new cable system.

III. Cables in Hazardous Settings

Submarine cables traversing the continental shelf are subject to frequent wave and current action that is capable of shifting sand and gravel,[25] to create conditions favouring cable abrasion and suspension; the latter exposing the cable to

---

22 W.J. Schmitz, "On the Interbasin-scale Thermohaline Circulation" (1995) 33(2) *Review of Geophysics* 151–173; and T. Whitworth *et al.*, "On the Deep Western-boundary Current in the Southwest Pacific Basin" (1999) 43 *Progress in Oceanography* 1–54.

23 C.D. Hollister and I.N. McCave, "Sedimentation Under Deep Sea Storms" (1984) 309 *Nature* 220–225.

24 W.W. Chadwick Jr. *et al.*, "Vertical Deformation Monitoring at Axial Seamount Since its 1998 Eruption Using Deep-sea Pressure Sensors" (2006) 150 *Journal of Volcanology and Geothermal Research* 313–327.

25 For a discussion of the characteristics and behavior of soil and sand particles on the seabed, the effects of water pressure, and the mobility of sea bedforms see generally P.G. Allan, "Cable Security in Sandwaves", Paper presented at the International Cable Protection Committee Plenary Meeting, Copenhagen, May 2000.

fatigue if a suspension moves or strums with the water motions. For example, in the near-shore zone off California, the coaxial Acoustic Thermometry of Ocean Climate (ATOC) cable suffered abrasion and fatigue as it moved repeatedly over a rocky substrate under wave action.[26] In a similar case, the first power cables installed in the tide-swept Cook Strait, New Zealand in 1961, were abraded by the daily flow of sand and gravel that locally removed the jute sheath and steel wire reinforcing to expose the inner core.[27] However, subsequent improvements in design and protective materials produced replacement cables that have effectively resisted abrasion since their deployment in 1991.

Storms

Hurricanes, typhoons and cyclones are major hazards to coastal and shelf cable infrastructure. Not only do the powerful winds increase wave and current action that enhance seabed erosion and sediment transport, but they can also generate storm surges with the potential to damage onshore infrastructure especially over flat lying areas such as deltas. This was the case with Typhoon Nargis in 2008, when a 4 m-high storm surge swept over the Irrawaddy Delta and damaged a submarine cable station.[28] Hurricane Katrina wreaked its own brand of havoc in the United States in 2005. Large waves and an 8 m-high storm surge helped destabilize exposed parts of the Mississippi Delta to form mudflows that swept into the Gulf of Mexico. The mudflows displaced and buried pipelines and other infrastructure associated with offshore hydrocarbon production platforms.[29] Telecommunications suffered a significant blow during Hurricane Sandy in 2012. A combination of strong winds (130 km/h), low atmospheric pressure (946 millibars) and the funneling effect of narrow coastal inlets produced a surge of 4 m. The combined effect of these occurrences, together with 180 mm of rain, resulted in widespread flooding and loss of power to lower Manhattan, New York.[30]

---

26 For a discussion of the Acoustic Thermometry of Ocean Climate (ATOC)/Pioneer cable experiment see Irina Kogan *et al.*, "ATOC/Pioneer Seamount Cable After 8 Years on the Seafloor: Observations, Environmental Impact" (2006) 26 *Continental Shelf Research* 771–787.

27 L. Carter, "Geological hazards and their impact on submarine structures in Cook Strait, New Zealand" Paper presented the Australasian Conference on Coastal and Ocean Engineering, Launceston, 30 November–4 December 1987, 410–414.

28 L.L. Ko, "Experience of Nargis Storm in Myanmar and Emergency Communications" presentation by Myanmar Ministry of Communications, Posts & Telegraphs, 9 July 2011. Available online at http://www.itu.int/ITU-D/asp/CMS/Events/2011/disastercomm/S4C-Myanmar.pdf (last accessed 1 June 2013).

29 M.C. Nodine *et al.*, "Impact of Hurricane-Induced Mudslides on Pipelines", Presentation at Offshore Technology Conference, 30 April–3 May 2007, Houston, Texas. Available at doi: 10.4043/18983–MS.

30 National Oceanic and Atmospheric Administration (NOAA), 2012. US Climate Extremes Index; NE USA. National Climate Data Center, NOAA, see http://www.ncdc.noaa.gov/extremes/cei/graph/ne/cei/01-12 (last accessed 1 June 2013).

At least one submarine cable was damaged, although the main effect on telecommunications, including the internet, was the closure of several large data-centers due to basement flooding and subsequent loss of mains and auxiliary electrical power. As a result up to 10 per cent of the New York networks went offline.[31] Despite this temporary loss, the overall network system continued to function, a testament to its resilience. Yet another major disturbance, Hurricane Iwa, in 1982, accelerated ocean current speeds of up to 7 km/hour which, together with large waves, triggered landslides and turbidity currents that swept down the continental slope off Oahu, Hawaii.[32] En route, these sediment flows reached 11 km/hour and broke six submarine cables. One cable section was unrecoverable due to burial by sediment, attesting to the size and power of these seabed-hugging sediment flows.

Despite the frequent occurrence of extreme storms over coasts and adjacent seas, their impact on cables is minor. Less than 10 per cent of faults were sustained through storm damage, which is small when compared with damage caused by shipping and fishing.[33]

Extreme floods also pose a threat to deep-ocean cables located hundreds of kilometers from land. This was demonstrated graphically during Typhoon Morakot in 2009, in the Strait of Luzon between Taiwan and the Philippines (see Figure 10.2). Morakot was the wettest tropical cyclone recorded over Taiwan. Almost 3 m of rain fell in three days to form floods that carried so much sediment the river discharge dived to the seabed (normally river discharge spreads over the sea surface as freshwater is less dense than seawater). In the case of southern Taiwan's Gaoping River, the discharge plunged into a submarine canyon and formed two turbidity currents; one during the peak flood and a second three days after the flood when flood sediment deposited in the canyon was remobilized, possibly by large waves.[34] This second, larger turbidity current passed down the Gaoping Canyon into the Manila Trench—a distance of 370 km. Current speeds reached 60 km/hour and broke at least six submarine cables in addition to the two cables damaged by the first turbidity current.

Earthquakes and Tsunamis

Cable-damaging submarine landslides and turbidity currents can also be set off by ground shaking associated with earthquakes. On 18 November 1929,

---

31 J. Cowie, Hurricane Sandy: outage animation, available online at www.renesys.com/blog/2012/10/hurricane-sandy-outage-animati.shtml (last accessed 1 June 2013).

32 A.T. Dengler *et al.*, "Turbidity Currents Generated by Hurricane IWA" (1984) 4(1) *Geo-marine Letters* 5–11.

33 Kordahi *et al.*, *supra* note 2.

34 L. Carter *et al.*, "Near-synchronous and Delayed Initiation of Long Run-out Submarine Sediment Flows from a Record-breaking River Flood, Offshore Taiwan" (2012) 39(12) *Geophysical Research Letters* L12603.

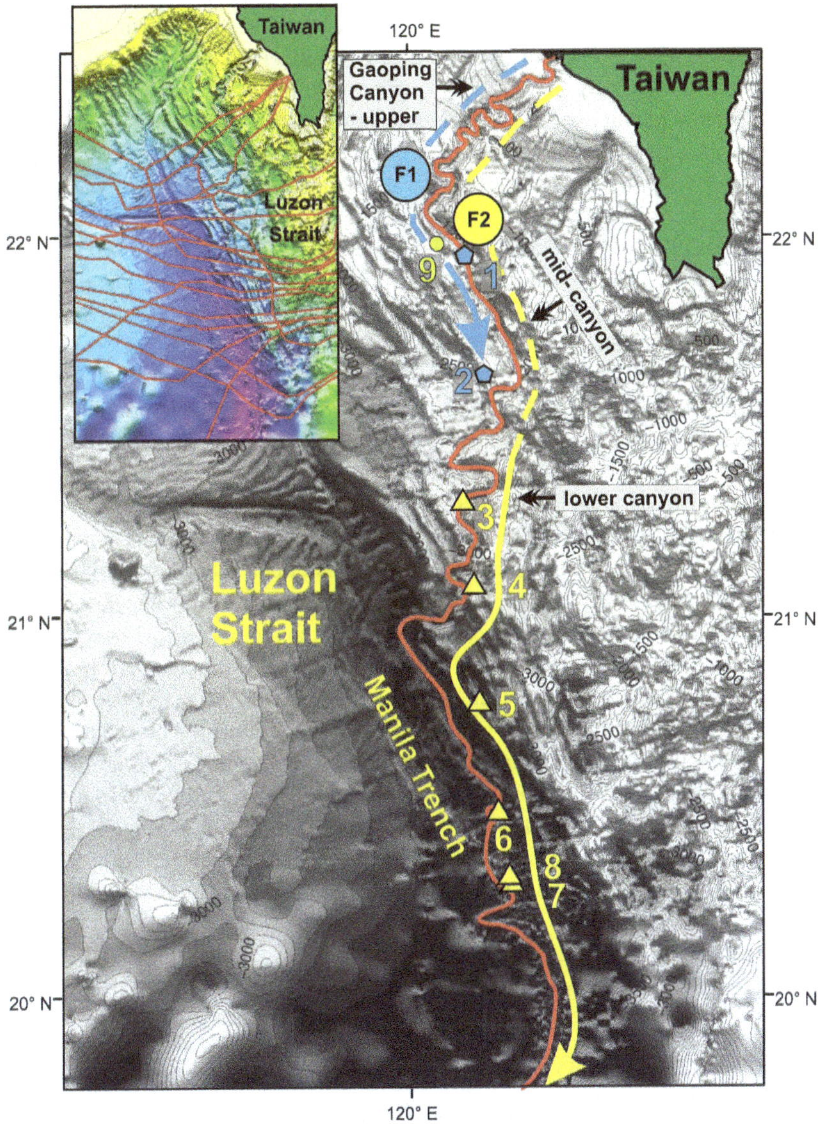

Figure 10.2 The Strait of Luzon off southern Taiwan, through which at least 18 cables pass (inset) connecting SE Asia with the rest of the world. The red line is the path of Gaoping Canyon and Manila Trench along which two flood-formed turbidity currents passed: F1 (blue) during the peak flood and F2 (yellow) three days after the flood. (Image courtesy of L. Carter, Victoria University of Wellington and the American Geophysical Union)

a magnitude M7.2 earthquake shook the seabed off the Grand Banks, Newfoundland. At least eight submarine cables were broken concomitant with the main shock.[35] Subsequent studies show these initial breaks resulted from a series of landslides triggered by severe ground shaking within a 100 km radius of the earthquake's epicenter.[36] By mixing with water, some landslides were transformed into a fast moving mud- and sand-laden turbidity current that moved down-slope breaking at least five additional submarine cables en route. From the timing of the cable breaks and their location it was possible to estimate current speeds, which reached 65 km/hour on a journey of over 650 km. Overall, the turbidity current carried about 200 km^3 of sediment into water depths over 4500 m.

The Grand Banks cable study remains a text-book example and was the forerunner of several papers that recorded other cable breaks related to earthquakes including Algeria 1954[37] and 2003,[38] Fiji[39] and Papua-New Guinea[40] amongst others. Most recent research relates to breaks associated with the M9.0 Great Tohoku Earthquake off northern Japan in 2011 (studies of which are still underway) and the M7.0 Hengchun earthquake off southern Taiwan in 2006.[41] In the case of Taiwan, the main shock and similar magnitude aftershocks had epicenters 20–55 km offshore, directly in the Strait of Luzon cable corridor (see Figure 10.2, inset), and resulted in strong ground shaking. The main shock was accompanied by the near-instantaneous failure of three cables under substantial landsliding. At least three turbidity currents followed at various times and appear to have been related to the large aftershocks as well as to the main shock. These currents passed rapidly down-slope to create an additional 19 cable faults, mainly within the Gaoping Canyon/Manila Trench. Most of the faults occurred sequentially over approximately nine hours as the first turbidity current—associated with the first shock—travelled at least 246 km from lower Gaoping Canyon to Manila Trench at depths over 4000 m. Times and distances between the breaks indicate average speeds of 46 km/hour along steep parts of the Gaoping Canyon, slowing to 20 km/hour along the gently sloping floor of the Manila Trench. Almost three years later, cables in the same canyon/trench system were disrupted by turbidity

---

35 B.C. Heezen and M. Ewing "Turbidity Currents and Submarine Slumps, and the 1929 Grand Banks Earthquake" (1952) 250 *American Journal of Science* 849–873.

36 D.J. Piper *et al.*, "Sediment Slides and Turbidity Currents on the Laurentian Fan: Sidescan Sonar Investigations Near the Epicentre of the 1929 Grand Banks Earthquake" (1985) 13 *Geology* 538–541.

37 B.C. Heezen and M. Ewing "Orleansville Earthquake and Turbidity Currents" (1955) 39(12) *American Association of Petroleum Geologists Bulletin* 2505–2514.

38 G. Dan *et al.*, "Mass Transport Deposits on the Algerian Margin (Algiers area): Morphology, Lithology and Sedimentary Processes" (2003) 28 *Advances in Natural and Technological Hazards Research* 527–539.

39 R.E. Houtz and H.W. Wellman, "Turbidity Current at Kandavu Passage, Fiji" (1962) 99 *Geological Magazine* 57–62.

40 D.C. Krause *et al.*, "Turbidity Currents and Cable Breaks in the Western New Britain Trench" (1970) 81 *Geological Society of America Bulletin* 2153–2160.

41 S.-K. Hsu *et al.*, *supra* note 6.

currents associated with Typhoon Morakot (Figure 10.2). In light of research into river discharge from earthquake ravaged Taiwan,[42] it is pertinent to ask if the record-breaking sediment discharge from the Gaoping River during Morakot—estimated to be around 150 million tonnes[43]—resulted not only from the exceptional rainfall but also destabilization of the landscape by the 2006 Hengchun earthquake. In 2010 another turbidity current swept down the Gaoping Canyon/Manila Trench and broke at least nine cables. On this occasion the trigger appears to have been a swarm of onshore earthquakes of magnitudes < M4.0 to M6.4, with epicenters 120 km north of the Strait of Luzon. Ground shaking in the Strait was very weak, which suggests the seabed sediment was only quasi-stable and could be disturbed by small ground motions. Again, a relevant question is whether such quasi-stable deposits related to the large sediment influx of Typhoon Morakot? From published and ongoing research, it appears that the hazard prone Taiwan seabed reflects the interaction between earthquakes and weather/climate; (i) earthquakes provide the means to destabilize landscapes making debris available for transport to the ocean, and are a mechanism to trigger submarine landslides (for example Hengchun, 2006) and (ii) intense rainfalls increase river discharge into the ocean thus providing conditions favourable for the formation of turbidity currents (for example Typhoon Morakot, 2009).

Offshore earthquakes may be accompanied by tsunamis, which in recent years have proven a major hazard for land-based telecommunications, cable landing stations and potentially cables laid on the continental shelf. The 2004 Indonesian earthquake was accompanied by a devastating tsunami that swept across the Indian Ocean to inundate low lying coastal areas, especially those close to the epicenter off Sumatra. Terrestrial communications were severely disrupted as far away as South Africa, where a submarine cable may have been damaged by tsunami debris carried offshore.[44]

As two tectonic plates snapped past one another to form the 2011 Japan earthquake, up to 5 m of vertical plate movement generated a major tsunami that swept across the Pacific causing widespread damage, which included breaking off 125 km^2 of an Antarctic ice shelf (Figure 10.3). Near the epicenter, located about 130 km offshore in the Japan Trench, tsunami waves quickly reached the coast and moved up to 5 km inland.[45] Maximum wave height varied with coastal topography but often exceeded 10 m and locally reached 19.5 m. As a result, telecommunications infrastructure, which included land-based mobile and fixed line

---

42 S.J. Dadson *et al.*, "Earthquake Triggered Increase in Sediment Delivery from an Active Mountain Belt" (2004) 32 *Geology* 733–736.

43 L. Carter *et al.*, *supra* note 34.

44 C. Strand and J. Masek "Sumatra-Andaman Islands Earthquake and Tsunami of December 26, 2004: Lifeline Performance" (2005) 29 *Monograph* American Society of Civil Engineers.

45 H. Tanaka *et al.*, "Coastal and Estuarine Morphology Changes Induced by the 2011 Great East Japan Earthquake Tsunami" (2012) 54(1) Coastal Engineering Journal available online at http://www.worldscientific.com/worldscinet/cej (last accessed 1 June 2013).

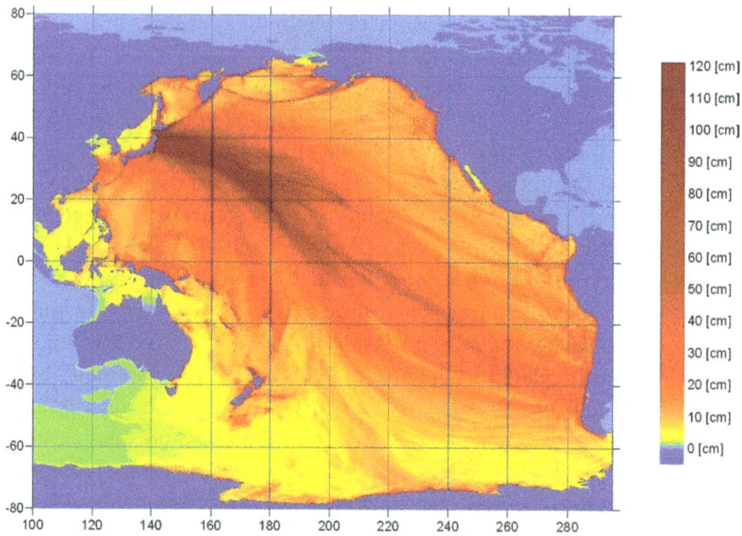

Figure 10.3 The displacement of seabed caused by the Tohoku earthquake, which pushed up over 15,000 square kilometers of seabed by up to 5 m—a vast displacement that generated a tsunami with deep water wave heights exceeding 120 cm near the earthquake epicenter but rising as the waves moved into shallow water and onshore where a maximum height of 19.5 m was recorded. (Image courtesy of National Oceanic and Atmospheric Administration (NOAA))

systems, together with at least one cable landing station, was severely damaged. Damage resulted as the waves moved onshore and then drained back into the ocean, especially where the returning water was funneled by the topography to form an eroding sluice. Whether moving onshore or offshore, the power of the tsunami was increased by the debris it carried. Around twenty regional and telecommunications submarine cables land near Tokyo and it was not surprising that they suffered some degree of damage.[46] The full extent and causes of the damage have yet to be finalized but: (i) the pronounced and prolonged shaking (2.5 minutes) that accompanied the main shock; (ii) the large and frequent aftershocks, e.g. 14 shocks > M6.0 within six hours of the main earthquake, and (iii) disturbance of the seabed by the advancing tsunami and associated retreating waters, collectively point to the triggering of submarine landslides and turbidity currents. Certainly, sediments in the Japan Trench contain turbidity current deposits from previous perturbations.[47]

---

[46] BBC, 2011 "Japan to repair damaged undersea cables" http://www.bbc.co.uk/news/technology-12777785 (last accessed 1 June 2013).

[47] S. Lallemand *et al.*, "Subduction of the Daiichi Kashima Seamount in the Japan Trench" (1989) 160 *Tectonophysics* 231–233, 237–242.

Volcanoes

Despite the hazardous potential of active volcanoes via explosive eruptions, lava and hot debris flows, seismic activity and landslides, their risk to cables is minor. This is because volcano locations are generally obvious and cable route planners are able to avoid them. However, cables do land in regions of active volcanism, including Hawaii, Lesser Antilles, Japan and New Zealand amongst others. With respect to the larger islands, cables can be routed through non-active zones. However, avoidance may not be possible with small islands, as was the case in 1902 when the volcanoes, La Soufrière, St Vincent, and Mont Pelée, Martinique erupted; an occurrence that was followed by the loss of cable connectivity.[48] Although there was no determinative finding as to the cause of the cable damage, it is reasonable to conclude that it resulted from an avalanche of volcanic debris from Mont Pelée.

Floating and Fixed Ice

Floating and fixed ice in polar regions can damage cables from coastal waters to water depths of 500 m and deeper where large icebergs can plough the seabed. In a case study from western Greenland, cable damage was confined mainly to coastal waters in water depths < 25 m and resulted from impacts of fixed and floating ice. That region is swept by icebergs that follow an anticlockwise path around the Labrador Sea and back down the eastern seaboard of Canada via the Labrador Current. Iceberg scouring is the scraping of an iceberg along the seabed and is capable of crushing and breaking cables. Other potential impacts of ice can result from the growth of pressure ridges, the formation of ice crystals on a cable potentially causing it to become buoyant, and the formation of scouring vortices where melt water or river water drain through holes in surface ice. Hazard assessment relating to ice impacts on the seabed is gaining interest in light of plans to develop two new cable routes through the Arctic Sea in order to connect Asia with western Europe (see next section).

IV. Climate Change

Disaster information gathered by the reinsurance industry for the past six decades reveals a marked increase in the number of natural catastrophes, particularly since

48 G. Pararas-Caryannis, "Eruptive Processes of Stratovolcanoes of the Lesser Antilles (Islands of Montserrat, Martinique, St. Vincent and Granada)—Mechanisms of Flank Failures and Tsunami Generation" (2004) 22(2) *Journal of Tsunami Hazards* excerpts available online at http://www.drgeorgepc.com/TsunamiVolcanicCaribErProc.html (last accessed 1 June 2013).

1980.[49] Much of this change is related to storms, floods and associated landslides, and extreme climatic episodes such as prolonged droughts. In contrast, the number of earthquakes, tsunamis and volcanic eruptions are contained within a reasonably restricted range of <~140 events/year. Against the backdrop of increasing climate and weather-related hazards, 2010 saw their geological counterparts comprise < 10 per cent of all hazards. Since 1980, the world's population has risen from 4.5 billion to 7.0 billion, so it may be argued that the upward trend may also relate to population expansion into more hazardous regions. Indeed this may be true, but that data reveal no marked increase in catastrophes associated with earthquakes and tsunami, suggesting that the upward trend is reflecting more hazardous climate and weather. This conclusion is consistent with the finding and projections of the Intergovernmental Panel on Climate Change (IPCC).[50]

The impact of weather and climate related catastrophes on lives, infrastructure and economies is immense. Even normal 'bad weather' such as snow storms, takes its toll. For example, in the United States the cost of normal bad weather is estimated to be 3.4 per cent of the gross domestic product.[51]

Submarine cables are not immune to the hazards associated with climate change. And although cable fault data are insufficient to definitively identify fault trends relating to climate, it can be noted that recent damage to submarine and coastal telecommunications, such as that caused by Typhoon Morakot and Hurricane Sandy, is at least consistent with climate change projections.

According to the IPCC[52] and more recently published research, the ocean and coastal zones are projected to respond to a warmer, more vigorous climate. Of those projections, the following are particularly relevant to cables and related coastal infrastructure.

Sea level rise. Warming and thermal expansion of the ocean, coupled with additional melt water from land-based ice, have caused sea level to rise at 3.2 mm/year. This is a global average rate and masks considerable variation caused by regional conditions, for instance the rise in sea level off the northeastern United States is faster than the global average, which may reflect the added effects of

---

49 Munich R.E., NatCatService, Great Natural Disasters Since 1950. Available online at http://www.munichre.com/en/reinsurance/business/non-life/georisks/natcatservice/default.aspx (last accessed 1 June 2013).

50 Intergovernmental Panel on Climate Change, 2007. Oceanic Climate Change and Sea Level in Climate Change 2007—The Physical Basis. Contribution of Working Group 1 to the Fourth Assessment Report of the IPCC (S. Solomon *et al.*, eds, Cambridge University Press, 2007) 385–432.

51 J. Lazo *et al.*, "U.S. Economic Sensitivity to Weather Variability" (2011) Bulletin of the American Meteorological Society. See also http://www2.ucar.edu/news/4810/economic-cost-weather-may-total-485-billion-us (last accessed 1 June 2013).

52 Intergovernmental Panel on Climate Change, *supra* note 50.

changes in ocean currents and land elevation.[53] In that context, cable operators should account for local conditions at a landing, as modest changes in sea level can significantly increase the hazard potential. Research into the relationship of sea level rise and flooding around Australia suggests that as a general rule, a 100 mm rise in sea level increases the frequency of flooding threefold.[54]

Increased frequency and/or intensity of major storms. A warmer ocean and more vigorous atmosphere are projected to lead to more frequent and/or intense hurricanes, cyclones and typhoons that will markedly increase wave and current activity in coastal and shelf waters, and strengthen storm surges. Hurricane Sandy and the related storm surge that caused so much damage to Manhattan datacenters may be a window into the future, as it is part of an observed trend towards a more extreme climate in the northeastern United States.[55]

Regional changes in rainfall. A warmer atmosphere and ocean are likely to change patterns of precipitation, thereby affecting coastal flooding and the delivery of sediment to the continental shelf and beyond. The most recent and relevant example is Typhoon Morakot whose flood waters were so voluminous and laden with debris that they formed turbidity currents, which broke submarine cables well over 200 km from the river mouth.[56] Morakot was the wettest typhoon on record to strike Taiwan, but in light of the short time span of local meteorological records, all than can be said is that its occurrence is consistent with a warmer atmosphere.

Changes in wind patterns. The core of strong westerly winds has moved towards the poles, resulting in changes in waves, currents and potentially storm surges, for example, the core of the westerly winds in the Southern Hemisphere has shifted around 5 degrees of latitude southwards to strengthen major ocean currents.[57] Likewise, model simulations are suggesting that Southern Ocean wave height will increase and because these waves travel across ocean basins, the change may

---

53 A.H. Sallenger Jr. *et al.*, "Hotspot of Accelerated Sea-level Rise on the Atlantic Coast of North America" (2012) 2 *Nature Climate Change* available at http://www.nature.com/nclimate/journal/v2/n12/full/nclimate1597.html (last visited 1 June 2013).

54 Sea-Level Rise 2012, ACE CRC Report Card, Australia, Antarctic Climate and Ecosystems, available online at http://www.acecrc.org.au/access/repository/resource/1c91bb6a-15f5-1030-998b-40404adc5e91/ACE%20SLR%20REPORT%20CARD.pdf (last accessed 1 June 2013).

55 National Oceanic and Atmospheric Administration (NOAA), 2012, "US Climate Extremes Index; Northeast USA", National Climate Data Center, see http://www.ncdc.noaa.gov/extremes/cei/graph/ne/cei/01-12 (last accessed 1 June 2013).

56 Carter *et al.*, *supra* note 34.

57 J.R. Toggweiler *et al.*, "Midlatitude Westerlies, Atmospheric CO_2 and Climate Change During the Ice Ages" (2006) 21 *Paleoceanography* PA2005, available at doi: 10.1029/2005PA001154.

have implications for coasts of both hemispheres, especially in the presence of rising sea level. However, the same simulations suggest about a quarter of the planet will experience reduced wave heights.[58]

Density-driven ocean currents projected to vary. The density of ocean water is dictated by temperature, salt content and pressure; as sea water cools and/or becomes more saline its density increases causing it to sink. In the polar and sub-polar regions, fresh melt water from ice sheets and glaciers is reducing coastal ocean salinity and hence its density. Models and limited observations are suggesting that melt water from Greenland is lowering the density of the upper ocean thus reducing the sinking of water that contributes to the *Atlantic Overturning Circulation* whereby heat lost from the North Atlantic surface currents together with increased salinity from evaporation cause those surface waters to sink north of Iceland and in the Labrador Sea and to flow southwards at depths > 1100 m. The long-term impact of this change would be (i) to reduce those deep southward currents and (ii) increase local sea level rise because sinking locally depresses the ocean surface.

Indirect impacts. Climate change can alter the activities of other seabed users and hence the risk they pose to cables. Offshore wind farms are expanding in European seas and elsewhere as nations attempt to reduce greenhouse gas emissions as well as improve the security of energy supplies and meet greater energy demand.[59] As a result of wind farm expansion, the choice of new cable routes may be restricted and the maintenance of existing cables jeopardized—a situation that should be eased by initiatives such as development of cable placement guidelines for the renewable energy and submarine cable sectors.[60]

Recent research also suggests that commercial fisheries are responding to climate change, although scientists are cautious in noting that changes caused by human influences on the ocean can be difficult to distinguish from natural variations.[61] Nevertheless, a warmer ocean has encouraged southern fish species to migrate north, for example, anchovies, previously prominent in the Mediterranean Sea, are now appearing in commercial quantities off the United Kingdom.

---

58 M.A. Hemer *et al.*, Projected Changes in Wave Climate from a Multi-model Ensemble" (2013) *Nature Climate Change* available at doi:10.1038/nclimate1791.

59 For additional information on competing uses of ocean space and off-shore energy, refer Chapters 11 and 16.

60 The Crown Estate "Publication of Offshore Proximity Guidance, 29 August 2012, see www.thecrownestate.co.uk/news-media/news/2012/collaboration-between-submarine-cable-and-offshore-wind-industries-ushered-in-with-publication-of-proximity-guidance/ (last accessed 1 June 2013).

61 M. Frost *et al.*, "Impacts of Climate Change on Fish, Fisheries and Aquaculture" (2012) 22 *Aquatic Conservation: Marine and Freshwater Ecosystems* 331–336. Available online at http://onlinelibrary.wiley.com/doi/10.1002/aqc.2230/pdf (last accessed 1 June 2013).

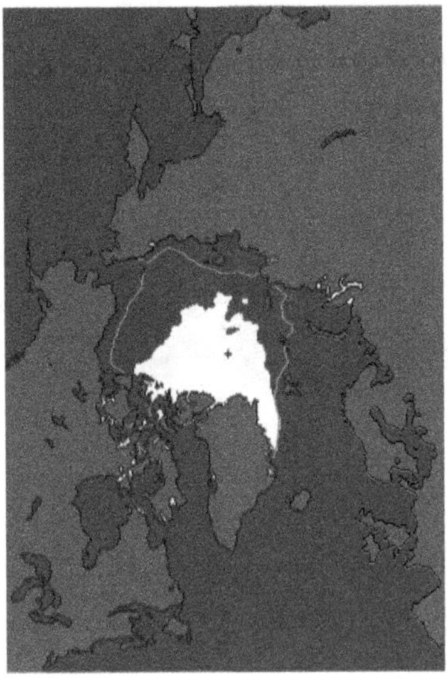

Figure 10.4 Sea ice observed in September 2012, when it reached its minimum extent since late 1978. Over that time, the extent has reduced by 45 per cent resulting in the summer opening of the Northwest and Northeast Passages off northern Canada and northern Russia respectively. The pink line is the median extent. (Image courtesy of National Snow and Ice Data Center)

Some deep-dwelling fish have increased their preferred depth of occurrence by 3.6 m/decade. If benthic fish are affected by this northward expansion, it may result in changes in bottom trawling.

New opportunities. Climate change will not only present more risk to submarine cables but will also provide opportunities. Over the past 35 years, the Arctic Ocean has witnessed a major loss of summer sea ice in response to a warmer ocean and increased storminess, the latter causing the ice pack to fragment making it more susceptible to melting (Figure 10.4).

In September, 2012 Arctic sea ice reached its minimum extent since satellite records began in late 1978.[62] If this trend continues, the summer Arctic could be ice-free within a decade or so—a change that has encouraged development of fiber optic links with remote Arctic communities and Europe (Figure 10.5). The more advanced proposal is the *Russian Optical Trans-Arctic Submarine Cable System* (ROTACS) intended to run from Tokyo through the Russian Arctic and terminate in London. En route branches will land in South Korea, China and

62 National Snow and Ice Data Center (NSIDC), 2012. "Arctic Sea Ice News and Analysis", see generally http://nsidc.org/arcticseaicenews/ (1 June 2013).

Figure 10.5 The Alcatel-Lucent cableship *Peter Faber* completes the Northwest Passage in September 2008 en route from the Pacific to the Atlantic to lay a cable from Greenland to Iceland. That voyage signifies the growing importance of the Arctic Ocean not only for vessel passage but the future development of submarine cables across the top of the world. (Photograph courtesy of Alcatel-Lucent)

northern Russia. A second system, *Arctic Fibre*, is initially planned to service northern Canada via the Northwest Passage that is part of a longer link to eventually connect Tokyo with London.

Conclusion

Despite the dominance of cable faults caused by human activities, faults resulting from natural hazards can disrupt multiple cables thus threatening telecommunications and internet traffic over large regions. This is especially the case where cables are forced to pass through constricted corridors that traverse earthquake and/or flood prone zones of the circum-Pacific Rim and the Mediterranean. The risk is heightened where cables cross submarine canyons where earthquake- and flood-triggered turbidity currents are focused within the canyon confines resulting in rapid flows capable of damaging cables over hundreds of kilometers. Although cable fault records are too short to definitively correlate faults with the present phase of climate change, impacts resulting from sea level rise, storm surge, storms and floods are at least consistent with climate change projections and observational trends as attested, for example by Hurricane Sandy, which is part of a trend towards more extreme weather events in the northeastern United States. However, climate change is also providing new opportunities for the cable industry with the opening up of routes through the Arctic Ocean—an opportunity that will be challenged by new hazards of an icy environment.

CHAPTER ELEVEN

PROTECTING SUBMARINE CABLES FROM COMPETING USES

Robert Wargo and Tara Davenport

Introduction

Nearly everyone has heard the story of the first submarine cable being pulled up by a fisherman who, thinking it was a new species of seaweed, cut a piece out and took it to a local university for study.[1] Indeed, damage to submarine cables by human activity in the oceans has been a consistent and inevitable threat to the safety and security of submarine cables. The vulnerability of submarine cables to damage from fishing, shipping and resource exploration activities has increased over the years, propelled by the ever-expanding ability of States to find new uses for ocean resources. The purpose of this Chapter is to highlight the main causes of cable faults arising from external human aggression, to examine the various legal and policy challenges faced by States in minimizing such faults, and to discuss best practices on the protection of cables by both governments and the cable industry. Hostile or intentional actions against submarine cable infrastructure by terrorists and pirates are addressed in Chapter 12.

I. The Causes of Cable Faults

External aggression is generally defined as damage to a cable caused by force or object external to the cable. It includes incidences of external human aggression (fishing gear, anchor dragging, dredging and other human activity) and natural occurrences, such as submarine earthquakes, landslides or abrasion (Chapter 10 examines damage caused to cables as a result of natural occurrences).

[1] D.P.F. Chisholm, "The International Cable Protection Committee" (1979) 46(1) *Telecommunications Journal* 29–32.

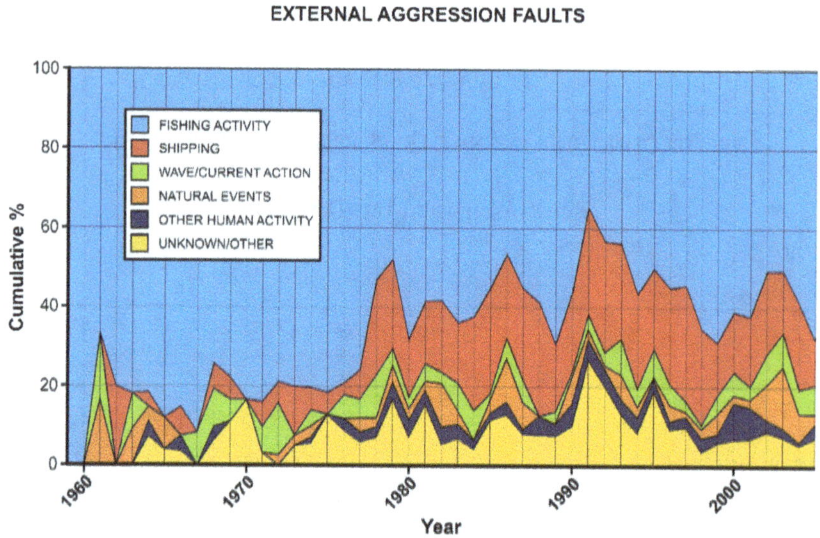

Figure 11.1 Historical percentage of cable fault causes. (Chart prepared by Wood and Carter, reproduced courtesy of the International Cable Protection Committee)

A cable fault can take one of several forms; damage to the outer insulation which results in seawater coming into contact with the power conductor, damage to the optical fibers such that they can no longer transmit light, damage to both the power conductor and the optical fiber and, finally, a complete break in the cable.

Analysis undertaken in a 2007 study indicates that between 72–86 per cent of all cable faults are caused by external aggression.[2] Eighty per cent of cable faults caused by external aggression are attributable to human activity, with fishing being responsible for more than 60 per cent of all external aggression faults.[3] Figure 11.1 provides a representation of the overall percentage of faults from 1960 until mid-2000 and also illustrates that the majority are caused by commercial fishing.

As indicated earlier, not all cable faults occur as a result of human activity. Faults have also been caused by various geologic processes, such as earthquakes and their resultant after-effects (tsunamis, undersea landslides, turbidity currents), abrasion of the cable on a rough seabed, and general equipment failure. Two of the more novel causes of cable failure in the past include incidents of weather observation buoys that have gone off station and become entangled in

---

[2] M.E. Kordahi *et al.*, "Trends in Submarine Cable System Faults" (2007) available online at http://www.suboptic.org/archives/suboptic-2007-archive/oral-and-poster-papers.aspx?catid=5 (last accessed 6 June 2013).

[3] Kordahi *et al.*, *ibid.*

submarine cables[4] and shark bite. The latter problem has since been eliminated by adding additional polyethylene and metal protective layers to the cable.[5]

Bottom contact fishing is, by far, the cause of the greatest number of faults worldwide. The equipment and practices employed in bottom contact fishing activities can be very detrimental to submarine cables. For example, clam dredges can dig into the seabed up to 6–8 inches and scallop dredges rake along the bottom.[6] Both of these gear types tend to work in shallow water and can damage cables that are not sufficiently buried. Otter and beam trawls skim along the seabed and are used to harvest bottom fish that dwell either in or on the seabed. Otter trawls can be used in water depths of 1000 m or more and can damage unburied cables. In areas where adequate burial can be achieved, generally in the order of 1 m burial depth, the threat from these types of commercial fishing practices is greatly reduced. Other detrimental fishing practices include the use of stow nets. Stow nets are deployed in tidal flows and their movement is slowed by large anchors. The anchors dig into soft sediment and have been responsible for causing numerous faults even to well-buried cables off the coast of China (see Figure 6.8). In areas where static gear such as pots or traps are used and sometimes lost, the grapnels used to recover them can also damage cables.

Anchors are the second leading cause of external aggression faults to cables, with damage typically occurring in two forms. The first form is damage occurring as a result of ships anchoring outside approved anchorages in areas near cable landings. For example, in the early 1990s overcrowding in the approved Caribbean anchorage area around the high shipping volume Christmas holiday season in the Panama Canal often resulted in cable breaks. This was eventually remedied by moving the cables further away from the anchorage. In 2007–2008, during the worldwide economic downturn, large numbers of unemployed cargo ships congregated off of Singapore, awaiting orders. To avoid crowded anchorages and fees, many of these ships anchored in areas where submarine cables were laid, causing 21 cable faults.[7] This situation was improved by the joint actions of

4 On 9 March 2008, TAT-13, a trans-Atlantic telecommunication cable linking the United States to Europe was damaged by NOAA ODAS weather buoy #44004, the anchor of which drifted across the cable.

5 L.J. Marra, "Sharkbite on the SL Submarine Lightwave Cable System: History, Causes and Resolution" (1989) *IEEE Journal of Oceanic Engineering* 14(3), 230–237. This will be addressed in further detail in Chapter 7.

6 E. Wagner *et al.*, "International Legal Issues in Submarine Cable Protection" SubOptic Conference 1993 (Table 1: External Aggression Fault Statistics, North Atlantic Cables, 1968–1992).

7 R. Beckman and T. Davenport, "Workshop on Submarine Cables and Law of the Sea" 14–15 December 2009 at 32, footnote 28, powerpoint presentation by Singapore Maritime Port Authority reporting 21 faults (21 January 2008–20 July 2009) established by Automatic Identification System (AIS) as caused by ship anchors in the vicinity of the Singapore Strait, http://cil.nus.edu.sg/wp/wp-content/uploads/2009/10/Workshop-Report-29-Jan-2010.pdf (last accessed 6 June 2013).

Indonesia, Malaysia and Singapore in securing from the International Maritime Organization (IMO) a circular warning of strict enforcement by the three States of non-anchoring restrictions over cables.[8]

The second form of anchor damage occurs as a result of an anchor being unintentionally dragged by a vessel underway.[9] An improperly stowed anchor can be dragged a substantial distance from a vessel underway without being discovered. For example, in 18 April 1986 the cargo ship M/V *Aconcagua* broke three of the then four trans-Atlantic cables off the coast of New Jersey a total distance of approximately five miles apart. When spotted by a patrol plane shortly after, in transit to New York harbor she was still dragging her anchor chain. The vessel had failed to properly secure its anchor for sea, and when the anchor paid itself out during a gale, it severed the submarine cables as it skipped along the seabed. One of the submarine cables that suffered damage was a direct 'hot line' between the governments of the US and the USSR.[10] The next day the US version of the Financial Times was not printed as it could not be transmitted across the Atlantic on the remaining cables that were still operating. The cable owners subsequently arrested the ship in New York harbor and recovered over USD2.2M in damages from the vessel's protection and indemnity club.

There are a number of other human activities that have the potential to damage submarine cables. These include dredging sand for beach reclamation and navigation channels, offshore energy extraction and its associated activities such as anchor handling. The installation of new cables may also cause damage to previously laid cables[11] and, as noted above, navigational and meteorological buoys have also caused breaks in cables.

The statistics[12] indicate that the majority of faults caused by external aggression occur in water depths of less than 100 m. When cables come ashore they necessarily transit shallow water. It is in these areas that cables are at their most vulnerable, because frequently these waters are heavily fished. Because cables tend to come ashore near large population centers they may also be close to shipping lanes and anchorages situated near busy port facilities. As noted above, bottom tending fishing gear such as trawls and dredges and ships' anchors are fre-

---

8 IMO Circular 282 Information Concerning Anchoring in the Traffic Separation Scheme in the Straits of Malacca and Singapore, 27 November 2009.

9 Loss Prevention Bulletin circulated by vessel protection and indemnity (P&I) clubs (18 March 2009), www.iscpc.org (located under 'Publications') reporting 21 cable faults (2006–2008) established by AIS caused by ship anchors in United Kingdom waters.

10 The 'hot line' was one of 4200 circuits in TAT-7, a commercial cable and underscores how vital international State communications are enmeshed in the commercial cable infrastructure. See Chapter 15 for a discussion of international treaties related to special State to State communications.

11 *AT&T Corp v. Tyco Telecomms (US), Inc* 255 F Supp 2d 294, 2004 AMC 1964 (S.D.N.Y. 2003); *Concert Global Network Services, Ltd v Tyco Telecomms (US) Inc*, Soc'y of Mar Arbs No 3770 (12 September 2002), available at 2002 WL 34461677.

12 Kordahi *et al.*, *supra* note 2.

quent causes of cable faults. Sand mining or dredging for beach reclamation can also occur nearby cable landings and have the potential to damage submarine cables or remove enough sediment that buried cables become unburied.

II. International Law on the Protection of Submarine Cables

Chapter 3 of this Handbook sets out in detail the applicable international law on the protection of submarine cables, as provided in the 1884 Convention for the Protection of Submarine Telegraph Cables (1884 Cable Convention)[13] and the 1982 United Nations Convention on the Law of Sea (UNCLOS).[14] For the purposes of this Chapter, salient provisions of UNCLOS (which incorporates the relevant provisions from the 1884 Cable Convention) will be briefly highlighted.

Within Territorial Waters (Internal Waters, Territorial Seas and Archipelagic Waters)

As discussed in Chapter 3, UNCLOS divided the oceans into different maritime zones where coastal States and other States have different rights and obligations. The 12 nm territorial sea[15] and archipelagic waters[16] are considered areas within the territorial sovereignty of the coastal/archipelagic State (territorial waters). In these waters the coastal/archipelagic State has the authority to regulate all activities, subject to the right afforded to foreign vessels of innocent passage in the territorial sea[17] and archipelagic sea lanes passage in archipelagic waters.[18]

Coastal States[19] and archipelagic States[20] have an express right to adopt laws and regulations relating to innocent passage/archipelagic sea lanes passage through their territorial sea and archipelagic waters in order to protect submarine cables. They also have a general competence to enact laws to protect submarine cables within such territorial waters. However, under UNCLOS there is no

13 Convention for the Protection of Submarine Telegraph Cables, adopted 14 March 1884, TS 380 (entered into force 1 May 1888) (1884 Cable Convention). The provisions of the Cable Convention are generally accepted as customary international law in the United States: See *Restatement of the Law (Third): The Foreign Relations Law of the United States*, Volume 2 (American Law Institute Publishers, 1987) § 521, comment f (1986). As at July 2013 there are 40 States Parties to the 1884 Cable Convention. A complete copy of the Convention is contained in the Appendix.

14 United Nations Convention on the Law of the Sea, adopted 10 December 1982, 1833 UNTS 3 (entered into force 16 November 1994) (UNCLOS). Select UNCLOS provisions are contained in the Appendix.

15 See generally UNCLOS Part II.

16 See generally UNCLOS Part IV.

17 UNCLOS Art 17.

18 UNCLOS Art 53.

19 UNCLOS Art 21(1)(c).

20 UNCLOS Art 52.

obligation on coastal States to adopt laws and regulations to protect submarine cables within territorial waters.

Outside Territorial Waters (EEZ, Continental Shelf and High Seas)

The exclusive economic zone (EEZ) and the continental shelf are maritime zones where coastal States are afforded sovereign rights over resources and certain jurisdictional competences[21] but where other States also have certain freedoms, including the freedom of navigation and the freedom to lay submarine cables and pipelines.[22] It is clear that the coastal State does not have sovereignty in these areas akin to the sovereignty that it has in the territorial sea. The high seas lie beyond the EEZ and the continental shelf. The high seas are open to all States and no State may validly purport to subject any part of the high seas to its sovereignty.[23]

Articles 113 to 115 of UNCLOS address the protection of submarine cables on the *high seas* and are based on three articles in the 1884 Cable Convention. They are also applicable to submarine cables laid in the EEZ[24] under Article 58(2) as well as to cables laid on the continental shelf.[25]

Article 113 of UNCLOS requires States to adopt laws and regulations to provide that the breaking or injury by a ship flying its flag or by a person subject to its jurisdiction of a submarine cable beneath the high seas done willfully[26] or through culpable negligence, is a punishable offence.[27] Such laws and regulations must also apply to conduct calculated or likely to result in such breaking or injury. However, it shall not apply to any break or injury caused by persons who acted to save lives or their ships, after having taken all necessary precautions to avoid such an occurrence. Article 113 essentially extends a State's criminal jurisdiction (usually limited to territory) over acts of breaking or injury to submarine cables done "willfully or through culpable negligence". This extension of jurisdiction only applies to ships flying the State's flag on the high seas or EEZ, or to its

21 UNCLOS Arts 56 and 77.

22 UNCLOS Arts 58, 78 and 79.

23 UNCLOS Arts 87 and 89.

24 UNCLOS Art 58(2).

25 See M. Nordquist *et al.*, (eds.), *United Nations Convention on the Law of the Sea 1982: A Commentary*, Volume III (Martinus Nijhoff, 1995) at 270, 273, and 278.

26 See generally *Peracomo et al., v Sociéte Telus Communications, Hydro-Québec, Bell Canada v Royal and Sun Alliance Insurance Company of Canada (the Realice)*, 2012 FCA 199 (29 June 2012), aff'g 2011 FC 494 (2011).

27 See generally, *The Government of the Netherlands, Post Office v G'T Mannetje-Van Dam [Fishing Cutter Go 4]*, File No 325/78 (District of Rotterdam, decision rendered 20 November 1978, aff'd sub nom, GT Mannetje Post Office No 69 R/81 (The Court at the Hague, Second Chamber, decision rendered 15 April 1983); *Alex Pleven*, 9 Whiteman, *Digest of International Law* at 948–951 (1961).

nationals who commit such acts, consistent with general principles of international law on the prescription of extra-territorial jurisdiction.

Article 114 of UNCLOS, which is based on Article IV of the 1884 Cable Convention, requires every State to adopt laws and regulations concerning the liability of owners of cables for the cost of repairs to existing cables which are damaged in the course of laying or repair operations.[28]

Article 115, which is based on Article VII of the 1884 Cable Convention, provides that every State should adopt laws and regulations to provide for an indemnity to be paid by cable owners to ship owners whose master sacrifices an anchor, a net or any other fishing gear in order to avoid injuring a submarine cable, provided that the ship owner has taken all reasonable precautionary measures beforehand.[29]

The practice of sacrificing fishing gear or an anchor to avoid injury to a submarine cable is considered a basic tenet of prudent seamanship.[30] Complementing this legal obligation is the customary practice in the cable industry to provide for procedures in line with Article 115 and Article VII, using toll-free 24/7 telephone numbers, cable warning charts and websites to provide mariners at sea immediate access to knowledgeable cable owner representatives if the mariners suspect they may have fouled a submarine cable or have questions on a cable's location. For industry it is far preferable to pay compensation for fishing gear or anchor sacrifice than meet the cost of repairing a damaged cable.

III. Law and Policy Challenges in the Protection of Cables from Competing Uses

Governments have the opportunity and ability to assist cable owners in the protection of this important critical infrastructure through effective legal and policy protections. Often, however, the laws and policies designed to protect submarine cables are inadequate or antiquated and have failed to keep pace with advances in the marine realm. This section will expound on the various legal and policy challenges that States encounter in attempting to protect submarine cables from competing uses in the oceans.

---

28 See generally, *AT&T Corp v Tyco Telecomms (US), Inc* 255 F Supp 2d 294, 2004 AMC 1964 (SDNY 2003); *Concert Global Network Services, Ltd v Tyco Telecomms (US) Inc*, Soc'y of Mar Arbs No 3770 (12 September 2002), available at 2002 WL 34461677.

29 See generally, *Agincourt Steamship Company Ltd v Eastern Extension, Australasia and China Telegraph Company Ltd*, 2 KB 305 (1907).

30 United Kingdom Hydrographic Office, *The Mariner's Handbook* at §§ 3.121–3.122, NP100 (ed. 9, 2009); *Société Telus Communication v Peracomo Inc.*, 2012 FCA 199 (29 June 2012).

Lack of Adequate National Laws on the Protection of Submarine Cables

As noted above, 72–86 per cent of all cable faults are caused by external aggression and of those, more than 80 per cent are attributable to human activities, with fishing causing the majority of the faults.[31] One of the major problems in the protection of submarine cables is the absence of adequate legislation to meaningfully penalize persons or entities who cause damage to submarine cables, either within or outside territorial waters.

Within territorial waters, many States either do not have legislation criminalizing damage to submarine cables (whether negligent or intentional),[32] or only impose minimum fines which are inadequate to compensate owners for repairs or to serve as a deterrent. For example, when American cable owners objected to the US law which imposed only US$5000 for willful injury to submarine cables and US$500 for negligent damage, it was observed that the "insignificant maximum penalty provides little incentive for enforcement authorities to assign full-time legal and investigative personnel to prosecute vessel owners caught damaging a submarine cable."[33]

While States are not obliged by UNCLOS to impose penalties for damage to submarine cables in territorial waters, coastal States have extensive powers to regulate all activities within such waters and it is clearly in their best interests to do so. This is especially because the majority of external aggression faults occur in water depths of less than 100 m. Adequate penalties would provide an incentive to owners/masters of fishing and other vessels to exercise more caution when engaged in their respective activities, thereby reducing the possibility of damage to submarine cables.[34]

---

31 Kordahi *et al.*, *supra* note 2.

32 For example, in a review of the national legislation of Southeast Asian States, none of the States had an express provision imposing penalties for negligent or willful damage to submarine cables. However, it warrants note that intentional (as opposed to negligent) damage to cables may be covered by general legislation criminalizing damage to telecommunications infrastructure within their territory: See, for example, Section 21 of Brunei's Telecommunications Order 2001, Section 41 of Singapore's Telecommunications Act and Sections 44, 72 and 73 of Thailand's Telecommunications Business Act 2001.

33 S. Coffen-Smout and G.J. Herbert, "Submarine Cables: A Challenge for Ocean Management" (2000) *Marine Policy* 24 at 444; D. Burnett, "Cable Vision" US Naval Institute Proceedings, August 2011 at 68–69, reporting lack of prosecution for damaging a US-Cuba cable in 1997.

34 This would especially be the case if Protection and Indemnity Clubs (P&I Clubs) put pressure on owners to ensure that their masters minimize risk of damage to submarine cables. Loss Prevention Bulletin circulated by vessel P&I clubs (18 March 2009), www.iscpc.org (located under 'Publications'); Owners of fishing vessels also respond to the deterrent value of civil and criminal sanctions as demonstrated in a instruction by one owner to his fishing fleet that stated "All Captains will be held responsible for each and every warning letter received regarding... vessels dangerously close

For damage to submarine cables that occurs outside territorial waters (i.e. in the EEZ, continental shelf and high seas), many States Parties to UNCLOS have not implemented their obligations on the protection of submarine cables, in particular, their obligation in Article 113 of UNCLOS to penalize damage to submarine cables. Indeed, this has been consistently recognized by the United Nations General Assembly, which has called upon States to implement their obligations under Article 113 of UNCLOS.[35] States that have enacted such legislation are commonly parties to the 1884 Cable Convention[36] and, in many cases, their legislation has not been updated and the penalties are so low that they provide neither an incentive for authorities to prosecute or for vessels to take minimum precautions to avoid damage to submarine cables.[37]

Lack of Understanding on Submarine Cables Leading to Ineffective Policies

It is a commonly held assumption on the part of governments that satellites provide the majority of their countries' international telecommunications needs. This highlights the general lack of knowledge that government officials have on the importance and nature of submarine cables and often results in policies that undermine the protection of submarine cables.

For example, governments will frequently request that cables be confined to cable corridors in an effort to reduce their overall footprint and appease other ocean users. However, this reduces the resiliency of the overall undersea network, as placing multiple cables in a concentrated location increases their

---

to . . . cables. Furthermore, responsible Captains will be fined $1000 for the first warning incident and immediately dismissed for the second warning incident." E. Wagner *et al.*, "International Legal Issues in Submarine Cable Protection" SubOptic Conference Paper (1993) at 4.

35 See General Assembly Resolution A/Res/66/231 dated 5 April 2012 at para 12; General Assembly Resolution A/67/L.21 dated 23 November 2012 at para 133.

36 See for example, Sections 21 and 22 of the US Submarine Cables Act, 29 February 1888, 47 US Code 21; Australia's Submarine Cables and Pipelines Protection Act 1963, Act No 61 of 1963 and Section 3 of the UK Submarine Telegraph Act 1885. It should be noted, however, that in 2005, Australia introduced a new regime on the protection of cables: See Part 2, Schedule 3A, Telecommunications Act 1997.

37 The penalty in the United States under Sections 20 and 21 of the 1888 Submarine Cables Act, 29 February 1888, is US$5000 for willfully breaking a submarine cable and US$500 for negligently breaking one. In Australia, although Australia has adopted extensive legislation on the protection of submarine cables through the establishment of cable protection zones (see Part 2, Schedule 3A of the Telecommunications Act 1997), the Submarine Cables and Pipelines Protection Act 1963 still applies to submarine cables beneath the high seas or EEZ which do not fall within established protection zones. The penalty for intentional breakage of submarine cables is AUS$2000 or imprisonment for 12 months and for negligent damage, the penalty is AUS$1000 or imprisonment for 3 months: See Section 7 of the Submarine Cables and Pipelines Protection Act, Act No. 61 of 1963. But see updated legislation at 272–274.

vulnerability and the potential impacts of a single incident such as a terrorist attack or an earthquake. This lack of understanding was demonstrated in the Eastern Scotian Shelf Integrated Ocean Management (ESSIM) Plan of Canada. In that instance the Canadian provincial government intended not only to group submarine cables into a corridor but also to include other utilities, such as pipelines and electrical cables, in the same location. Cable owners worked collaboratively with the Canadian federal government to draw attention to the inconsistencies of the ESSIM plan with UNCLOS and Canadian law, and to educate regulators on the importance of submarine cables as critical infrastructure. The cable industry ultimately prevailed in having the cable utility corridor concept removed from the ESSIM plan.

The lack of understanding and knowledge on submarine cables is also discernible in the tendency of governments to ignore submarine cables when planning or regulating activities in maritime zones under their jurisdiction. This is notably evident when these activities incorporate new technologies *vis-à-vis* submarine cables, one of the oldest technologies in the ocean. For example, governments are currently very interested in 'green' (renewable) technologies such as offshore wind and wave power. These technologies offer the promise of reducing States' dependence on carbon based energy and foreign oil imports. However, in some instances the eagerness of governments to approve these projects has resulted in them ignoring the fact that submarine cables were already in place and needed to be protected and maintained. In practical terms, the spacing of wind turbines needed to allow for cableships to enter the wind farm, carry out repairs on the telecommunications cable, and exit after completion, without an unreasonable risk of collision with the wind turbines during these processes. This problem was demonstrated in the early rounds of wind farm leases released by the UK when several proposed leases overlapped with areas in which active submarine cables were located. Consultation with the cable owners prior to the issuance of the leases would have avoided this issue. Cable owners in the UK (represented by the Regional Protection Committee, Subsea Cables UK) engaged with the government and the renewable energy industry operating in the UK to work on guidelines for siting renewable energy structures in the vicinity of submarine cables to avoid this issue in the future.[38]

Similarly, the general drive by governments to adopt Marine Spatial Planning schemes runs afoul of the same issues. It would be unacceptable to site a Marine Protected Area, a sand borrow area or an artificial reef in the area of a submarine cable if the effect is that the cable is damaged, protection of the cable is reduced or the ability to maintain the cable is compromised. (Chapter 7.)

---

[38] See ICPC Recommendation 13, Issue 1B (27 September 2010) 'Proximity of Wind Farm Developments and Submarine Cables'. Also refer to Chapter 16 at 372 on Submarine Cables and Offshore Energy.

Lack of Systems and Resources to Update Maps and Charts on the Location of Cable Routes

Maritime law and safety depends heavily upon navigational charts and Notices to Mariners produced by government-operated hydrographic offices. Due to constant fiscal pressure on the part of governments to reduce expenditure, these hydrographic offices are frequently under-resourced and do not have sufficient staff to produce and distribute charts and notices in a timely manner in order to keep pace with new installations, including submarine cables. If a recently laid cable is not depicted on a chart and no Notice to Mariners is issued identifying its route, the cable is at risk because those who have the potential to damage it do not know its location and will not take steps to avoid it. In addition, if the entity responsible for the cable break is identified, there may be no means to recover damages from them as they were not notified of the location of the cable. This occurred recently with a cable break off Trinidad. When a cableship arrived to repair the damaged cable, it found an oil rig in the vicinity of the fault. Discussions with the rig and support vessels indicated that one of the rig's anchors had likely broken the cable. Follow-up investigations indicated that the cable had never been added to charts by the local cable owner, and it became evident that there was no system in place for updating such charts. Efforts to recover damages were abandoned once this fact came to light. (See 268 n.42.) It should also be noted that even if a Notice to Mariners is issued in a timely manner by the responsible authorities so that all mariners have received constructive notice, lax mariners may not keep up to date and edit their charts with the revised data, further endangering the cable.

IV. Best Practices for Cable Protection by the Cable Industry

The cable industry takes various steps to avoid or mitigate damage to its cables. In this regard, the industry has developed a standard set of protection strategies that operate worldwide, though they may need to be tailored to certain geographic areas or to address specific types of threats. These can be thought of as an industry 'best practice' and span the life cycle of the cable, from concept through to retirement.

Proper Routing, Cable Engineering and Installation

In the early stages of planning for a new cable, prior to the route survey being conducted, a desktop study (DTS) should be performed. The ultimate aim of the DTS is to locate the safest and most technically viable route for the cable.[39] The

---

[39] This is examined in Chapter 4. Also note ICPC Recommendation 9, 'Minimum Technical Requirements for a Desktop Study' (also known as Cable Route Study) available

DTS will use published and grey literature, as well as previous experience of other cables along the route, to identify the majority of the threats that exist for the proposed cable. Navigational charts will be reviewed to identify shipping lanes, anchorages, wrecks, protected areas, munitions dumps or sand borrow areas to ensure that such hazards will be avoided. Local fishing, maritime, coast guard and environmental organizations should be consulted within every State in which the cable will land. A properly conducted DTS will provide vital information for use during the route survey (see Chapter 4). A review of Automatic Identification System (AIS) or Vessel Monitoring System (VMS) data at this time will assist in clarifying the risks presented by commercial fishing or anchoring in the area.[40]

Once a route is identified via the DTS, the Cable Route Survey can be performed to further refine decisions regarding the route and the type of burial and armoring that will be required for the cable. The Cable Route Survey will provide precise detail on the proposed route which will enable cable engineers to avoid obstructions and features not identified in the DTS. Previously unknown or uncharted features, such as wrecks and seamounts, will be identified and avoided. Areas that exhibit trawl scars may require changes to be made to the armoring or burial previously specified in the DTS. Bottom sediment will also be identified, and this information will be used to plan for the appropriate cable burial method; whether by plow burial or by remotely operated vehicle (ROV). All of these items originally identified in the DTS are further refined by the results of the Cable Route Survey.

As a result of the DTS and Cable Route Survey, a cable will be designed and properly engineered to meet the rigors of the route selected. Information regarding specific threats will identify the areas along the route where cable armor and burial will be required. Although burial may be recommended throughout specific areas, bottom conditions such as hard seabed or steep slope may not allow for burial. In these cases armoring the cable will be the best option (see Chapter 5).

The route the cable will follow is inevitably a compromise between risk, budget and time. The safest route for any cable is the route that minimizes the risk over the lifespan of the cable within the budget and timetable of the cable owner. In many areas earlier laid cables followed the safest routes, and as a result there may be crowded landings and limited options for route variation. In some developing areas the last cable to land in a country may be a telegraph cable from the 1800s and local knowledge about how a cable may fare at a particular landing point may be incomplete.

---

at http://www.iscpc.org/Recs/Recommendation_09_Iss_04.pdf (last accessed 6 June 2013).

40 A. Palmer-Felgate, "New Methodologies for Desk Top Study Cable Route Planning" (2010) SubOptic Conference, available at http://www.suboptic.org/archives/suboptic-2010-archive/papers-and-presentations-posters.aspx?catid=36 (last accessed 20 June 2013).

Post Installation Cable Awareness Programs Are Essential

Once the cable has been installed the cable owner should take steps to advise other maritime users of its presence. The cable route should be provided to the leading charting authorities so they can publish the cable locations on navigational charts and issue Notices to Mariners. At a minimum the three major producers of worldwide international chart systems,[41] as well as the local national charting authority, should confirm receipt of the cable as laid position lists. These steps are essential not only for the physical safety of the cable but also for its legal protection. A prudent mariner should be navigating using the most up to date information available for an area. Publishing the route ensures that, in the event that the cable is broken and those responsible for its breakage are identified, the cable owner has done what is required by providing notice and will be in a better position to recover damages from those at fault.

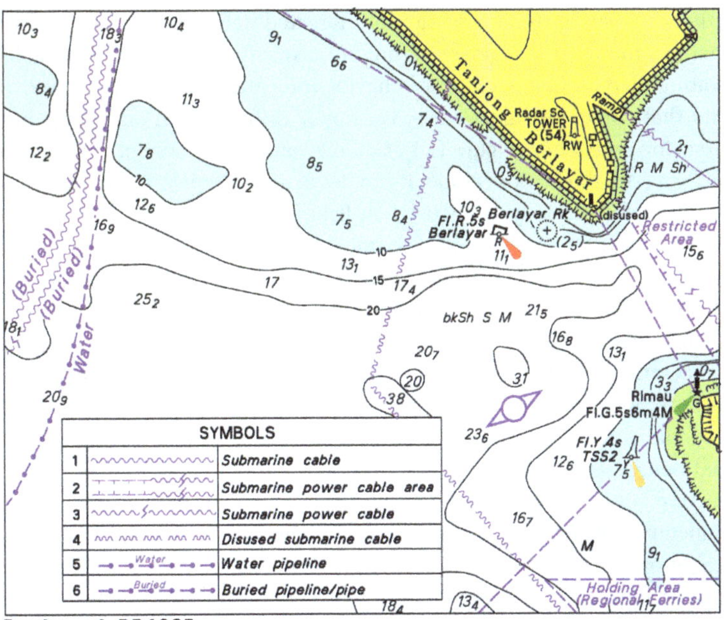

Portion of DB4035

Figure 11.2 Navigational chart extract showing symbols for submarine cables, both in service and out-of-service, a submarine cable area and a submarine power cable. The charted magenta symbols are based on the 'Chart Specifications of the International Hydrographic Organization'. These symbols alert mariners to the presence of submarine cables. (Image courtesy of the Maritime and Port Authority of Singapore)

41 UK Hydrographic Office (United Kingdom), NOAA and DMA (United States), and Service Hydrographique et Océanographique de la Marine (SHOM) (France).

It should be borne in mind that if there is a breaking or injury to a cable but its presence "has not been adequately marked [on a chart], there can be no question of 'culpable negligence' on the part of the navigators" as required under Article 113 of UNCLOS.[42]

Most cable owners go far beyond the minimal notice to advise other maritime users of the presence of a cable and employ active Marine Liaison and Cable Awareness/Protection programs. These programs target specific marine user groups with the capacity to damage cables and/or a history of cable damage. The programs are conducted to increase marine user's knowledge and regard for the importance of avoiding damage to cables. Liaison programs should be uniquely tailored to the risks identified and the local audience. Programs should be presented in the local language and be conducted at a technical level appropriate to the audience, paying heed to local laws, culture, customs and other factors.

New technologies such as the AIS and VMS, and other maritime monitoring programs can sometimes identify the vessels, companies or specific ports involved in activities creating risk, enabling more targeted liaison before damage occurs. Typical AIS monitoring involves setting up a warning zone around a cable and generating alerts when a vessel's behavior indicates it may be involved in an activity that could damage a cable, i.e. slowly drifting in an anchorage indicating dragging an anchor (Figure 11.4). Cable awareness personnel can then either notify the vessel directly or contact its owner in an attempt to stop damage before it occurs. Since VMS is typically used on fishing vessels and the location of their fishing activities may be viewed as a trade secret, this data is often bundled by authorities and, rather than targeting specific vessels, cable awareness personnel may visit specific ports to target the appropriate audience.

In each area the cable owner should carefully weigh the benefits of joining a liaison committee with other ocean user groups. The committees may be related to offshore renewable energy or petroleum, but are most commonly related to commercial fisheries. Liaison committees meet regularly to discuss topics of mutual benefit and facilitate interaction between both user groups. While this interaction is generally positive and offers numerous benefits, it is often funded by monetary contributions from the cable owner and is sometimes forced upon the cable owner by the local permitting agency. Cable owners should be in a position to join liaison committees as a matter of choice not because installation permits are withheld until the cable owners agrees to join the committee.

---

[42] See Commentary to Article 62 of the 1956 Draft Articles on the Law of the Sea at 294, *Yearbook of the International Law Commission*, Volume II, UN Doc. A/3159 (1956).

Cable Awareness programs may employ a variety of tools to communicate with their intended audience. Tools such as charts or cable data, which can be loaded directly into a vessel's navigation systems, are often provided free of charge to groups such as commercial fishermen and marine construction companies. Cable awareness staff may also conduct port visits prior to a cable being installed, or immediately after installation, so as to ensure that the most recent information is provided in a timely manner. During port visits small scale 'chartlets' with the planned or installed route of the cable are often provided. It is important to provide this information in the local language spoken in the port otherwise the message is unlikely to be disseminated to the parties who most need to be aware of it. In the event that a vessel is unmanned during a visit, cable personnel frequently leave information for the master to ensure that she/he has information on the cable route or is aware of how to contact cable liaison personnel with any questions. Often cable awareness personnel will conduct meetings in a port with groups of fishermen; indeed, liaison committees provide an excellent means of reaching important maritime users such as these.

Pursuing Civil Actions against Vessels Causing Damage to Submarine Cables

In the event that a cable is damaged and the culprit can be identified, the cable owner has the option to pursue legal action against the person/entity responsible for the break. In these cases a civil claim will be made for the full value of the repair. With most repairs costing in excess of $1M, the successful prosecution of a claim against the responsible party acts as an effective deterrent against future damage when word of the legal action spreads.

Figure 11.3 A submarine cable damaged by an anchor. (Photograph courtesy of D. Burnett)

To recover in a civil case against a vessel or party that injures a submarine cable, five elements must be proven to establish a maritime tort:

- The cable must have been damaged by mechanical (man-made) means.
- The vessel must have had either actual notice[43] of the position of the cable or constructive notice[44] as a result of the cable's position being shown on a nautical chart.
- The vessel must have been at the cable fault location at the approximate time of the fault.
- The vessel must have been engaged in an activity capable of causing the damage (such as fishing, anchoring or dredging).
- No other vessels must have been in the same area at the approximate time of the fault.

With these elements established, the burden shifts to the vessel to prove it was not involved in the damage. AIS data is ideal for establishing the third and fourth elements. (Figure 11.4.)

Jurisdiction for a civil suit lies in the following venues:

- Any State where the cable lands.
- The flag State of the culprit vessel.
- Any port where the ship can be physically arrested.

It is recommended that the vessel be arrested as soon as possible after it is identified in order to obtain and preserve testimony and evidence and to obtain security for the cable owner's damage claim in an amount sufficient to cover the damage estimates. Damage claims include the cable repair ship hire, composed of standing charges and running costs, spare cable and equipment used in the repair, and restoration costs. Because of the special maritime nature of a cable damage case, it is recommended that the cable owner retain experienced admiralty counsel as soon as it is known that a culprit vessel is suspected of causing the fault.

---

[43] *United States v North German Lloyd*, 239 F. 587,589 (SDNY 1917); *The Elsie*, 288 F. 575 (ND Cal 1923); *Alex Pleven*, 9 Whiteman, *Digest of International Law*, 948–951 (1961).

[44] Novorossiisk, Department of State Bulletin, Volume XL, No 1034, 555–558 (29 April 1959); *Peracomo et al., v Sociéte Telus Communications, Hydro-Québec, Bell Canada, v Royal and Sun Alliance Insurance Company of Canada*, 2012 FCA 199 (29 June 2012), aff'g 2011 FC 494 (2011).

V. Protection of Submarine Cables by Governments—Best Practices for Domestic Legislation and Enforcement

Around the world governments are coming to the realization that submarine cables represent critical infrastructure, on par with power generation and food and water distribution.[45] Governments are also realizing that they have a cooperative role to play with cable owners in order to protect these vital assets.[46] This section identifies some of the more important initiatives that assist cable owners in protecting submarine cables.

Establish Sufficient Penalties for Damage to Submarine Cables

What can governments do to aid cable owners in the protection of their cables? First, governments can enact legislation that makes it an offence (with sufficient penalties) to damage cables in their territorial waters, regardless of whether the acts are intentional or negligent. Although there is no obligation under UNCLOS for States to enact such legislation, it is critical to do so, given that many of the cable faults occur within territorial waters and are attributable to the negligence of vessels. It would also be preferable that such legislation sets out the offence as specifically as possible to address both negligent and intentional damage caused to submarine cables in territorial waters rather than subsume damage to submarine cables under offences relating to general damage to telecommunications infrastructure.[47] This would send the signal that submarine cables are a critical component of a State's infrastructure. For enforcement purposes it would also minimize difficulties in identifying an offence that will cover the act.

Second, with regard to damage to cables outside of territorial waters, States that are parties to UNCLOS should implement their treaty obligations on the protection of cables and, in particular, the obligations contained in Article 113.

---

[45] See for example, General Assembly Resolution A/Res/65/37 dated 17 March 2011, General Assembly Resolution A/Res/66/231 dated 5 April 2012 and General Assembly Resolution A/67/L.21 dated 23 November 2012, all of which have recognized that submarine cables transmit most of the world's data and communications and are vitally important to the global economy and the national security of all States.

[46] See for example, the Press Release of the Asia-Pacific Economic Co-operation (APEC) Committee on Trade and Investment which called for the increased harmonization of submarine telecommunication cable protection and recognized the importance of cooperating on initiatives that can help to create a more coherent regulatory environment to facilitate vital submarine cable protection, repair and maintenance. See APEC Press Release issued by the APEC Committee on Trade and Investment on "Submarine Cable Resilience Critical to Connectivity" 6 February 2013, Jakarta, Indonesia; APEC Workshop on Enhancing Supply Chain Connectivity: Submarine Telecommunication Cable Resilience in the Asia Pacific, 15–16 October 2013, Bali, Indonesia. The collaborative partnership between industry and APEC stands out as a positive model which should be emulated by other organizations.

[47] As noted in footnote 32.

States that are not parties to UNCLOS should also consider adopting legislation that gives effect to Article 113 to ensure that vessels registered under their flag and their nationals are held accountable for damage to cables in the high seas or in the EEZ. As discussed in Chapter 3, the provisions on the protection of cables in UNCLOS are customary international law and are therefore binding on non-parties to UNCLOS as well as to States Parties.[48] If States adopt legislation to criminalize damage to cables and impose penalties commensurate with the importance of the cables, such measures will, at a minimum, discourage conduct likely to damage cables and will serve as an important acknowledgement that submarine cables are critical infrastructure.

Establish Cable Protection Zones

Another option available to States is to enact cable protection zone legislation. This approach has been adopted in Australia and New Zealand. The legislation of both these countries has been described as examples of an "integrated approach to the management of competing ocean uses through zoning,"[49] and there have been calls for other States to adopt similar cable protection legislation.[50] Colombia[51] and Uruguay[52] have enacted the use of protection zones and established meaningful penalties to protect cables landing in those countries. A brief overview of the cable protection legislation of Australia and New Zealand is set out below.

Australian Cable Protection Legislation

The Australian government enacted Schedule 3A to the Telecommunications Act of 1997 in 2005 under which submarine cables are installed and protected in Australian waters. The objectives of the legislation were *inter alia* to "provide security and reliability for the submarine cable component of Australia's national

---

48 The provisions on the protection of submarine cables in the 1958 Geneva Convention on the High Seas which were incorporated into UNCLOS, purportedly to codify existing customary international law at the time. See R. Churchill and A.V. Lowe, *The Law of the Sea* (3rd ed, Juris Publishing, 1999) at 24. Also see discussion in Chapter 3.

49 Y. Takei, "Law and Policy for International Submarine Cables in the Asia-Pacific Region" paper presented at the 2nd National University of Singapore-Asian Society of International Law Young Scholars Workshop, Singapore, October 2010, available at http://asiansil.org/publications/2010-13%20-%20Yoshinobu%20Takei.pdf (last visited 6 June 2013).

50 M. Scwhartz, "Legal Protection of Submarine Cables in South America" 12 August 2009, Developing Telecoms, available at www.developingtelecoms.com/legal-protection-of-submarine-cables-in-latin-america.html (last accessed 6 June 2013).

51 Resolution No. 204 of the Director General, Maritime Authority of Colombia (19 April 2012).

52 Penal Code, Book 2, Title 6, Crimes Against Public Security, Chapter 4, Articles 128, 148, and 217 (22 February 2011).

information infrastructure", "clarify the liability for compensation" and "provide increased consistency and clarity in the Commonwealth telecommunications regulatory regime."[53]

Schedule 3A allows the Australian Communications and Media Authority (ACMA) to "declare protection zones over submarine cables of national significance, to grant permits to install submarine cables in Australian waters and to set significant penalties for non-compliance."[54] *Australian waters* are defined as the waters of the territorial sea, exclusive economic zone and the sea above the part of the continental shelf of Australia that is beyond the limits of its EEZ.[55] Before establishing a cable protection zone, ACMA is required to undertake several steps, including developing a proposal for the protection zone,[56] ensuring that the cable is nationally significant,[57] publishing the proposal for public and other consultation purposes[58] and considering the impact of a new cable installation on the marine environment.[59] Schedule 3A also prescribes the processes for granting a permit to a cable owner to install or lay a submarine cable within a protection zone (known as a protection zone permit) or outside a protection zone within Australian waters (known as a non-protection zone permit).[60]

A range of activities can be prohibited in the protection zone, including the use of certain fishing gear, certain fishing activities and any activity that involves a serious risk that an object will connect with the seabed and result in damage to the cable.[61] Other activities can be significantly restricted, including using nets that are above the seabed at all times, towing certain equipment, and installation

---

53 Telecommunications and Other Legislation Amendment (Protection of Submarine Cables and Other Measures) Bill 2005—Explanatory Memorandum, available at http://www.acma.gov.au/webwr/_assets/main/lib100668/expl%20memo%20submarine%20cable%20bill%20pdf.pdf (last visited 21 February 2013).

54 The Australia Communications and Media Authority (ACMA), "Report on the Operation of the Submarine Cable Protection Regime: A Report on Five Years' Operation of Schedule 3A of the Telecommunications Act 1997, the Submarine Cable Protection Regime" September 2010 at 5, available online at http://www.acma.gov.au/scripts/nc.dll?WEB/STANDARD/1001/pc=PC_311993 (last visited 14 May 2013).

55 See Clause 2, Schedule 3A, Telecommunications Act 1997. It should be noted that the designation of the "sea above the part of the continental shelf of Australia that is beyond the limits of its EEZ" as Australian waters is technically inconsistent with UNCLOS. Article 78 states that the "rights of the coastal State over the continental shelf do not affect the legal status of the superjacent waters" which in the case of extended continental shelf, would be high seas, and would not be subject to the jurisdiction of Australia.

56 Clause 15, Schedule 3A of the Telecommunications Act 1997.

57 Clause 18, Schedule 3A of the Telecommunications Act 1997.

58 Clause 17, Schedule 3A of the Telecommunications Act 1997.

59 Clauses 20–21, Schedule 3A of the Telecommunications Act 1997.

60 See generally Part 3 of Schedule 3A of the Telecommunications Act 1997.

61 See Clause 10, of Schedule 3A of the Telecommunications Act 1997.

of oil or gas pipelines.[62] The legislation provides for significant criminal penalties for both intentional[63] and negligent[64] damage to cables in a protection zone, as well as for engaging in prohibited or restricted activities within a protection zone.[65]

To date, ACMA has declared three protection zones. There is a protection zone off the coast of Perth around the SEA-ME-WE3 cable which links Australia to Southeast Asia, the Middle East and Western Europe. The Perth Protection Zone extends 51 nm from Perth's coast and covers an area up to 1 nm on either side of the SEA-ME-WE3 cable.[66]

The other two protection zones are located off the coast of Sydney. The Northern Sydney Protection Zone is around the Southern Cross Cable which links Australia's communications network with New Zealand, Fiji and the United States and extends 40 nm off Sydney's coast.[67] The Southern Sydney Protection Zone is around the Australia-Japan Cable which links Australia with Guam, Japan and Australia. It extends 30 nm offshore.[68]

New Zealand Legislation

In 1996 New Zealand enacted the Submarine Cables and Pipelines Protection Act.[69] The legislation allows for the establishment of cable protection zones within the internal waters, territorial sea and EEZ of New Zealand.[70] Fishing, shipping activities and anchoring are prohibited within the zone.[71] The legislation provides for specific offences in relation to the protected areas, with significant penalties,[72] and specific enforcement powers relating to offences within the protected zones.[73] To date, eleven cable protection areas (known as cable protection zones) have

62 Clause 11, Schedule 3A of the Telecommunications Act 1997.
63 Clause 36, Schedule 3A of the Telecommunications Act 1997.
64 Clause 37, Schedule 3A of the Telecommunications Act 1997.
65 Clauses 40–44, Schedule 3A of the Telecommunications Act 1997.
66 See ACMA Website at http://www.acma.gov.au/WEB/STANDARD/pc=PC_100223 (last visited 21 February 2013).
67 *Ibid.*
68 *Ibid.*
69 Submarine Cables and Pipelines Protection Act 1996, Public Act 1996 No. 22, Date of assent, 16 May 1996. Reprinted 1 October 2008. The Act is administered by the Ministry of Transport and is available online at http://www.legislation.govt.nz/act/public/1996/0022/latest/DLM375803.html (last visited 21 February 2013).
70 *Ibid.*, Sections 12 and 15.
71 *Ibid.*, Section 13.
72 *Ibid.*, Section 15.
73 *Ibid.*, Sections 16–23.

been established.[74] The area of the cable protection zones appear to be much less extensive than that adopted by Australia.[75]

For surveillance and enforcement purposes, protection officers are established under the Act.[76] Protection officers have authority to undertake numerous enforcement activities, including gathering evidence for prosecution purposes, ordering a ship to leave a cable protection area, seizure of ships and fishing equipment, and requiring that the master or ship owner produce documents and information.[77]

The New Zealand legislation expressly provides that nothing in the Act affects the civil liability for damages that may arise from damage to a cable.[78] A person who damages a cable may therefore be subject to both criminal prosecution and be liable for civil damages.

Cable Protection Zones Consistent with UNCLOS?

Both Australian and New Zealand legislation on cable protection zones allow the relevant authorities to establish cable protection zones in areas outside of territorial waters. Indeed, the cable protection zones established by Australia extend beyond the 12 nm territorial sea of Australia and into its EEZ. This raises issues as to whether the cable protection legislation is consistent with UNCLOS.[79] Under the Australian legislation, a permit is required for any submarine cable being installed in 'Australian waters', an area which extends to areas beyond the 12 nm

---

74 These are around Great Barrier Island, Hauruki Gulf, Kawau Island, Whangaparoa Peninsula, Muruwai Beach Takaroa, Cook Strait, Oaonui, Hawke's Bay, Maui A and B and Pohokura Protection Area: See New Zealand Ministry of Transport website at http://www.transport.govt.nz/ourwork/sea/protectingunderseacables/ (last accessed 6 June 2013).

75 For example, the Cable Protection Zone in the Cook Strait of New Zealand is approximately seven kilometers wide. It serves to protect a number of electricity and telecommunication cables that lie unburied on the seabed between the North Island and South Island of New Zealand. See 'Cook Strait Submarine Cable Protection Zone' (February 2011) at 3, available online at https://www.transpower.co.nz/sites/default/files/publications/resources/cook-strait-booklet-2011.pdf (last visited 6 June 2013).

76 Section 16, New Zealand Submarine Cables and Pipelines Protection Act 1996. Section 2 of the Act also refers to enforcement officers, the latter being members of the New Zealand police force or officers of the New Zealand naval force.

77 *Ibid.*, Sections 17, 18, 20 and 21.

78 *Ibid.*, Section 6.

79 It is of note that the idea of establishing protection zones around submarine cables was mooted by the International Law Commission (ILC) during the preparation of the Draft Articles on the law of the sea in 1956 but was rejected as being inconsistent with the freedom of navigation: See *Yearbook of the International Law Commission*, Volume II, Doc. A/CN.4/97 (1956) at 12.

territorial sea.[80] It has been a specific complaint of cable companies that any requirement for permits in the EEZ is inconsistent with the freedom of all States (and their nationals) to lay, repair and maintain cables in this maritime zone. An international cable which traverses the Australian EEZ but does not land in Australia can, under UNCLOS, cross the cables in the protection zone and is entitled to be repaired by a non-Australian flag cable repair ship with no need for permits. If Australia were to prevent these activities on the basis of its domestic law, there would be a conflict with the interests of other State(s) whose nationals owned the transiting cable. A simple fix to this issue is for Australia to amend its domestic law to simply recognize that it will be implemented in accordance with international law.

A second inconsistency relates to the fact that the cable protection legislation restricts activities otherwise permitted in the EEZ. It has been noted that

> [a] protection zone for a submarine cable outside the territorial sea could be validly asserted by a State, provided the basis of jurisdiction was tied to one that could be claimed under the regime for the EEZ or continental shelf. That is to say, protection over a cable could be achieved by restricting activities that could be validly regulated in the EEZ or continental shelf.[81]

To the extent that cable protection zones prohibit or restrict activities such as fishing, resource exploration and marine scientific research, they are arguably consistent with a coastal State's rights in the EEZ or continental shelf. However, the situation is less clear when it comes to other activities unrelated to a coastal State's rights in its EEZ/continental shelf. Anchoring is part of the freedom of navigation allowed to other States in the EEZ and is not covered under the competences given to coastal States in their EEZs. Arguably, any designation of a no-anchorage area in cable protection zones would have to be done under the auspices of the International Maritime Organization.[82]

Regardless of this potential inconsistency (the effects of which can be easily mitigated) cable protection zones are valuable tools in the protection of cables and should be carefully evaluated by States seeking to enhance the protection of critical communications infrastructure.

---

[80] See for example, submissions of two cable operators, Telstra and Southern Cross, in Report of the ACMA, *supra* note 54 at 22–23.

[81] S. Kaye, "The Protection of Platforms, Pipelines and Submarine Cables Under Australian and New Zealand Law" in *Maritime Security: International Law and Policy Perspectives from Australia and New Zealand*, Natalie Klein *et al.*, eds, (Routledge, 2010) at 192.

[82] According to the IMO, it is recognized as the only international body for developing guidelines, criteria and regulations on an international level for ships' routing systems, which include the designation of no anchorage areas. See IMO website at www.imo.org/safety/mainframe.asp?topic_id=770; and IMO Resolution A.572 (14), Amendments to the General Provisions on Ships Routeing.

Enforcement of Legislation

Cable protection legislation should be accompanied by robust enforcement measures. A vital aspect of enforcement is compliance monitoring through patrols in areas where critical cables are laid. For example, it has been observed in relation to Australia's cable protection legislation, that the legislation "would not prevent a major submarine cable outage either through deliberate or inadvertent breaches without adequate monitoring. Monitoring needs a review and a solution utilizing patrols by existing law enforcement agencies or technical systems such as Automatic Identification System (AIS)."[83] Cable patrols by local naval or coast guard forces in the normal course of their duties can be an effective deterrent. In addition to compliance monitoring, those caught breaking a cable or involved in activities that could break a cable should be actively prosecuted (in addition to being liable to the cable owner for damages).

In New Zealand, the responsibility for arranging and paying for the vessel and aerial patrols is borne by the cable companies. If a vessel or prohibited activity is detected in a cable protection zone, marine police are summoned to enforce the law. From 2005 to 2011 there were six successful prosecutions by New Zealand law enforcement agencies and no cable damage.[84] Staff employed by or contracted to a cable company in New Zealand may be appointed as protection officers by the Ministry of Transport. For example, Transpower New Zealand Limited owns and operates cable links laid in the Cook Strait Cable Protection Zone. Transpower monitors the protection zone using both patrol vessels and helicopters, and members of these crews have been appointed as protection officers.[85]

Better Balancing of Competing Activities in Areas where Submarine Cables are Located

As offshore activities expand and diversify, governments must continue to consider cables as critical infrastructure. The installation, repair and maintenance of cables require physical clearance from other installations such as those used for petroleum and renewable energy. The vessels that repair submarine cables are generally in excess of 100 m in length, often tow grapnels behind them or may need room to manoeuver behind an ROV. All of these factors must be taken into consideration when siting renewable activities in areas where submarine cables have been laid. Export cables or pipelines that cross an active cable should do so as close to 90 degrees as practicable given local conditions. Government

83 See ACMA Report, *supra* note 54 at 15–16.

84 This is based on reports shared on the ICPC Members' Only Database (personal copies with authors).

85 See, 'Cook Strait Submarine Cable Protection Zone' *supra* note 75.

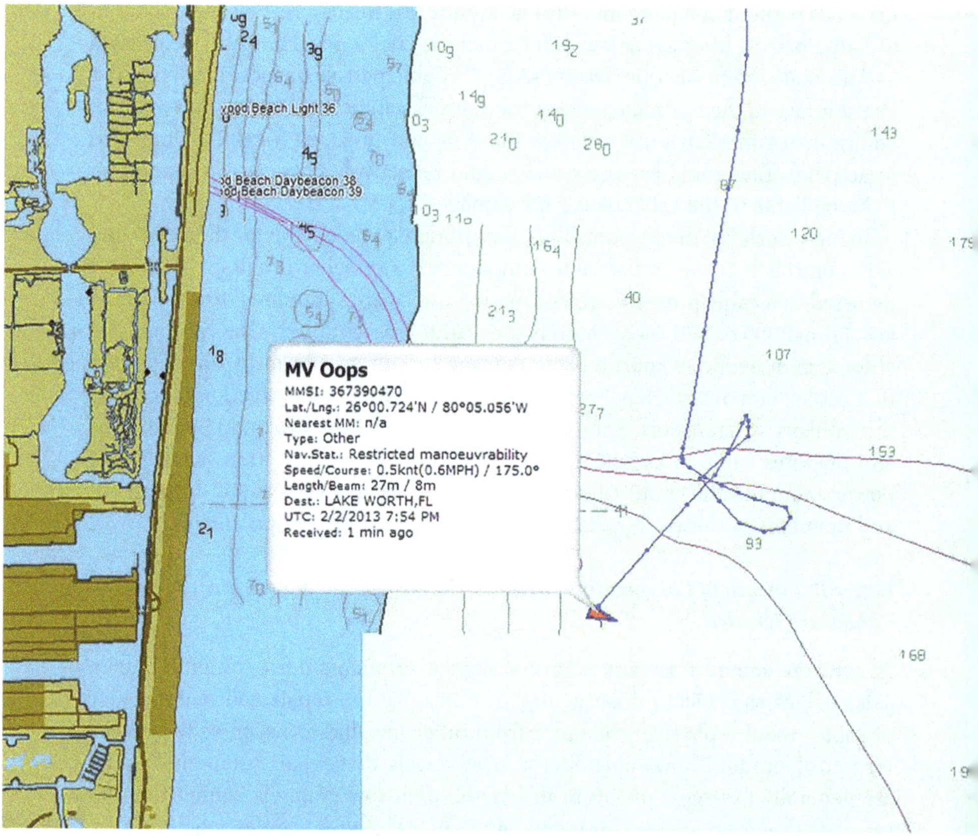

Figure 11.4 An Automatic Identification System (AIS) image showing a vessel's suspicious plotted movements over three international cables off the coast of Florida. With this 24/7 real time information, the vessel identified by AIS can be contacted by the cable owners and/or coast guard or naval authorities to be warned away from the cables or investigated if there is a fault. (Image provided courtesy of Sea Risk Solutions LLC)

supported Marine Spatial Planning activities should include consultation with affected cable owners.

Assist Cable Owners in the Identification of Parties that Damage Cables

AIS and VMS data identify the location of vessels engaged in their normal activities. In the event that those activities damage a submarine cable, location data can be valuable in identifying the parties responsible (Figure 11.4). Governments should consider allowing the release of fishing VMS data and other maritime monitoring information in support of cable protection, including identification of parties that damage cables. British Telecom has done pioneering work with the UK coast guard in this respect and has been able to prevent many cable faults and save millions of dollars. In essence both the company and the coast guard work closely and informally for rapid responses to threatening situations by vessels near cables.

Conclusion

Balancing competing uses of the oceans has been a perennial problem for States. While shipping, fishing and resource exploration and exploitation are all vital economic activities essential to a State's survival, submarine cables have become equally as important to the economy and security of States. Accordingly, both governments and the cable industry must continue to collaborate and co-operate in the protection of submarine cables. The reality is that most cable faults are preventable and both cable owners and States have every incentive to do all that they can to prevent them, from the time a cable is conceived and throughout its useful life. Cables must co-exist with, and be protected from, the very activities that have the potential to break or damage them.

CHAPTER TWELVE

PROTECTING SUBMARINE CABLES FROM INTENTIONAL DAMAGE—THE SECURITY GAP

Robert Beckman

Introduction

Submarine communications cables route approximately 97 per cent of the world's data traffic and provide essential services such as the internet, phone and banking services. They have undoubtedly become critical communications infrastructure. An intentional attack on international submarine cables could cause devastating damage to the world's economy and to the security of many States. Yet a significant number of States do not have legislation in place making it a crime to intentionally damage submarine cables for personal or political gain. Further, while there are currently global conventions that make it an 'international crime' to intentionally damage or destroy airports, international aviation facilities,[1] lighthouses and other aids to international maritime navigation,[2] there is no global convention that makes it an international crime to intentionally damage or destroy international submarine cables.

The purpose of this Chapter is to examine the gaps and loopholes in the current legal regime governing the protection of submarine cables from terrorist attacks and other intentional acts that damage or destroy cables. It calls on States to amend their domestic legislation to make it a criminal offence to intentionally destroy or damage international submarine cables, and for States to support a global convention to establish a cooperative regime to combat intentional acts

1 See for example, the Convention for the Suppression of Unlawful Acts Against the Safety of Civil Aviation, adopted 23 September 1971, 974 UNTS 177 (entered into force 26 January 1973) [Montreal Convention] and Protocol for the Suppression of Unlawful Acts of Violence at Airports Serving International Civil Aviation, supplementary to the Convention for the Suppression of Unlawful Acts against the Safety of Civil Aviation, adopted 24 February 1988, 1589 UNTS 474 (entered into force 6 August 1989).

2 Convention for the Suppression of Unlawful Acts against the Safety of Maritime Navigation, adopted 10 March 1988, 1678 UNTS 201 (entered into force 6 March 1992) [SUA Convention].

against submarine cable networks. Part I will examine the incidents of intentional damage to submarine cables and Part II will discuss general principles of criminal jurisdiction, as many attacks against submarine cables will occur outside the territory of any State. Part III will explore the adequacy of the legal regime established in the 1982 UN Convention on the Law of the Sea (UNCLOS)[3] and Part IV sets out recommendations on what steps can be taken to protect submarine cables from intentional damage. Part V discusses recent developments and increasing recognition of the threat to submarine cables and Part VI sets out some conclusions.

I. Incidents of Intentional Damage to Submarine Cables

The large majority of cable breaks are caused by negligence resulting from fishing and shipping activities or from natural hazards such as earthquakes and typhoons. There have, however, been several isolated incidents of intentional damage being caused to submarine cables. For example, on 23 March 2007, at least two vessels were involved in hostile activities on the high seas against the TVH cable system involving the removal of 98 km of cable, and against the APCN cable system involving the removal of 79 km of cable, including critical optical amplifiers. A cable repair vessel arrived at the scene and photographed one of the vessels, which was registered in Vietnam, in the act of removing the cable. The extensive nature of the damage was such that repairs could not be completed for three months because new amplifiers had to be built at a factory.[4] In November 2007, there was a report of intentional sabotage of a cable in Bangladesh, which resulted in a total loss of communications for at least one week.[5] In addition, there have also been reports of cable theft in Jamaica in 2008,[6] and a 2010 attack by separatists against the beach manhole connection of a submarine cable system linking the Philippines with Japan.[7] In March 2013, it was reported that 16 tons of submarine cables laid on the seabed between Bangka Island and the

---

[3] United Nations Convention on the Law of the Sea, adopted 10 December 1982, UNTS 1833 (entered into force 16 November 1994) [UNCLOS].

[4] M. Green and D. Burnett, "Security of International Submarine Cable Infrastructure: Time to Rethink?" in M. Nordquist *et al.*, (eds), *Legal Challenges in Maritime Security* (Martinus Nijhoff Publishers, 2008) at 559–563.

[5] M. Sechrist, "Cyberspace in Deep Water: Protecting Undersea Communications Cables by Creating an International Public-Private Partnership" (Harvard Kennedy School, 2010) at 38 available online at http://live.belfercenter.org/publication/20710/cyberspace_in_deep_water.html?breadcrumb=%2Fexperts%2F2223%2Fmichael_sechrist (last accessed 9 June 2013).

[6] *Ibid.*

[7] A. Dalizon, "Reds Bomb Cagayan Globe Site, Disarm Cop, Guards" *People's Journal*, 11 June 2010.

Riau Islands in Indonesia were stolen.[8] In the same month, there were also reports that three men were arrested by the Egyptian coastguard after an attempt to cut the SEA-ME-WE-4 cable, although the motives of such an act remain unknown.[9] On 8 July 2013 PT Indosat filed a report that 31.7 km of a critical cable linking Indonesia to Singapore had been removed through an act of theft.[10]

To date, there has been no large-scale terrorist attack against submarine cables. However, it has been observed that at some point in the future submarine cables could be targeted by terrorist groups.[11] This potential threat was highlighted in December 2010 when the US State Department labelled a WikiLeaks disclosure of critical infrastructure around the world, including submarine cables and cable landings, as tantamount to giving terrorists a target list.[12] The risk of a deliberate attack against submarine cables with the intention of crippling the world's telecommunications system is not as remote a possibility as is commonly perceived. Even more alarming, the asymmetrical act could be as simple as a vessel dropping an anchor and dragging it several kilometers in an area where there is a network of cables laid, causing a multitude of cable faults and maximum damage to international telecommunications systems. Alone or in combination with car bomb attacks on terminal landing stations or against cableships and their spares depots, significant damage to critical international infrastructure would quickly impact economies. Cable system owners and cableship operators work hard to maximize security for their systems and ships, but they require help from national governments on an international scale in order to meaningfully reduce these threats by preventing hostile actions, especially outside of territorial seas and, in the event of a successful attack, restoring communications and quickly repairing damaged cable infrastructure. This Chapter examines the current legal basis for actions by national governments to protect submarine cable systems.

---

8 A. Panji, "Duh, Kabel Internet Indonesia Sering Dicuri untuk Besi Tua" Kompas, 27 March 2013, available online at http://tekno.kompas.com/read/2013/03/27/09324795/duh.kabel.internet.indonesia.sering.dicuri.untuk.besi.tua (last accessed 7 June 2013).

9 "Three Cable-Breaking Incidents Affect the Internet" ICPC Environment Update, Issue 127, 3 April 2013.

10 Jakarta Post, "Ministry Sets up Submersible Cable Task Force" 8 July 2013, see http://www.thejakartapost.com/news/2013/07/08/ministry-sets-submersible-cable-task-force.html.

11 S. Kaye, "International Measures to Protect Oil Platforms, Pipelines and Submarine Cables from Attack" (2006–2007) 31 *Tulane Maritime Law Journal* 337–423, at 418.

12 See M. Clayton, "WikiLeaks List of Critical Sites: Is It a Menu for Terrorists?" *Christian Science Monitor*, 6 December 2010, available at http://www.csmonitor.com/USA/Foreign-Policy/2010/1206/WikiLeaks-list-of-critical-sites-Is-it-a-menu-for-terrorists (last accessed 7 June 2013).

II. Criminal Jurisdiction Over Acts at Sea

Intentional acts of damage to submarine cables will occur either within areas under the territorial sovereignty of a coastal State (such as internal waters, the territorial sea, archipelagic waters and international straits) or in areas outside the territorial sovereignty of a coastal State (the EEZ, continental shelf, high seas and deep seabed area).[13] A State may only prosecute the perpetrators of such intentional damage to cables in a manner consistent with the general principles of criminal jurisdiction. It is therefore important to understand what these general principles are and how they operate in international law.

Jurisdiction refers to the:

> [P]ower of a State under international law to govern persons and property by its municipal law. It includes both the power to prescribe rules (prescriptive jurisdiction) and the power to enforce them (enforcement jurisdiction). The latter includes both executive and judicial powers of enforcement.[14]

As established above, there are two types of jurisdiction. 'Prescriptive jurisdiction' generally refers to the authority of a State to *prescribe* laws and make them applicable to persons or circumstances.[15] 'Enforcement jurisdiction' describes the "authority of a State to take action to enforce those laws through, for example, arresting, detaining, prosecuting, convicting, sentencing and punishing persons for breaking those laws".[16]

Under international law, all States have the right to exercise prescriptive and enforcement jurisdiction over events occurring and persons (whether nationals or foreigners) present in their territory. This is known as the 'principle of territoriality'.[17] This also applies to maritime areas within the territorial sovereignty of a coastal State, namely, internal waters, territorial seas, archipelagic waters and international straits. [18] In these areas, a coastal State is entitled to apply its domestic laws, i.e. it has prescriptive jurisdiction.[19] The coastal State is also entitled to *enforce* these domestic laws, either by measures such as board-

---

13 For an overview of the various maritime zones recognized under international law, please refer to Chapter 3 on "Overview of the International Legal Regime Governing Submarine Cables."

14 D. Harris, *Cases and Materials on International Law* (7th ed, Sweet and Maxwell, 2010) at 227.

15 See International Bar Association, *Report of the Task Force on Extraterritorial Jurisdiction*, July 2008, at 7–8 available at http://tinyurl.com/taskforce-etj-pdf (last accessed 7 June 2013) [IBA Report].

16 *Ibid.*

17 Harris, *supra* note 13, at 228.

18 Since archipelagic waters are also within the territorial sovereignty of coastal States, the criminal laws would also apply in the archipelagic waters of States like Indonesia and the Philippines.

19 D. Nelson, "Maritime Jurisdiction" in Max Planck Encyclopedia of Public International Law, at 1 available online at http://www.mpepil.com/ViewPdf/epil/entries/law-9780199231690-e1195.pdf?stylesheet=EPIL-display-full.xsl (last accessed 7 June 2013).

ing, searching and arresting, or by measures imposed by courts, such as fines and imprisonment, i.e. enforcement jurisdiction.[20]

However, if the criminal act takes place outside the territory of a State, it is required to exercise *extra-territorial jurisdiction.* An exercise of jurisdiction is extra-territorial "where it provides that acts taking place abroad may be offences within the local jurisdiction and individuals may be subject to local courts in respect of those acts".[21] International law generally recognizes five bases of extra-territorial jurisdiction, which will be briefly summarized for the purposes of this Chapter.[22]

First, the 'nationality principle' allows States to exercise jurisdiction over its nationals for crimes committed anywhere in the world. Second, the 'passive personality principle' allows States to assert jurisdiction over a foreigner for a crime committed outside its territory against one of its own nationals. The 'protective principle' allows States to exercise jurisdiction over a limited range of crimes committed by foreigners outside its territory where the crime prejudices the State's vital interests. The 'effects doctrine' allows States to exercise jurisdiction over certain conduct by foreigners outside its territory where the conduct has a certain effect within the State. Finally, the 'universality principle' enables States to assert jurisdiction over certain crimes committed by foreigners outside the State's territory and having no connection to or impact on a prosecuting State. The exercise of universal jurisdiction is usually confined to war crimes, crimes against humanity, genocide, torture and piracy.[23]

For crimes that are committed in maritime areas outside of territorial sovereignty, such as in the EEZ, on the continental shelf, or on the high seas, the situation is more complicated. Criminal acts which take place in the EEZ and continental shelf are in principle outside the territory of a coastal State, but occur in maritime zones where certain rights and jurisdiction of the coastal State are recognized. If such acts take place on the high seas, they are also outside the territory of any State. Generally, in such maritime zones outside the territorial sovereignty of a coastal State, coastal States are permitted to exercise prescriptive jurisdiction provided that such prescriptive jurisdiction is recognized in UNCLOS.[24] For example, because UNCLOS gives coastal States sovereign rights over fisheries resources,[25] the coastal State is also empowered to pass national legislation implementing such rights, as well as to create offences in respect of violations of these fishery laws.

---

20 *Ibid.*

21 M. Dixon, *International Law* (6th ed, Oxford University Press, 2007) at 147.

22 *Ibid.*

23 IBA Report, *supra* note 15 at 147.

24 This is consistent with the general view that a State is not able to extend its prescriptive jurisdiction outside its territory unless permissive rules support such an exercise: See V. Lowe, "Jurisdiction" in M. Evans, (ed) *International Law* (2nd ed, Oxford University Press, 2006) at 335.

25 UNCLOS Art 56.

However, with regard to *enforcement jurisdiction*, the general (and most important principle) is that only the flag State can exercise enforcement jurisdiction over vessels in the EEZ and on the high seas.[26] Ships in these maritime zones may not be boarded without the express consent of the flag State or the master.

There are limited exceptions to the principle of exclusive enforcement jurisdiction of the flag State in the EEZ and on the high seas. First, warships or ships on government service of all States[27] may board and arrest pirate ships in areas outside the territorial sovereignty of any State i.e. in the EEZ and on the high seas.[28] A ship is considered a pirate ship if it is intended by the persons in dominant control of the ship to be used for the purpose of committing any of the acts of piracy referred to in Article 101 of UNCLOS.[29]

Second, a warship may board another ship in the EEZ of another State or on the high seas if there are reasonable grounds for suspecting that the ship is engaged in piracy, the slave trade, or unauthorized broadcasting, the ship is without nationality, or the ship is the same flag as the warship. This is known as the right of visit under UNCLOS.[30]

Third, UNCLOS also recognizes that additional reasons for exercising the right to board foreign flagged ships may be established by treaty.[31]

Fourth, it should also be noted that a State has enforcement jurisdiction under UNCLOS in relation to certain other matters. UNCLOS gives coastal States the power to enforce their fishing laws and regulations in their EEZ, including the power to board, inspect and arrest ships violating their fisheries laws and regulations,[32] as well as limited enforcement jurisdiction to enforce their laws governing marine scientific research and pollution of the marine environment.[33]

These are exceptions to the principle of exclusive flag State jurisdiction in areas outside the sovereignty of any State and do not apply to warships and government ships owned and operated by States and used only on government non-

---

26 UNCLOS Art 89 provides that "(n)o State may validly purport to subject any part of the high seas to its sovereignty." Article 94 sets out the duties of the flag State over vessels flying its flag on the high seas. Articles 88 to 115 on the high seas apply to the exclusive economic zone in so far as they are not incompatible with Part V on the exclusive economic zone (UNCLOS Art 58 (2)).

27 UNCLOS Art 107.

28 See UNCLOS Art 105 which allows all States to seize a pirate ship or a ship taken by piracy and under the control of pirates (piracy is defined in Art 101), and arrest the persons and seize the property on board. Article 105 applies in the EEZ by virtue of Art 58(2) of UNCLOS.

29 UNCLOS Art 103.

30 See UNCLOS Art 110. This would apply in the EEZ by virtue of Art 58(2).

31 Article 110 of UNCLOS provides that "except where acts of interference derive from powers conferred by treaty". For example, Article X of the 1884 Cable Convention.

32 UNCLOS Art 73.

33 Article 56(b) of UNCLOS gives the coastal State jurisdiction over marine scientific research and the protection and preservation of the marine environment in the EEZ.

commercial service. Such ships have complete immunity from the jurisdiction of any State other than the flag State.[34]

III. Gaps in the Current Legal Regime Governing Submarine Cables

This Part examines the reasons why the current legal regime governing submarine cables,[35] consisting of the 1884 Convention for the Protection of Submarine Telegraph Cables (1884 Cable Convention)[36] and UNCLOS may not adequately protect submarine cables from intentional damage.

In Areas within the Territorial Sovereignty

While coastal States have the right to adopt laws and regulations to criminalize acts of intentional damage to submarine cables within their territorial sea and archipelagic waters pursuant to their general sovereignty over these areas, there is no obligation for them to do so under UNCLOS or otherwise. The drafters of the Convention appear to have assumed that coastal States would recognize that that they have an interest in protecting submarine cables which land in their territory or which pass through maritime zones under their sovereignty, and that they would adopt laws and regulations to protect submarine cables. However, very few States have express provisions criminalizing damage to submarine cables in their territorial waters,[37] although in some cases offences against submarine cables in the territorial sea will be covered by general legislation criminalizing damage to installations used for telecommunications.[38] Even when States do have legislation criminalizing intentional damage to cables, the penalties imposed are not sufficient to present an adequate deterrent to would-be perpetrators.[39]

---

34 UNCLOS Arts 95 and 96.

35 For a more detailed overview of the legal regime governing submarine cables, please see Chapter 3.

36 Convention for the Protection of Submarine Telegraph Cables, adopted 14 March 1884, TS 380 (entered into force 1 May 1888) [1884 Cable Convention].

37 For example, in the review of the national legislation of Southeast Asian States, none of the States had an express provision criminalizing intentional or willful damage to submarine fiber optic cables.

38 See, for example, Section 21 of Brunei's Telecommunications Order 2001, Section 41 of Singapore's Telecommunications Act, Sections 44, 72 and 73 of Thailand's Telecommunications Business Act (2001).

39 This was a specific complaint of U.S. Cable Owners about their national law, which imposed only a maximum penalty of US$5000 for wilful injury to submarine cables (47 U.S.C, § 21). This "insignificant maximum criminal penalty provides little incentive for enforcement authorities to assign full-time legal and investigative personnel to prosecute vessel owners caught damaging a submarine." See S. Coffen-Smout and G.J. Herbert, "Submarine Cables: A Challenge for Ocean Management" (2000)

In Areas Outside of Territorial Sovereignty

As discussed in Chapter 11, UNCLOS does provide some protection for submarine cables from intentional damage. Article 113 of UNCLOS, which is based on Article II of the 1884 Cable Convention, requires States to "adopt the laws and regulations necessary to provide that the breaking or injury by a ship flying its flag or by a person subject to its jurisdiction of a submarine cable *beneath the high seas* done wilfully or through culpable negligence... be a punishable offence" (emphasis added). Article 113 also applies to cables laid in a State's EEZ or on their continental shelf.[40] In other words, it applies to acts committed against cables beyond the limits of the territorial sea of the coastal States.

Arguably, 'wilfully' covers intentional acts of damage to submarine cables. Nonetheless, the protections afforded to cables under Article 113 are inadequate for several reasons. First, most States have not adopted laws or regulations incorporating this provision into their national laws. The practical effect being that when a submarine cable beneath high seas or EEZ is broken or damaged by intentional or reckless conduct, in many cases no crime has been committed because the act is not an offence under the laws of any State.

Second, and more importantly, Article 113 is inadequate because it does not apply to *foreign nationals* who intentionally break or damage cables, it only applies to nationals of that State or ships flying its flag (i.e. registered in the State). If a person intentionally breaks or damages a submarine cable that lands in a coastal State, it could seriously injure the economic and security interests of every State in which the cable lands. Therefore, landing States have a sufficient interest to justify them making such acts criminal offences under their laws, even if the acts are committed by foreign nationals outside their territory.

In fact, all States that depend upon the internet and submarine cables have a common interest in taking measures to ensure that any person who intentionally destroys or damages a submarine cable is arrested and charged with a criminal offence, wherever the act took place, whatever the nationality of the perpetrator, and regardless of their motive or purpose. There are global conventions on other matters, discussed below, which do exactly this, and the same model could be employed to protect submarine cables.

Third, Article 113 only addresses prescriptive jurisdiction and not enforcement jurisdiction over perpetrators who intentionally damage submarine cables. Indeed, under UNCLOS, the right to board[41] and arrest[42] vessels in the EEZ and

---

24 *Marine Policy* at 444; D. Burnett, "Cable Vision" U.S. Naval Institute Proceedings, August 2011 at 68–69.

40 UNCLOS Art 58(2).

41 UNCLOS Art 110.

42 For piracy (Art 105 of UNCLOS) and for unauthorized broadcasting from the high seas (Art 109).

high seas is limited to certain specified circumstances. In contrast to UNCLOS, Article X of the 1884 Cable Convention, allows warships to require the master of a vessel suspected of having broken a cable to provide documentation to show the ship's nationality and thereafter to make a report to the flag State. However, Article X is only binding on States Parties (which to date, numbers at 40) and States Parties have not often exercised their rights under this provision.[43]

It can be argued that the intentional theft or destruction of a submarine cable outside the territorial sea of any State can be piracy as defined in Article 101 of UNCLOS if such theft or destruction is an act of depredation and if a submarine cable on the seabed outside the territorial sea is property in a place outside the jurisdiction of any State. Under Article 101 piracy consists of "any act of depredation, committed for private ends by the crew or the passengers of a private ship . . . and directed . . . against property in a place outside the jurisdiction of any State." If the theft of submarine cables were deemed "piracy" under UNCLOS, this would give any warship the right to board and arrest perpetrators under Article 105 of UNCLOS. This argument has been made by commentators.[44] However, it is a creative interpretation of Article 101, and most States Parties to UNCLOS may be reluctant to arrest such persons as "pirates." Generally, States have objected to a right to board without the consent of the flag State, even for the suppression of serious crimes,[45] and will likely have the same concerns about intentional damage to submarine cables.

It is evident from the above analysis that there are serious security gaps in the current legal regime and that neither the 1884 Cable Convention nor UNCLOS adequately address the issue of intentional damage caused to submarine cables. Given that submarine cables have become vital communications infrastructure, States, the International Cable Protection Committee (ICPC) and the cable industry should work together to address this gap. The next Part of this Chapter sets out recommendations on the steps that can be taken to protect submarine cables from intentional damage.

---

43 There is only one reported case of a State Party relying on Article X of the 1884 Convention and this was the boarding and inspection of log books of the Soviet Trawler *Novorossik* by officers of the US naval vessel *Roy O Hale* in 1959 after the US naval vessel had grounds to suspect that the trawler was responsible for cable breaks in the area: See "US and USSR Exchange Notes on Damage to Submarine Cables" *The Novorossiisk*, Dept of State Bull Vol XL, No 1034 at 555 (20 April 1959).

44 Green and Burnett, *supra* note 4, at 573–574.

45 R. Beckman, "International Responses to Maritime Terrorism" in V. Ramraj, *et al.*, *Global Anti-Terrorism Law and Policy* (Cambridge University Press, 2005) at 268.

IV. The Way Forward—Recommendations

The Adoption of National Legislation

The first step that all States should take is to adopt national legislation making it an offence to damage submarine cables (and related infrastructure) within territorial waters. Second, States should implement Article 113 of UNCLOS. Third, they should adopt national legislation for acts of intentional damage to cables in the EEZ or high seas, provided that such cables land in their territory or service their telecommunications system. This is on the basis that such damage committed outside the territory of the State has effects within the territory of that State (based on the effects doctrine) or that it prejudices the vital interests of the State (pursuant to the protective principle).

Given the importance of submarine cables to the world economy and security, it appears axiomatic that States should adopt such legislation. However, many States appear to lack the political will, interest or resources to do so. This is attributable to several reasons. First, there is a lack of awareness of the importance of submarine cables on the part of governments, although this situation appears to be gradually improving. Second, in many governments, there is no lead ageny responsible for submarine cables and hence, nobody to champion their cause. In many cases, there are multiple agencies within a government that are partially responsible for submarine cable regulatory governance, but there is no agency in charge for emergency security issues internationally. Third, to date there has not been a serious incident involving intentional damage to submarine cables, and governments have a tendency to act in response to crisis or serious events rather than pre-emptively. Lastly, there appears to be a general reluctance to adopt legislation which may have far-reaching extra-territorial effects. Accordingly, ensuring that States adopt such legislation may be easier said than done.

Adoption of a Global Convention on the Protection of Critical Communications Infrastructure

Perhaps the most effective way to ensure that States adopt national legislation on a particular issue is to adopt a global convention. The international community has developed a series of global conventions designed to enhance cooperation to protect critical communications infrastructure. The conventions are part of a series of United Nations conventions commonly referred to as the 'UN terrorism conventions'. The first of these conventions was the Convention for the Suppression of Unlawful Seizure of Aircraft, which was signed at the Hague on 16 December 1970 and referred to as the 1970 Hague Convention. The UN Terrorism conventions include the following:

- Convention for the Suppression of Unlawful Acts against the Safety of Civil Aviation, signed at Montreal on 23 September 1971.[46]
- Protocol for the Suppression of Unlawful Acts of Violence at Airports Serving International Civil Aviation, supplementary to the Convention for the Suppression of Unlawful Acts against the Safety of Civil Aviation, signed at Montreal on 24 February 1988.[47]
- Convention for the Suppression of Unlawful Acts against the Safety of Maritime Navigation, done at Rome on 10 March 1988.[48]
- Protocol to the 1988 Convention for the Suppression of Unlawful Acts against the Safety of Maritime Navigation, 2005.[49]

The objective of these terrorism conventions is to establish a comprehensive cooperative regime to ensure that persons who intentionally destroy or damage critical infrastructure for international civil aviation or for international maritime navigation are punished for their actions as criminals. The conventions are designed to create a cooperative mechanism among State parties to ensure that persons who commit the offences defined in the convention will be arrested and prosecuted, no matter what their nationality, and no matter where the act took place. The scheme of all the conventions is as follows:

1. State Parties are obligated to make the offences defined in the convention a crime under their national laws, punishable by severe penalties.
2. State Parties are required to establish jurisdiction over the offences defined in the convention when they have a link or connection to the offence because the act took place within their territory, or was committed by their national or from a ship flying their flag.
3. State Parties are also required to establish jurisdiction over the offence when the alleged offender is 'present in their territory' and the State chooses not to extradite them. This requirement in effect makes the offence an 'international crime' among the States that are party to the convention. It requires States to enact legislation giving their courts jurisdiction to try the offender, even though the offence was committed by a foreign national outside their territory, so long as the offender is physically present in their territory.

---

46 *Supra* note 1.

47 *Supra* note 1.

48 *Supra* note 2.

49 Protocol of 2005 to the Convention for the Suppression of Unlawful Acts Against the Safety of Maritime Navigation, adopted 14 October 2005, IMO Doc. LEG/CONF.15/21, (entered into force 28 July 2010) [SUA 2005].

4. If the alleged offender is present in their territory, State Parties have a legal obligation to take them into custody and to ensure their presence in their territory. The State then has only two choices—it must either extradite the alleged offender or prosecute them. The State can extradite the alleged offender to another State Party that has jurisdiction, such as the State of nationality of the offender, the State in whose territory the offence was committed, or the State on whose ship the offence was committed. If they elect not to extradite the alleged offender, the State's only option is to prosecute the offender in their courts.
5. The conventions include provisions which make it possible to extradite alleged offenders to other State Parties, even in the absence of an extraction treaty between the two countries.
6. The conventions contain provisions that require State Parties to provide mutual legal assistance to assist the State where the alleged offenders are prosecuted. The legal assistance would include matters such as providing evidence or witnesses.

It should be noted that although the UN conventions are commonly referred to as the 'UN terrorism conventions' there are no provisions in most of the conventions requiring that the offence be committed by a terrorist group or with a terrorist motive or purpose. The conventions create offences for the destruction or damage of infrastructure, such as airports or air navigation facilities[50] when such acts are done 'unlawfully and intentionally'. The motive or purpose of the offender is not relevant.

The one UN terrorism convention which could apply if a submarine cable was subjected to a terrorist attack is the International Convention for the Suppression of Terrorist Bombings,[51] which was adopted by the General Assembly of the United Nations on 15 December 1997. Under that Convention, it is an offence to unlawfully and intentionally use an explosive or lethal device against an infrastructure facility with the intent to cause extensive destruction of such facility or where such destruction results in or is likely to result in major economic loss. Under the Convention, an 'infrastructure facility' means any publicly or privately owned facility providing or distributing services for the benefit of the public, such as communications. However, this Convention would only apply if submarine cables were destroyed as a result of bombing, that is, by use of an explosive or

---

50 For example, Art 1(d) International Convention for the Suppression of Terrorist Bombings, and Art II (1)(b) of the 1988 Protocol for the Suppression of Unlawful Acts of Violence at Airports Serving International Civil Aviation, Supplementary to the Convention for the Suppression of Unlawful Acts Against the Safety of Civil Aviation.

51 International Convention for the Suppression of Terrorist Bombings, adopted 15 December 1997, 2149 UNTS 256 (entered into force 23 May 2001). As at 14 May 2013 there are 165 States Parties.

other lethal device. It would not apply if a cable were cut or destroyed by other means likely to be employed in an ocean environment.

If there are global conventions that make it an 'international crime' to intentionally damage airport infrastructure or navigational infrastructure, why is there no convention that makes it a crime to intentionally destroy or damage submarine cables? There could be several reasons.

One reason why the legal regime governing submarine cables has been neglected is that in many States there is no government ministry or agency responsible for international law and policy issues relating to submarine cables. The corollary being that there is often a low level of knowledge and understanding about the cables themselves and the activities undertaken by cable companies. Therefore, there has been no systematic review by national governments of the adequacy of the legal regime governing submarine cables.

A second reason why the legal regime has been neglected lies with the nature of the cable industry and the development of submarine cables. Historically, the interaction between the cable industry and States has been very limited. Since the advent of broadband the cable industry has invested billions of dollars in new fiber optic submarine cables, however, for the most part it has done so by using members of consortiums or shipping agents to obtain the necessary permits and approvals from national governments. Industry has cooperated amongst itself through organizations such as the ICPC. However, States have only been admitted as members of the ICPC since 2010 and are often not represented at its meetings.[52] Also, the ICPC has no observer status with any of the UN agencies that have an interest in submarine cables.

A third reason why the legal regime governing submarine cables has been neglected is that there is no international organization or agency responsible for protecting submarine cables. The International Civil Aviation Organization (ICAO) is responsible for the safety and security of international civil aviation, and it has adopted global conventions to protect airports and other infrastructure for international civil aviation. The International Maritime Organization (IMO) is responsible for the safety and security of international maritime navigation, and has adopted global conventions to protect infrastructure for international shipping. Unfortunately, there is no international organization or agency responsible for submarine cables. Therefore, there is no UN agency that is taking the initiative to protect submarine cables.

The International Telecommunication Union is the UN specialized agency responsible for regulating information and communication technologies, but principally deals with telecommunications satellites rather than with submarine cables. It appears to have little interest in championing a new global convention

---

52 To date Australia, Malta, New Zealand, Singapore and the United Kingdom have joined as government members. The US Navy enjoys membership status as a cable owner.

to combat terrorist attacks on international submarine cables. The UN Division of Ocean Affairs and Law of the Sea is aware of the issues, but is not in a position to champion a new convention.

The international agency that may be the most interested in supporting an effort to adopt a global convention for the suppression of unlawful acts against the safety of international submarine telecommunications cables is the UN Office on Drugs and Crime (UNODC). It is the UN body that is responsible for overseeing the UN conventions to combat terrorism. Its officials are most likely to understand that international submarine cables are critically important communications infrastructure, and that a convention similar to the others they oversee should be adopted to make it an international crime to intentionally destroy or damage international submarine cables.

The critical issue may be how to get the issue of submarine cables on the agenda of UNODC, and whether the UNODC has the political influence within the UN system necessary to generate momentum to draft and adopt a new global convention. This is unlikely to happen unless a group of member States of the UN take the lead to work with the ICPC to actively promote the issue at the UN in general and with the UNODC in particular. Among the States that have shown the greatest interest in the protection of submarine cables are Australia, Singapore and the United States.

Other Measures to Protect Submarine Cables

Given that the intentional cutting of submarine cables by terrorists is a serious potential threat to the economy and security of the coastal State, governments should have contingency plans in place to deal with a terrorist attack on a submarine cable network within their territorial waters. Part of this planning would be a standard procedure whereby the cable industry would immediately notify the relevant government agencies, preferably through a designated lead agency, whenever there is a cable break or when suspicious activity is observed. The coastal State would then conduct a risk assessment to determine the likelihood of a possible hostile action or whether an actual break or fault could be due to a terrorist attack or an intentional act.

If a terrorist attack occurred against multiple submarine cables outside the territorial sea of any State, it is not clear which State would have the legal authority under international law to respond to the attack. The attack is likely to have been committed by persons onboard a ship, but if the ship is outside the territorial sea of any State, the flag State has exclusive jurisdiction over the ship, and it cannot be boarded without the consent of the flag State. Furthermore, the act of intentionally damaging or destroying submarine cables may be a criminal offence under the law of the State of nationality of the offender, or under the law of the flag State. However, as noted earlier, many States have not enacted legislation to fulfil their obligations under Article 113 of UNCLOS.

In addition, navies and coast guards of most States are not likely to have any procedures permitting them to exercise jurisdiction against persons on board ships who have intentionally damaged or destroyed cables in an area outside the territorial sovereignty of any State. They may not even be authorized under their national laws and legislation to take any action against vessels or persons who have intentionally damaged or destroyed submarine cables.

Therefore, what is required is a regional agreement among States served by a cable which authorizes the navy or coast guard authorities of any member State to take action against persons onboard any ship who are suspected of having intentionally damaged or destroyed submarine cables. The navy and coast guard vessels should be authorized to seek permission from the flag State to board suspect vessels. If the suspect vessel is flying the flag of a State which is a party to the regional agreement, the flag State should be obliged to grant permission for the suspect vessel to be boarded.

What is also required is an information-sharing arrangement among all the States Parties to the regional agreement to share information on suspected attacks on submarine cables and to fully cooperate in the event of an attack on submarine cables in the region but outside the territorial sea of any State.

Another recommendation is for State Parties and the cable industry to carry out joint international exercises to plan and develop protocols and practical responses that will serve the international community in preparing for possible disruptions to submarine cable infrastructure and its repair. Too often a State Party may conduct an exercise domestically with no or limited involvement from its domestic telecommunications companies and without considering the reality of international submarine cable infrastructure. A State may have superb domestic defences and repair plans for the infrastructure within its territory, but it remains vulnerable to hostile actions conducted against other State in whose territory the international cable lands if that State has lax defences and the international contracts and requirements for repair of cables are not understood.

V. Recent Developments—Increased Recognition of the Potential Threat to Submarine Cables

In the past few years, there has been an increasing recognition of the importance of protecting cables. For example, officials and scholars have begun to write on the problem of protecting submarine communications cables.[53] In 2010 the

[53] See, for example, Sechrist, *supra* note 5; M. Matis, "The Protection of Undersea Cables: A Global Security Threat" Strategy Research Paper, United States Army War College, 3 July 2012, available online at http://oai.dtic.mil/oai/oai?verb=getRecord&metadataPrefix=html&identifier=ADA561426; Burnett, Cable Vision, *supra* note 39 at 66; APEC:

EastWest Institute and the IEEE Communications Society published *The Reliability of Global Undersea Cable Communications Infrastructure (ROGUCCI) Study and Global Summit Report.*[54] The ROGUCCI Report makes ten recommendations for actions by industry and governments. The ICPC also works with these institutions to establish industry practices and improve transparency in line with these common goals.

For the past three years the protection of submarine cables has been recognized by the international community in the annual resolution of the United Nations General Assembly on Oceans and Law of the Sea. For example, Resolution 66/231 on Oceans and the Law of the Sea was adopted the United Nations General Assembly at its 93rd plenary meeting on 24 December 2011.[55] The preamble of the resolution contains the following paragraph on submarine cables:

> *Recognizing* that fibre-optic submarine cables transmit most of the world's data and communications and, hence, are vitally important to the global economy and the national security of all States, conscious that these cables are susceptible to intentional and accidental damage from shipping and other activities, and that the maintenance, including the repair, of these cables is important, noting that these matters have been brought to the attention of States at various workshops and seminars, and conscious of the need for States to adopt national laws and regulations to protect submarine cables and render their wilful damage or damage by culpable negligence punishable offences[.]

The operative paragraphs of the resolution contain the following paragraphs on submarine cables:

> 123. *Also calls upon* States to take measures to protect fibre-optic submarine cables and to fully address issues relating to these cables, in accordance with international law, as reflected in the Convention;
> 124. *Encourages* greater dialogue and cooperation among States and the relevant regional and global organizations through workshops and seminars on the protection and maintenance of fibre-optic submarine cables to promote the security of such critical communications infrastructure;
> 125. *Encourages* the adoption by States of laws and regulations addressing the breaking or injury of submarine cables or pipelines beneath the high seas done wilfully or through culpable negligence by a ship flying its flag or by a person subject to its jurisdiction, in accordance with international law, as reflected in the Convention;
> 126. *Affirms* the importance of maintenance, including the repair, of submarine cables, undertaken in conformity with international law, as reflected in the Convention;

Submarine Cable Resilience Critical to Connectivity, *APEC Committee on Trade and Investment*, News Release 6 February 2013.

54 K. Rauscher, ROCUGGI The Report, IEEE Communications Society and EastWest Institute, 2010, available online at http://www.ieee-rogucci.org/ (last accessed 9 June 2013).

55 Available at http://daccess-dds-ny.un.org/doc/UNDOC/GEN/N11/472/68/PDF/N1147268.pdf?OpenElement (last accessed 9 June 2013).

The resolution demonstrates that the international community is beginning to recognize that submarine cables are vitally important to the global economy and the national security of all States, and that measures should be taken to protect them. However, the resolution fails to call for any measures beyond those already set out in Article 113 of UNCLOS. It fails to recognize that a terrorist attack on submarine cables could cause incalculable damage to the world's economy and to the security of many States. It also fails to call for a global convention to establish a cooperative regime to combat intentional acts against submarine cable networks around the world. Accordingly, while it is most certainly a positive development, there remains a long way to go.

Conclusion

The international community has recognized that terrorists pose a threat to international transportation and communications networks, but the potential terrorism threat posed to submarine cables has not been recognized or addressed. Therefore, the legal regime governing submarine cables has been neglected, and is in need of review.

A high priority should be given to drafting and adopting a global convention to make the act of intentionally damaging or destroying international submarine cables an 'international crime'. This convention can follow the same scheme as the other UN terrorism conventions. States who understand the problem should take the lead in pushing for the adoption of a global convention through the UNODC. In the meantime, efforts should be made to promote greater cooperation at the regional level to protect international submarine cables and to ensure that States enact legislation to make the act of damaging or destroying a cable a crime under their national laws punishable by severe penalties. Governments should also cooperate with the cable industry to enable them to react quickly in response to a terrorist attack on international cables outside their territory.

PART V

SPECIAL PURPOSE SUBMARINE CABLES

CHAPTER THIRTEEN

SUBMARINE POWER CABLES

Malcolm Eccles, Joska Ferencz and Douglas Burnett

Introduction

Submarine power cables are cables used to transmit electrical power from one location to another. Advances in power cable technology are allowing power cables to conquer major ocean distances and depths between States. The cables themselves, at a basic level, comprise solid copper conductor cores, insulation and armoring. The physical properties of the copper core significantly increase the size and weight of the cables. This results in additional challenges for cable owners seeking to lay, repair and maintain the cables, not least of which is the need to procure the specialized vessels and equipment necessary to conduct these activities.

Each power cable system is unique. Unlike fiber optic cables, which often have standardized parts, each power cable system is custom made to specifications by the manufacturer and there are no universal jointing kits that can be used for repairs. Their unique construction compounds many of the issues that cable owners encounter when cables are damaged. This Chapter seeks to expound on some of these challenges. It first charts the development of power cables, examines the different types of cables and their uses and then discusses the ways in which the activities associated with laying and repairing power cables differ from those employed for fiber optic cables.

I. History of Submarine Power Cables

Submarine power cables have traditionally been installed for two reasons: the integration of domestic island electricity networks with mainland electricity networks and the interconnection of energy markets. By joining two markets together surplus energy can be traded, thereby providing consumers with an efficient energy option as generation outputs can be optimized across the two markets.

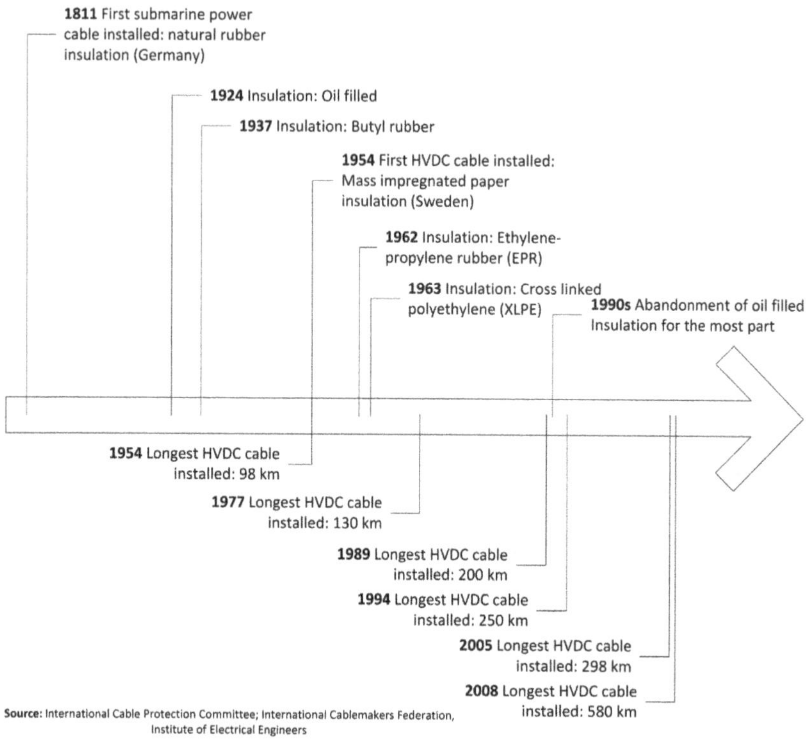

Figure 13.1 High Voltage Direct Current (HVDC) Evolutionary Timeline. (Image courtesy of J. Ferencz, Basslink)

An example of this optimisation is evident in the BritNed HVDC link[1] that connects the United Kingdom with energy markets in the Netherlands. The Executive Director of National Grid noted:

> This ability we now have to move power across national borders means we can use the full potential of...energy..., making it easier to import when...not available and export when there is a surplus.[2]

The first submarine power cable was installed in 1811. It was insulated with natural rubber and was laid across the Isar River in Bavaria, Germany.[3] In the 200 years that have elapsed since this first cable was laid, there have been significant developments in cable technology and construction. Primary drivers for the metamorphosis in cable design can be summarized in four simple terms: length

---

[1] HVDC refers to high voltage direct current.

[2] D. Carrington, "BritNed Power Cable Boosts Hopes for European Supergrid" *The Guardian* (11 April 2011). Available online at http://www.guardian.co.uk/environment/2011/apr/11/uk-netherlands-power-cable-britned (last accessed 20 May 2013).

[3] R. Kandiyoti, "Under the Sea" (2009) 4(14) *Engineering & Technology* at 26.

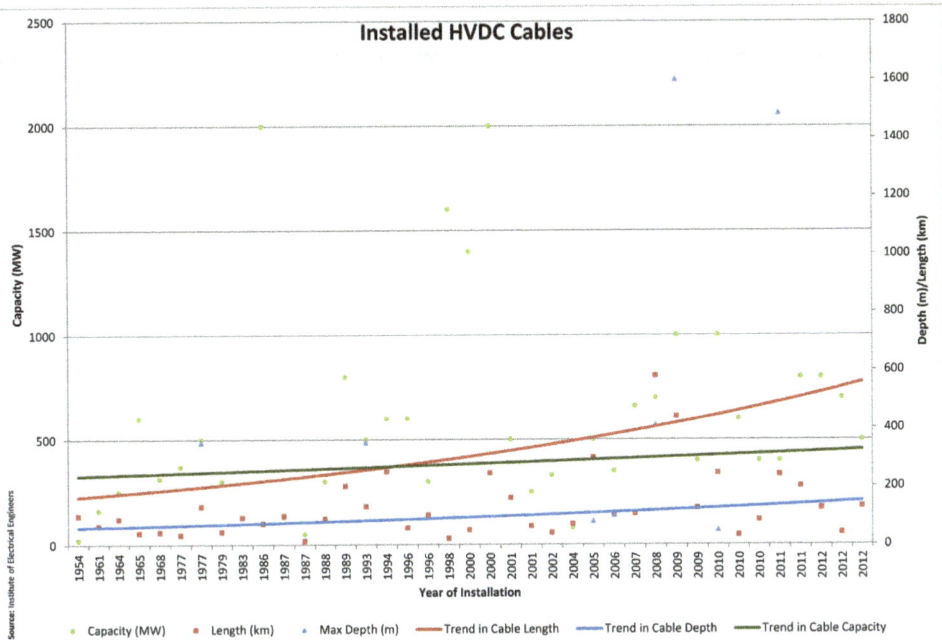

Figure 13.2 Installed High Voltage Direct Current (HVDC) cable systems showing depth, length and capacity. (Image courtesy of J. Ferencz, Basslink)

of lay, depth of lay, volume of electricity and system reliability. These drivers have in turn affected the basic construction and design of power cables through innovative use of materials, conductor design and manufacture.

Historical trends indicate that the lengths of cables are increasing, for example the first High Voltage Direct Current (HVDC) power cable Gotland 1 was installed in 1954 at a length of 98 km from Gotland Island to the Swedish mainland. Nearly 50 years later in 2005 Basslink was installed at a length of 298 km. Basslink, which connects the Australian states of Tasmania and Victoria, became the world's longest power cable but only retained this title for three years before being overtaken by NordNed in 2008. NordNed, which connects Norway to the Netherlands, has an installed length of 580 km, almost doubling the previous cable length. Proposed future installations are longer again with cables mooted between Borneo and the Malaysian peninsular (670 km)[4] and even between Iceland and the United Kingdom (900 km).[5]

---

4 R. Moses, "Sime to power up mega Bakun project" *New Straits Times* (Malaysia, 2 February 2007).

5 E.B. Hreinsson, "Renewable Energy Resources in Iceland—Environmental Policy and Economic Value" *Nordic Conference on Production and Use of Renewable Energy*

Power cable projects have become very ambitious in recent years and are linking regions of increasing distance; projects that would have been considered unfeasible decades earlier.[6] The depths of waters being traversed are also increasing. The deepest laid cable system to date is SAPEI, which was laid in 2008 at a maximum depth of 1600 m. SAPEI connects Sardinia to the Italian mainland and is 400 km in length. At these depths the quality of cable design and construction is paramount. At a depth of 1600 m the pressure exerted on a cable is approximately 160 times that of sea level. Any imperfections created during the manufacture of a cable would create a weakness and be crushed by the pressure, leading to premature electrical and/or mechanical failure. Increased water depth places an additional burden on the parties who manufacture, install and repair the cables. Hypothetically if a cable is in need of repair at a water depth of 1600 m the repairing vessel would potentially need to lift around 4000 m of cable, no easy feat given that the conductor for the cable is made of solid copper and weighs somewhere between 50 to 70 kg per meter. These considerations require installation, protection, and maintenance plans that are markedly different from those employed for fiber optic submarine cable systems.

II. Types of Submarine Power Cables

There are two forms of submarine power cables, Alternating Current (AC) and Direct Current (DC). Practically all submarine power cables, whether AC or DC, transport High Voltage (HV)[7] energy. AC and DC represent the form or syntax of the electricity transmitted. It is important to understand the difference between these two electricity formats, as they define the basic design assumptions of the entire cable system. It is also important to note that integrating AC with DC or *vice versa* requires specialist power electronics and ultra-high speed control systems (microseconds).

Each cable system is installed to rated maximum voltage. The voltage rating specifies the electrical insulation that is required and is as fundamental to the cable design as the copper conductor. Modern submarine power cables are rated in ranges up to 500 kV or 500,000 volts, at present this represents the currently

(9–11 of July 2008) Vaasa, Finland. See https://notendur.hi.is//~egill/rit/vaasa_2008_hreinsson_renewable_energy.pdf (last accessed 4 April 2013).

6 It is worth noting that Basslink was first mooted as a concept in the 1960s by the then Hydro Electricity Corporation of Tasmania see http://www.siemens.com.au/news/basslink, (last accessed 4 April 2013) but was not constructed until 2005 when improvements in technology made it feasible.

7 HV is defined as any voltage exceeding 1 kilovolt (kV) AC or 1.5 kilovolts DC (International Standard IEC 60038).

installed voltage ceiling.[8] The following sections will discuss the various attributes of the two electrical mediums with respect to their physics, use and limitations.

AC Systems

AC is inherent to all modern terrestrial electricity systems and is also commonly known as 'three-phase AC'. It is produced by generators and transported to industrial and residential customers via networks of towers, poles, wires and cables. AC interchanges the direction of flow of current at regular intervals (milliseconds), which means that the flow of current is sent in one direction then the other. Three-phase AC is characterized by three overlaid alternating waveforms or phases, as shown in Figure 13.3. Each colour represents the cyclic nature of one current interchange (or phase) and, as can be seen, each current phase is offset or delayed with respect to time, thus giving it its characteristic waveform. Arguably, the domination of AC in terrestrial networks was influenced by the well-documented historical rivalry between Edison and Tesla as much as it was the result of any economic or technological drivers.[9]

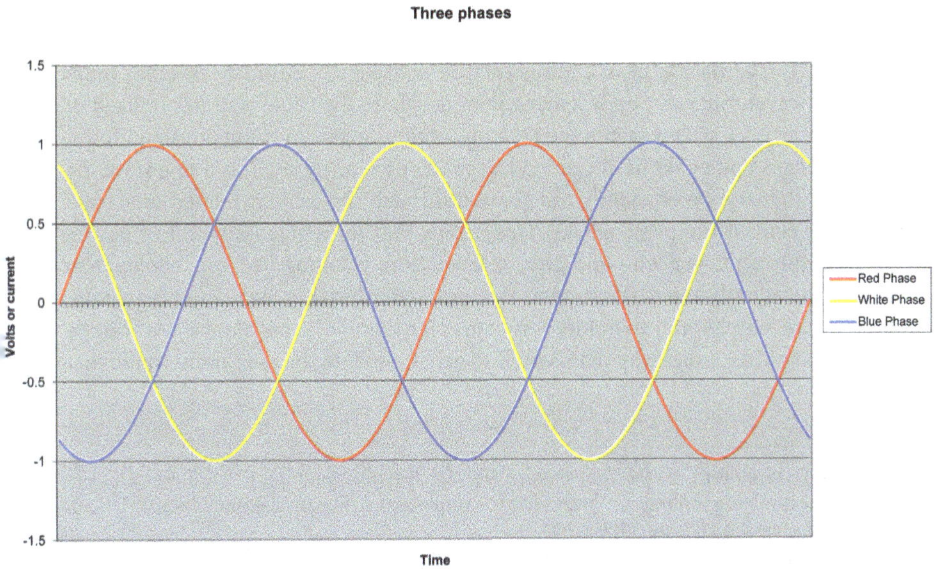

Figure 13.3 Typical three-phase Alternating Current (AC) waveforms. (Image courtesy of J. Ferencz, Basslink)

---

[8] Western HVDC link that will connect the Scottish and English electricity networks will be rated at 600 kV, making it the highest rated DC submarine power cable in the world. See http://prysmiangroup.com/en/corporate/about/special_projects/western-HVDC-link/index.html (last 4 April 2013).

[9] T. McNichol, *AC/DC: The Savage Tale of the First Standards War* (1st ed, 2006, Jossey-Bass).

In practical terms an AC cable system requires one conductor per phase, with a total of three conductors for a complete circuit. The three conductors may be located in one cable or they may be three individual cables, depending on the project's installation and logistics requirements. Each phase generates its own electro magnetic field (EMF), and this EMF interacts with the fields produced by the other two phases, causing inefficiencies in the transfer of energy. The effect is exacerbated by the close proximity of the cables and the magnitude of the field that is produced. This inefficiency presents a limitation in the length of AC cable that can be effectively installed, as an AC submarine cable can only be used to transfer energy between electricity networks at a critical distance of around 50 km.[10] There are some exceptions to this general rule, such as the Manx HVAC Interconnector that links the Isle of Man to England (104 km).[11]

When interconnecting three-phase power systems together it is absolutely crucial that each system is synchronized.[12] This means that the voltage magnitude and frequency of interchange cycle must be identical or catastrophic failure will occur to equipment within either system. This is important because in some regions of the world different network frequencies are present, being either 50 or 60 Hz, and commonly these differences are circumscribed by national boundaries. However this is not always the case. For example, within the Japanese electricity network 50 and 60 Hz share equal prominence for historical reasons.[13]

Due to the nature of AC, infrastructure is designed to satisfy the maximum of the waveform, essentially leaving periods where the cable is under-utilized or over-utilized, in other words the area under the waveform is not constant, forming troughs and peaks in the energy output. For instance, as AC is a sine wave the power that can be efficiently transmitted through the cable is related to around 70 per cent of the peak value.[14] This means that for a given amount of power, AC requires thicker wire and greater insulation, resulting in larger cables than the equivalently rated DC system. However, transformers and switching equipment are simpler and therefore less expensive than DC equivalents.[15] AC can be thought of like a highway that is built to accommodate the maximum traffic that

---

[10] See D. Woodford, *HVDC Transmission* (18 March 1998, Manitoba HVDC Research Centre), available online at https://hvdc.ca/uploads/ck/files/BasisPrinciplesofHVDC.pdf (last accessed 20 May 2013).

[11] See Manx Electricity Authority, http://www.gov.im/mea/projects/ (last accessed 4 April 2013).

[12] See J. Arrillaga *et al.*, *Flexible Power Transmission: The HVDC Options* (2007, Wiley).

[13] See P. Fairley, *IEEE Spectrum: Why Japan's Fragmented Grid Can't Cope* (6 April 2011), see http://spectrum.ieee.org/energy/the-smarter-grid/why-japans-fragmented-grid-cant-cope (last accessed 20 May 2013).

[14] See A. Haddad and D. Warne, *Advances in High Voltage Engineering* (2004, The Institution of Electrical Engineers).

[15] See J. Arrillaga, *High Voltage Direct Current Transmission* (2nd ed, 1998, The Institution of Electrical Engineers).

occurs at rush hour yet the rest of the time during the day it is relatively empty, whereas DC is the same highway with the same amount of traffic yet that traffic is evenly distributed throughout the day without peaks. In short, the road does not have to be overly wide to facilitate the same amount of traffic.

DC Systems

The main difference between DC and AC is that DC does not cyclically change direction and that it is only necessary to have two conductors in order to complete a DC circuit; with an HV Cable and a low voltage or return cable being the usual configuration of a single DC circuit. In some projects this return cable is sometimes replaced by a sea return where the return current is transported via submerged electrodes. Sea returns are used less frequently because the risk of electrolysis requires existing submarine asset owners to be more cautious.

For example Basslink was first designed using a sea return to significantly reduce project capital costs in comparison with a two cable system configuration.[16] Westerweller and Price state that the system configuration was changed:

> ...to eliminate perceived environmental impacts caused by the return current flowing through the electrodes and sea [and as a result] a medium-voltage return cable was introduced...

The addition of the return cable for Basslink mitigated anxieties surrounding claims of EMF, electrolysis and corrosion of third-party submarine assets.[17] Environmental issues such as these will be discussed later in this Chapter as well as in Chapter 7.

In comparison to AC, DC does not constitute a sinusoidal waveform. This means that the cable can be fully utilized up to its maximum rating. Simply put, there is no wasted area under the waveform, as explained in the highway analogy. The key to determining the use of AC or DC in the submarine application is essentially one of distance. The physical attributes of the different types of electricity, as briefly highlighted above, is the fundamental determinative factor. At a basic level, the interaction between the individual phases in the three-phase AC waveform create electrical inefficiencies which means that beyond the critical distance discussed above the amount of energy received is insignificant. The electrical nature of DC mitigates this effect, meaning that large volumes of energy can be transported long distances with significantly fewer losses compared to an equal length AC transmission system.

As noted, a core requirement of AC systems is that they operate at a uniform frequency. Due to its physics, DC does not change direction and therefore

---

16 See T. Westerweller and J.J. Price, "*Basslink HVDC Interconnector—System Design Considerations*" 8th International Conference on AC-DC Power Transmission (2006, Institution of Electrical Engineers, London).

17 *Ibid.*

does not have a frequency. This effectively isolates the AC electricity network's frequency from the others so they need not be synchronized nor even use the same frequency, which is the case in Japan where 50 and 60 Hz are used.

III. System Design

When planning submarine power cable projects there are a number of basic design recommendations that should be satisfied to ensure a high level of reliability and longevity of the power cable. The design and commercial life of a submarine power cable is between 40 and 65 years, with the possibility of further life extensions. This section seeks to briefly set out some of the most influential factors used to determine the final design of a power cable system. The overview is not intended to capture all the factors that influence design or to be totally inclusive with regard to the factors discussed.

Identification of a business need is generally the initial point for deciding to electrically link one source of energy to another. This will usually occur even before the first marine survey has been conducted. The structure of the terrestrial electrical transmission networks is a key factor in determining the points where a submarine power cable will interconnect. Augmentation to the terrestrial electrical transmission network is difficult because of network strength and it can greatly increase overall project cost for a submarine power cable. Network strength relates directly to the impacts that faults have on system reliability. The greater the strength, the better the system will be able to ride out system interruptions. Hence submarine power cables are commonly interconnected to some of the larger pre-existing substations in the terrestrial transmission network. For example Basslink has its terrestrial transmission interconnections at the substations of Georgetown in the state of Tasmania and Loy Yang in the state of Victoria, which are both significant substations for the respective transmission networks. The Basslink route does not represent the shortest distance from Tasmania to Victoria. If the shortest route were selected it would be unlikely that enough energy could be either transported or used effectively due to capacity constraints in the terrestrial transmission networks.

After proposed preliminary routes have been selected, detailed seabed and environmental conditions are investigated. These investigations will take account of landing points and shore side connections. The value of engagement with local users (local community, fishermen and indigenous peoples) should not be underestimated during this process, as these groups may have a local knowledge that is not readily evident in a single survey campaign. A survey is simply a snapshot in time, and it is sometimes only with history and long term engagement that the true picture of the installation area becomes apparent.

Like building a house, the foundations for a power cable design are extremely important. Shaky foundations based on unreliable assumptions will invariably lead to a poorly constructed cable. Environmental and seabed considerations are factors that have a substantial influence on the design. The final cable route, cable design and the decision to increase protection for the cable may be influenced by these considerations. To illustrate this interaction consider that a cable route proposes crossing a sensitive environmental feature. Re-routing may not be economically or physically feasible and cutting through the feature may not be environmentally permissible. Routing the cable over the feature can give rise to a number of challenges, such as abrasive interaction between the feature and the cable. This interaction may cause damage to both the feature and the cable, and the cable designer may therefore be tasked with improving abrasion resilience of the cable, or retrofitted armor may be applied to the cable in the form of articulated pipe (Figure 13.5) or the like.

Surveys

As described in detail in Chapter 4, surveys are undertaken in order to understand and determine the suitability of the seabed for the proposed cable route or routes.[18] These surveys must take into account the different physical characteristics of power cables and their maintenance requirements compared to fiber optic cables. As with land, the seabed topology is variable and may take many forms from sand to rock, mountain to valley and everything else in between. In addition, there may be historically and culturally significant sites such as shipwrecks or underwater archaeological sites or even traditional burial regions that must be taken into consideration. There may also be significant threats to flora and fauna. The primary object of the survey is to accurately determine the best route for the cable that minimizes hazards.

The composition and topology of the seabed is a key factor in determining if and how the cable is to be protected. A soft muddy or sandy seafloor means that a protective burial may be feasible, whilst a hard rocky bottom may require some other form of protective covering.

The data obtained during the survey assists in determining the final route and cable design. For example, a cable route that traverses deep waters will require a cable design capable of supporting the weight of the cable when suspended from the lay ship to the seafloor. Alternatively a route laid through shallow waters may require burial or additional armoring to offer protection from third party damage.[19] Route deviation may be required so as to avoid crossing another cable or pipeline or to avoid a shipwreck of cultural significance and the human activities that are associated with it, such as activities of divers and salvors. These types of impediments to the passage of a cable may not be immediately apparent when the project is first proposed and their subsequent discovery may be a determinative

18 A comprehensive overview of cable route surveys is provided in Chapter 4.

19 *Third-Party Damage to Underground and Submarine Cables*, Technical Brochure 398, Cigre Working Group B1.21, December 2009.

factor for the cable owner in deciding whether to proceed with the project. Any significant variation to a cable route or design can result in substantial increases in projected installation costs. It is important to recall that power cables are physically much larger than fiber optic cables and a lot of material is required for their construction and installation. Increasing the length of a cable by tens of kilometers can increase the materials required by hundreds of tonnes. The bulk of a power cable's weight is found in the conductor, which is usually made from solid copper. The soaring world commodity prices for copper makes only too clear the impact that a few extra tens of kilometers of material would have on the value of a power cable installation, not to mention the additional manufacturing costs.

Environmental Factors

Electromagnetic fields (EMF), discussed in the section on AC above, are apparent in all cables regardless of whether they are AC or DC. Permitting authorities of coastal States take into account the potential effects of EMF on the marine environment when assessing the environmental impact of the cables.[20] It should be noted that the EMF of a DC cable is generally greater than that of an AC cable.[21]

Some marine fauna use geomagnetic fields for navigation and these fields occur naturally as variations in the Earth's magnetic field. Other marine fauna use the bioelectrical fields generated by all species for other purposes, such as detecting prey and mates. Concern has been expressed that submarine power cables could significantly increase or change the naturally occurring background EMF and have a deleterious effect on marine fauna.[22] The actual observations and studies to date of submarine power HVDC systems, however, indicate that these impacts do not negatively affect the species involved. (Chapter 7.)

Faugstad *et al.*[23] in their conference paper to the Cigre 2002 Paris session titled *Experience from the Licensing Process for the North Sea HVDC Interconnectors. Possible Consequences for Future Submarine Links*, commented that the magnetic field generated by HVDC cables is:

> a static magnetic field with the same magnitude as the natural earth field. Magneto-sensitive fish, such as Atlantic salmon and eel, has been observed based on the

---

20 Further discussion of the relationship between EMF and the marine environment is provided in Chapter 7.

21 See T. Olsson *et al.*, "Impact of Electric and Magnetic Fields from Submarine Cables on Marine Organisms; the Current State of Knowledge" 12 November 2010, available online at http://www.seai.ie/Renewables/Ocean_Energy/Foreshore_Lease_Consultation/Appendix_4_-_Impact_of_electric_and_magnetic_fields.pdf (last accessed 4 April 2013).

22 *Ibid.*

23 See K. Faugstad *et al.*, "Experience from the Licencing Process for North Sea HVDC Interconnectors. Possible Consequences for Future Submarine Links" Conference Paper 14-111, Cigre Session 2002.

> fear that their east-west migration could be disturbed. The Baltic Cable crosses the migration path of eel and salmon in and out of the Baltic Sea. Field experiments on eel showed just minor deviation, but no barrier to migration.[24] Atlantic salmon continue to migrate in and out of the Baltic Sea across the Baltic, Kontek, Kontiskan and Skagerrak subsea HVDC cables, and no migration problem seems to exist. Viking Cable had a laboratory test done on baby-eel to reveal any barrier effects from the magnetic fields. No effects were found.[25]

The concerns may be unfounded as a literature study conducted in 2010 by Vattenfall Power Consultant[26] regarding the current state of knowledge on the impact of the field from submarine cables on marine organisms notes that "[t]he main conclusion of the literature study was that the current amount of information on the subject is very limited. Still, no research results were found that suggested that present sub-sea power cables posed as a threat to marine environment due to EMF". This is consistent with the results of an EMF study carried out on the Basslink Interconnector which found that "[t]he magnetic field measurements have shown that at 5m above the cable, the measured magnetic field was virtually indistinguishable from background (less than 1% of the total background field)..."[27]

Field studies such as the Basslink study that are undertaken by independent university researchers are significant since they are based on actual observations and measurements in the marine environment. Furthermore, their observations support modelled calculations of the EMF from Basslink that provides confidence when estimating the extent and strength of the field associated with cables. They should be accorded more weight than theoretical and non-peer reviewed papers by writers with no practical field experience that opine about possible impacts without attempting to take into account the large amount of information and data available from power cable systems that have been installed in the world's oceans for many years. Additional information of environmental interactions of submarine power cables and the environment are covered in Chapter 7.

---

24 See H. Westerberg and M.L. Begout-Anras, "Orientation of Silver Eel (*Anguilla anguilla*) in a Disturbed Geomagnetic Field" National Board of Fisheries, Proceedings of the Third Conference on Fish Telemetry in Europe, Sweden, June 1999.

25 See H. Fock *et al.*, Schwimmverhalten von Jungaalen in einem experimentell geänderten geomagnetischen Feld. (2000) Germany, Reference 12 from paper by Faugstad *et al.*, *supra* note 23.

26 See T. Olsson *et al., supra* note 21.

27 D. Strong, Supplementary Technical Addendum, Matters arising from Minutes of Meeting Friday 31 August 2007. Report prepared by Coffey Natural Systems, CR 898_BSERC_Supplement, April 2008 for the Bass Strait Environmental Review Committee; J. Sherwood, "What's Going On Down There? Addressing Community Concerns about the Basslink Cable, SE Australia" ICPC Plenary, Miami, Florida, 22 May 2013, (presentation available upon request from the ICPC).

When a sea return path is used for an HVDC installation, increased corrosion may impact subsea metallic installations as an effect of the field set up by the sea electrode. The process, which is commonly known as electrolysis, uses the sea water as an electrolyte path to aid the displacement of atoms from the anode material, in this case the metallic structure to the cathode (the HVDC electrode).

The corrosion effect can be estimated and modelled so that plans can be developed to alleviate its effects.[28] This was successfully undertaken during the planning and construction phases of the Konteck and Balstic HVDC projects and further experience has shown that this issue can be mitigated economically if considered during the cable's planning stages.[29]

A second environmental issue that has been raised in the past relates to oil insulated submarine power cables. As indicated in Figure 13.1, which illustrates the chronological development of submarine cables, oil insulated submarine power cables ceased being used in the 1990s due to environmental and maintenance concerns. Insulation for modern submarine power cables is from mass impregnated paper or the much more prevalent *cross-linked polyethylene* (XLPE). Both provide better insulation properties without the environmental concerns associated with the now obsolete use of oil in submarine power cables.

IV. Power Cable Installation

Requirements

Submarine power cables are significant social and economic connectors. Consider that the traditional uses for submarine power cables have been to interconnect island communities to mainland energy networks, to reduce the reliance of island communities on power generated by fossil fuels or to transfer remote abundant under-utilized energy to energy starved regions. If the use of this infrastructure is lost it can have a catastrophic impact on interconnected communities, and repair campaigns can take many months depending on location. This is very different from repairs to a fiber optic telecommunication system as discussed in Chapter 6. Imagine no electricity to your home for several months; a very real prospect when a submarine power cable is damaged. Power cables, due to the costs involved, are generally not redundant geographically. Unlike fiber optic telecommunication cables, restoration of service by other oceanic power cables is not available in the case of monopole HVDC and AC cables. It is therefore extremely important that power cables are correctly installed and protected as soundly as possible. The challenges to making timely and cost efficient repairs are significant. Power cables are therefore often installed with more robust protection against third party damage than that found in fiber optic telecommunications systems.

---

[28] Faugstad *et al.*, *supra* note 23.

[29] *Ibid.*

Figure 13.4 Power cable being loaded on shipboard carousel. (Photograph courtesy of Global Marine Systems)

Vessels

The types of vessels required to install power cables are very different from the vessels used to lay fiber optic cables. There are a small number of power cable vessels in the world (approximately three to five vessels), the largest of which are controlled by power cable manufacturers. These vessels with their large cable bending radius and lifting requirements are generally not available for hire on a spot basis and chartering such vessels may require months or even years of advance planning. Figure 13.4 shows the cable carousel which is distinctive to this type of vessel. This is in sharp contrast to the approximately 38–42 cableships available for repairs and laying of fiber optic telecom cables which can be hired on a spot basis.

Burial

The required burial depth for a power cable will generally be agreed between the power cable owner, its insurers and the installation contractor.[30] In areas where increased sediment is apparent, a deeper burial may be attempted. Conversely, in areas where there is a reduction in the sediment cover there may be no need for burial passes and alternative protection may be required instead. Unlike telecommunication cables which undertake burial work in the shallow water closer to land, some submarine power cables are known to be buried the entire length

30 With regard to insurance it should be noted that submarine power cables are generally insured. This is in contrast to the vast majority of submarine telecommunication cables which are self-insured (i.e. no external insurance policy is purchased).

of lay. Not all submarine power cables are fully buried and most are installed at shorter distances than their fiber optic cousins and, to the most part, at shallower depths. If power cables were installed at greater lengths and to greater depths then it may not be economically viable to conduct burial for the entire length of the cable, despite the risk to the cable.

Protection

Given the cost of repairing power cables and the length of time repairs take, adequate protection at the installation stage is given a very high priority. There are three main methods used for power cable protection, these being direct burial, articulated pipes, and rock trenching. The goal for all three methods is to protect the power cable so that it is unlikely to ever require a marine repair.

Direct burial, as the name suggests, is the burial of the cable under the seabed. The target depth is generally set by the project requirements and the seabed conditions. Ideal depth of coverage is between 0.5 and 2 meters. Burial is achieved by water jetting a trench into soft seabeds. Rock trenching occurs in circumstances where the seabed is too hard. A trench is ground into the rock layer, with the cable laid into the trench and backfilled with softer material. Articulated pipes are used as protective covering where necessary, such as on reef crossings where both the reef and cable co-exist and the fragility of the reef makes it inappropriate to trench.

As discussed above, the utilization of a specific protection method will depend on the type of seabed on which the cable is laid. Post repair protection is likely to utilize the same methods employed during installation, with standard concrete mattressing being used where appropriate. A concrete mattress is a method of

Figure 13.5 An installed articulated pipe on the seabed. (Photograph courtesy of Basslink)

Figure 13.6 Illustration of a concrete mattress installed over a power cable.

protection and application of weight commonly used in the offshore oil and gas industry. The installation requirements and methodologies are well understood, manufacture is simple and the installation vessel can rapidly deploy mattresses for quick protection over short sections of exposed cable.

The need for concrete mattresses is likely to be identified during the cable deburial exercise, at which time seabed material is removed and it becomes apparent that the cable will require additional protection. This ensures manufacture of the mattress can commence in a timely manner, ready for installation immediately after the final joint (or bight) has been laid. Inspection videos are made after the deployment of the final bight and also after the burial runs. Prior to commencing cable protection activities the videos will be reviewed and evaluated to ensure that the cable is not sitting in a position that could result in damage if a mattress were applied, for example if the cable is laid over rocks or hard ground with suspensions.

In addition to these protection methods, rock dumping and grout bags can be used in place of mattressing for short lengths of exposed cable. Repair vessels cannot be used for rock dumping and a different vessel is required. The more labour intensive grout bags are installed by divers. When the bags are cut grout flows either side of the cable, filling suspensions and providing a protective capping for the cable.

V. Power Cable Repairs

During the lifespan of a submarine power cable, it may be necessary to undertake repairs. The reasons for repairs echo those of fiber optic cables, i.e. electrical or mechanical failure. Aside from manufacturing issues, the primary cause of damage suffered by power cables occurs as a result of third party damage, principally from anchors or fishing activities.[31] Faults can manifest in different forms, ranging from straightforward severing though to the more complex damage that can occur when the outer protective layers or insulation of a cable are compromised. What sets the repair of power cables apart from that of fiber optic cables is the time required to undertake the repair.

The repair and jointing of power cables takes days rather than hours to complete.[32] It is a highly specialized process and the critical requirements for a repair hinge upon a number of factors, not least of which is the repair vessel or barge. The vessel must be able to maintain position during this time, which invariably requires a suitable weather window. The repair vessel or barge must also have enough deck space to undertake the repair of (possibly) a number of cables at one time. With several jointing enclosures each being around 20 m in length and a cable bending radius being approximately 3 m, this can add up to a lot of working deck space.

Another fundamentally important requirement for undertaking submarine power cables repairs is the need for highly skilled and qualified cable jointers. This requirement should not be underestimated. There are no standard submarine power cable designs. In general each power cable is designed to meet the specifications of each individual project. Although common materials are used to construct power cables, there may be notable differences in the way in which the materials are used. The number of submarine cable manufacturers is few, especially at the leading edge of the HVDC market.[33] These manufacturers are the authorities of the submarine power cable industry with their own vessels and jointers. From start to finish each submarine power cable repair is a custom operation.

Due to the unique construction, size and relatively sparse global distribution of power cables in comparison to fiber optic cables, purpose built power cableships are very rare, already committed to contracts, and are essentially limited to these manufacturers and some exclusive operators. Unlike telecommunication cables, there are no maintenance zones or other shared repair assets. Each system is distinctive and as such each system owner will have unique requirements. The ideal repair would require a contract or charter with a power cableship owner. Other vessels with suitable characteristics could possibly be chartered and modified. It should be stressed, however, that this may increase the potential for problems due to uncertainty and cost during the repair process; especially if the cable is installed at a significant distance from a ready supply of vessels.

---

31 See "*Third-Party Damage to Underground and Submarine Cables*" Technical Brochure 398, Cigre Working Group B1.21, December 2009, available online at http://www.landsnet.is/uploads/1067.pdf (last accessed 4 April 2013).

32 This example only considers jointing time and not the time it takes to mobilize the repair platform or jointers, which may take many months.

33 400 kV or larger.

A review of the approach to power cable repairs in relation to 'best practice' indicates that while shorter repair times are the norm in Northern Europe, the Mediterranean and North America (all locations where the resources are on hand) longer repair times are more common elsewhere. Northern Europe and the Mediterranean are within two weeks sailing from the North Sea and there is a concentration of large high performance Dynamic Position (DP) equipped offshore vessels working in the oil and gas sector there. The same synergy can be seen in North America with its proximity to the Gulf of Mexico oil and gas region. Cable system owners in these regions may be more inclined to reduce operating costs and not have a dedicated repair platform in light of the greater selection of vessels on the open market. In contrast, owners of more remotely placed cables may have dedicated repair ships or barges.

As noted above, each power cable is unique. This also means that each joint is unique and any repair therefore requires specialized jointing. In contrast to fiber optic telecommunication cables, there is no Universal Joint Kit (UJC) for power cables that allows different systems to be connected and spliced from any cableship. In the case of power cables, splicing will generally require a team from the power cable manufacturer who will have to be contracted and transported to the fault location.

The basic process of repairs for power cables is similar to that of fiber optic cables. The differences arise due to the size and weight differentials of the cable types. The fundamental process once a fault occurs is as follows:

- Mobilizing the repair vessel and/or equipment;
- Determining and locating the fault;
- Recovering, repairing and testing; and
- Recommissioning the cable into service.

At this juncture it is important to note that the strategies behind submarine power cable disaster recovery planning are unique to each and every cable owner. Important factors will include insurance, customer requirements, geographic location, history and risk appetite. The repair strategy is derived from examining the probability of a submarine cable fault occurring. That is a function of threat versus protection afforded to the cable. Valid strategy considerations are as follows:

- No advanced planning and address a problem when an incident occurs;
- Procure specialized equipment for mobilizing onto a vessel of opportunity chartered in the event of a repair or standby vessel;
- Standby contract with a submarine cable repair company;
- Maintenance agreement with a submarine cable repair company;
- Procure vessel and equipment and maintain on permanent standby.

Vessel and Equipment

Given the specific requirements of power cables, vessel selection is immensely important. Vessel selection can severely limit the suitability of a proposed repair platform. Important factors include:

- *Size*

In the submarine power cable world the size of a cable dictates the suitability of a vessel. Submarine power cables are very heavy, with some AC cables weighing up to 100 kg per meter. Spare cable used in the repair has to be loaded, stored and deployed from the vessel. This means that the vessel, aside from having the physical deck space to lay out cable for jointing and handling, must also have sufficient structural strength to withstand any distribution of loading.

- *Jointing*

As discussed previously submarine power cable jointers are limited in general on larger cables to the manufacturers and they are highly specialized and have long lead times for their services. Third party jointers not affiliated with the original manufacturer are not an option. Joints are the potential weak spot of any cable and it is important that these joints are done to a high standard. The vessel must have sufficient space for jointing to take place under controlled conditions.

- *Bending radius*

The bending radius is the minimum radius that a cable can be safely bent without damage. Compromising the bending radius can distort the integrity of the cable construction and cause failure of both the electrical and mechanical properties of the cable. Submarine power cables have bending radii in the range of a few meters, which means they require a larger deck space than fiber cables.

- *Lifting capacity*

Power cables are heavy and the vessels must be capable of lifting the cable safely and reliably. The depth of water in which the repair is to take place will greatly influence this requirement.

- *Storage of cable*

Again, due to size and weight of power cables the storage of the spare cable on the vessel must be taken into consideration. This is important for vessel stability and safety. Issues of stability may prevent a vessel loading or departing for a repair and may necessitate the use of a naval architect or vessel augmentation.

- *Remotely Operated Vehicles/Divers*

When recovering a damaged section of cable it will be necessary to physically interact with the cable. This may take the form of de-burial, de-bundling or simply applying haulage wires for surface recovery. This requires underwater intervention through the use of divers or Remotely Operated Vehicles (ROVs). The depth of water generally determines whether divers or ROVs are necessary. The latter are used in deeper water.

Fault Finding

Determining the location of submarine power cable faults can be difficult. The design of the cable can have an impact upon fault locating tools, as can the type of fault itself, for example a complete severing compared to more complex insulation damage. As noted above each cable is unique in construction. In contrast fiber is a tube of glass, the quality of glass may differ but fundamentally it is still a tube of glass. Techniques used to locate faults on fiber cable are therefore more precise than they are for power cables. In practice a number of techniques would be used to increase the determination of the faults location.

A Time Division Reflectometer (TDR) may help determine the location of a fault. The TDR in simple terms detects the electrical reflections caused by inconsistencies in the cable by way of a pulsed test signal. These inconsistencies can occur at joints and at faults. Longer cables present a challenge to this technique as the time it takes a reflected signal to be received increases, thus increasing the exposure of the test equipment to false or misleading reflections.

While the TDR method is used from the shore, electroding is a method that requires underwater intervention. Electroding involves the use of a low frequency tone that is applied to the cable or armor at the landward end of the cable. A detector is then used subsea either by divers or by an ROV. The signal can be followed by the detector until it is lost, which signifies a possible fault. Problems arise, however, if the cable is not fully parted, as the signal will be transferred through the fault.

Another method of fault location is to look for bubbles. Absurd as it sounds, faults have been detected in this way. This method is only used when the cable systems are able to operate. The difficulty with this method is that the bubbles can be very small and infrequent, and given that cables can be hundreds of kilometers in length this makes for a highly inefficient means of locating a fault.

De-burial

Over time buried cables will tend to sink deeper into loose silty, sandy seabeds. On one hand this is good for cable protection, but on the other hand if a fault does occur the cable will need to be exposed. The repair of a cable requires a lateral distance from the damaged cable of approximately two and half times water depth exposed to recover and enact a repair by the repair vessel. In areas where the sea bottom is silty ROVs or divers will need to remove the covering layers of silt. Removing silt can be a very slow process and can be frustrated by re-covering that occurs through natural wave action well before the required amount of cable is exposed.

Repair

The speed of repair of a power cable is paramount; while most modern submarine power cables are constructed using water blockers, the water pressure at installation depth can force water into the cables layers. Water ingress under pressure, if unchecked, can cause greater damage to the cable than the original fault. In some cases to maintain cable integrity it may be necessary to undertake a cut and cap operation whereby the cable is cut, recovered to the surface for a waterproof cap to be applied to each exposed cable end, then buoyed off and laid on the seabed for recovery and repair at a time when resources or weather allow.

To repair the cable, one end is recovered and new cable stock is jointed to the existing cable and the joint is then laid on the seabed. The repair vessel then recovers the second cable end and joints the free end of the new cable stock to the second end. This joint is laid on the seabed. This repair process is similar to that employed for fiber optic cables, except that the jointing is far more complex and specialized.

VI. The International Legal Regime Governing Power Cables

An overview of the international legal regime governing submarine cables has been covered in Chapter 3 of this Handbook, and specific legal issues with regard to cable operations and the protection of cables have been covered in Parts IV and V. For present purposes, it suffices to say that the 1958 Geneva Convention on the High Seas[34] and the 1982 United Nations Convention on the Law of the Sea[35] do not differentiate between different types of submarine cables and hence, also apply to power cables.[36] It should be noted, however, that the 1884 Convention

---

[34] 1958 Convention on the High Seas, adopted 29 April 1958, 450 UNTS 11 (entered into force 30 September 1962). The United Nations Convention on the Law of the Sea (see below note 35) supersedes this treaty for States that are parties to both.

[35] United Nations Convention on the Law of the Sea, adopted 10 December 1982, UNTS 1833 (entered into force 16 November 1994).

[36] Indeed, Art 113 refers to the breaking or injury of "a submarine cable" on the one hand and the breaking or injury of "a submarine pipeline and high-voltage power cable" on the other.

for the Protection of Submarine Telegraph Cables[37] does not cover power cables and only applies to submarine cables used for telecommunications.

VII. Future Development

Deregulation

The deregulation and commoditization of electricity has, in some regions, created an opportunity to trade between markets. Deregulation has meant that traditionally government owned transmission network owners and operators have been privatized and afforded the opportunity to extend their market power beyond the traditional national boundaries. Where the privatization of government networks has not occurred the private sector has nevertheless interconnected networks via entrepreneurial submarine cable facilities. Funding for these projects can occur from straight utilization, capacity auctions and contracts for participation in each of the interconnected markets taking advantage of the arbitrage value of the energy.

Transmission Constraints

There are already many examples of submarine HVDC cables being used to manage issues caused by transmission constraints in terrestrial electricity networks. Many AC transmission corridors are simply filled to capacity. As cities and regions expand, demand for electricity increases. As housing and industry come to occupy greater expanses of land, this in turn encroaches upon the amount of land that would traditionally have been available for AC transmission corridors. Neptune RTS and Hudson are examples of submarine cables being used to alleviate transmission constraints in the United States. Neptune RTS links New Jersey with Long Island via HVDC cable system and Hudson links New Jersey to Manhattan via an HVAC cable system. Similarly, Western HVDC link, currently in the planning stage, will be used to mitigate the transmission constraints between the English and Scottish electricity networks.

---

[37] Convention for the Protection of Submarine Telegraph Cables, adopted 14 March 1884, TS 380 (entered into force 1 May 1888) (1884 Cable Convention). The provisions of the Cable Convention are generally accepted as customary international law, see *Restatement of the Law (Third): The Foreign Relations Law of the United States Vol 2* (American Law Institute Publishers, 1987) § 521, comment f (1986). As at 2 April 2013 there are 40 State Parties to the Cable Convention. A complete copy of the 1884 Cable Convention is contained in Appendix 3.

Concluding Remarks

Submarine power cables share many similarities with fiber optic cables yet they also share many differences. The relative size and individuality of designs provides the power cable owner with many challenges from construction through to operation and repair. When these cables are interrupted communities and governments can be affected beyond the impact to the cable system owners. While comparatively rare as compared to fiber optic cables, more and more power cables are being installed subsea and this trend seems likely to continue with these cables being laid longer, deeper and with greater capacity.

CHAPTER FOURTEEN

MARINE SCIENTIFIC RESEARCH CABLES

Lionel Carter and Alfred H.A. Soons

Introduction

Marine science has a fruitful association with the submarine cable industry that began symbolically with the first trans-oceanic cable between Newfoundland and Ireland. To assist in defining a route for this trans-Atlantic link, a cable consortium led by Cyrus Field turned to the United States Navy, which duly supplied information on ocean depths and the nature of the seabed.[1] These data were collected in 1853 and gave an early glimpse of the complexity of the North Atlantic crossing. Far from revealing a featureless substrate, the route was shown to traverse a remarkable topography of submarine shelves, plateaux, abyssal plains and the mid-Atlantic Ridge—a mountain chain that rivaled anything on land.

With the laying of the first cables came failures, but these provided unexpected surprises for science. Cables recovered for repairs from water depths of several thousand meters were often encrusted with marine life or covered with sediment containing marine organisms.[2] These discoveries were made during a time of intense debate as to whether marine life existed in water depths greater than 500 fathoms (914 m). Some scientists argued that the darkness, cold and high pressures were too extreme to support life, but the recovered cables showed otherwise. The famous biologist, Thomas Huxley, suggested that mud from a cable retrieved in 1858 contained slimy matter called *Bathybius*—a substance that was regarded as the foundation of life.[3] This so-called discovery was made in 1868, just a year before the HMS *Porcupine* returned to port

[1] J.S. Gordon, *A Thread Across the Ocean: The Heroic Story of the Transatlantic Cable* (Simon & Schuster, 2002) at 239.

[2] *Treasures of the Natural History Museum* (Natural History Museum, London, 2008) at 256.

[3] P.F. Rehbock, "Huxley, Haeckel, and the Oceanographers: the Case of Bathybius haeckelii" (1975) 66(4) *ISIS* 504–533.

with sponges and other animals dredged from 2435 fathoms (4453 m) on the Porcupine Abyssal Plain, west of the United Kingdom. The scene was set to resolve the debate once and for all. Accordingly, the British ship, HMS *Challenger*, was commissioned to undertake the world's first oceanographic survey that involved depth sounding, sampling of ocean waters and seabed and, critically, the marine life. The three year voyage between December, 1872 and May, 1876 revolutionized our view of the ocean. Many new life forms, unknown to science, were discovered in all ocean depths sampled. As for *Bathybius*, an ocean chemist onboard the HMS *Challenger* showed it was nothing more than a precipitate of calcium sulphate that formed when sediment samples were preserved in alcohol. When notified, Thomas Huxley quickly and graciously admitted he was in error. Nonetheless, he was correct in asserting that the deep ocean supported life—a remarkable paradigm shift that was in part due to the discovery of organisms on submarine telegraphic cables. From the telegraph era onwards,[4] recovered cables became an important source of marine organisms for researchers and museums, providing specimens from parts of the ocean rarely frequented by oceanographers.[5] In the same vein, other information collected during cable route surveys also contributed to improving knowledge of the deep sea.

In 1929, submarine telegraphic cables took on a new role—that of discovering major displacements of the seabed. On 18 November, a magnitude M7.2 earthquake shook the seabed off the Grand Banks, Newfoundland. At least eight cables broke concurrently with the main shock.[6] Subsequent studies showed these breaks resulted from a series of landslides triggered by severe ground shaking within an approximately 100 km radius of the earthquake epicenter.[7] By mixing with water, some landslides were transformed into a fast moving mud and sand-laden flow or *turbidity current*. This moved down-slope breaking more cables en route. From the timing of the cable breaks and their location it was possible to estimate current speeds, which reached 65 km/hour on a journey of over 650 km. Overall, the turbidity current carried about 200 km^3 of sediment into water depths over 4500 m. This event demonstrated for the first time that the deep ocean was far from quiet and stable, especially in regions subject to seismic activity.

---

[4] The telegraph era approximately spanned 1850–1950. For more information refer Chapter 1.

[5] See C.D. Levings and N.G. McDaniel, "A Unique Collection of Baseline Biological Data: Benthic Invertebrates from an Under-water Cable Across the Strait of Georgia" (1974) Technical Report No 441 Fisheries Research Board of Canada at 19; see also P.M. Ralph and D.F. Squires "The Extant Scleractinian Corals of New Zealand" (1962) 29 Zoology Publications, University of Wellington 1–19.

[6] B.C. Heezen and M. Ewing, "Turbidity Currents and Submarine Slumps, and the 1929 Grand Banks Earthquake" (1952) 250 *American Journal of Science* 849–873.

[7] D.J. Piper *et al.*, "Sediment Slides and Turbidity Currents on the Laurentian Fan: Side-Scan Sonar Investigations Near the Epicentre of the 1929 Grand Banks Earthquake" (1985) 13 *Geology* 538–541.

The Grand Banks event heralded a plethora of scientific research that continues today using cable fault data to provide information on the frequency, size, extent and causes of submarine landslides and turbidity currents (see Chapter 10). Such knowledge is helping to lessen the risk posed by these natural hazards.

I. Submarine Cables for Scientific Research

As early as 1948 research was underway to investigate the potential of submarine cables to measure ocean currents.[8] Ocean water is a conductor and resides within Earth's magnetic field. Thus the flow of sea water produces an electrical current that a submarine cable should be capable of detecting. This simple premise was tested in the Strait of Florida where the northward-flowing Florida Current passes over several disused coaxial and active fiber optic telecommunications cables connecting the United States with Caribbean islands. By measuring the voltages across a cable, it is possible to determine the transport of water passing through the Straits. By 1982, reliable measurements could be made and these continue today to produce a near-continuous record of the Florida Current, which is a key component of the North Atlantic Ocean circulation.[9]

Another scientific use of cables relates to the Acoustic Thermometry of Ocean Climate (ATOC) project. This was designed to study the thermal structure of the ocean using the speed of sound, which is known to vary with water temperature.[10] Sound transmitters were placed on the seabed off Hawaii and California in association with a series of sound receivers positioned around the North Pacific rim. The California sound source plus a hydrophone system were connected to shore by coaxial cable that supplied power and communications.[11] ATOC commenced operations in 1995 and continued for approximately a decade. However, even in the cable's out-of-service state it has supplied new insights regarding its physical

---

8 H. Stommel, "The Theory of the Electrical Field Induced by Deep Ocean Currents" (1948) 15 *Journal of Marine Research* 679–693.

9 M.O. Baringer and J.C. Larsen, "Sixteen Years of Florida Current Transport at 270°N" (2001) 28 *Geophysical Research Letters* 3179–3182; see also, H.R. Longworth *et al.*, "Historical Variability in Atlantic Meridional Baroclinic Transport at 26.5°N from Boundary Dynamic Height Observations" (2011) 58 *Deep-Sea Research II* 1754–1767 available at http://www.aoml.noaa.gov/phod/docs/2011_DSRII_Longworth_etal.pdf (last accessed 1 June 2013).

10 B.M. Howe, "Expanding the ATOC/NPAL North Pacific Array using the TPC-4 Submarine Cable" Proceedings paper from the Earthquake Research Institute of Tokyo, 8–9 November 2004 available online at http://www.ieee-jp.org/japancouncil/chapter/OE-22/Bruce.pdf (last accessed 1 June 2013).

11 I. Kogan *et al.*, "ATOC/Pioneer Seamount Cable After 8 Years on the Seafloor: Observations, Environmental Impact" (2006) 26 *Continental Shelf Research* 771–787, available online at http://www.mbari.org/staff/svonthun/ATOCcable_survey2006.pdf (last accessed 1 June 2013).

Figure 14.1 Marine life growing on the now out-of-service ATOC coaxial cable off California, which was the subject of a survey regarding its interaction with the seabed environment. Here the cable provides a substrate for the anemone, Metridium farcimen. The 3.2 cm-diameter cable was photographed at 140 m water depth. (Photograph courtesy of Monterey Bay Aquarium Research Institute)

interaction with the seabed environment, as documented for the continental margin off California (see Figure 14.1).[12]

A more recent development has been the use of cables to support sensors for the detection of natural hazards, particularly submarine earthquakes and tsunami.[13] Currently, detection of Pacific Ocean tsunami is based on a combination of well-established wave theory and computer models, together with observations from various sea level recorders, tsunami meters and satellites such as *Topex Poseidon*. The Pacific recording sites, together with those of the western Atlantic and European seas, number around 1170.[14] One prominent detection system is the DART buoy (Deep-ocean Assessment and Reporting of Tsunamis). These surface buoys are powered by solar- and wind-generated energy and gather data from tethered pressure and temperature sensors located on the seabed. That information is transmitted to shore via satellite. Following the Indonesian earthquake and giant tsunami of 2004, detection buoys, including DART, were deployed in the Indian Ocean—a region with sparse warning systems. A combination of poor

---

12 *Ibid.*

13 S. Ingle *et al.*, "The Next Wave in Tsunami Detection" (2012) 46 *Marine Technology Social Journal* 67–73.

14 See the National Oceanic and Atmospheric Administration's National Data Buoy Center, www.ndbc.noaa.gov (last accessed 1 June 2013).

maintenance, vandalism and other factors subsequently downgraded the buoys' warning capability.[15]

More robust systems comprised of seabed sensors linked to shore via fiber optic cable, are now coming to the fore, especially for tsunami detection close to land. For example the recently developed Seismic Tsunami Early Warning System (STEWS) located at 1360 m water depth off Oman, has a seabed package comprised of a seismometer, accelerometer, pressure gauge and pressure recorder. Because the supporting cable provides continuous communications and power, the seabed sensors deliver a stream of data in real time.[16]

The need for improved detection of distant deep-water tsunami in areas of the ocean with few monitoring sites, plus a growing requirement for better information on the deep ocean's response to global warming, have led researchers to seek new solutions. One possibility under consideration is the incorporation of environmental sensors in the repeaters of fiber optic telecommunications cable systems.[17] The so-called 'green repeaters' could potentially accommodate sensors to monitor ocean temperature and salinity, pressure and acceleration. Initial discussions between United Nations agencies [International Telecommunication Union (ITU), World Meteorological Organization (WMO), United Nations Educational Scientific and Cultural Organization (UNESCO) and International Oceanographic Commission (IOC)], together with the submarine cable industry including the International Cable Protection Committee (ICPC), the science community and legal experts, have highlighted important issues associated with the proposal. All parties recognize the exceptional strategic, societal and economic value of the submarine cable network that underpins the internet, communications and the global transfer of data.[18] Any modification to existing cable systems must be made with due care and in full consultation with the major parties. This caution reflects considerable legal, commercial and funding issues, for example, who is liable if a sensor(s) causes a cable failure, bearing in mind the high costs of cable repairs and loss of commercial revenue. Currently, there are several suggestions up for discussion that include:

---

15 P. Folger, "U.S. Tsunami Programs: a Brief Overview" Congressional Research Service, 14 March 2011, US Report R41686 at 8, available at http://fpc.state.gov/documents/organization/158601.pdf (last accessed 1 June 2013).

16 Ingle *supra* note 13.

17 Y. You, "Harnessing Telecoms Cables for Science" (2010) 466 *Nature* at 690–691; see also R. Butler "Using Submarine Cables for Climate Monitoring and Disaster Warning" (2012) International Telecommunications Union Report at 23.

18 L. Carter *et al.*, "Submarine Cables and the Oceans: Connecting the World" (2009) Report of the United Nations Environment Program and the International Cable Protection Committee (UNEP-WCMC-ICPC) at 64, available online at http://www.unep-wcmc.org/medialibrary/2010/09/10/352bd1d8/ICPC_UNEP_Cables.pdf (last accessed 1 June 2013).

- Sensors could be installed in repeaters to record temperature, pressure and acceleration. This is a 'simple approach' using sensors that are robust, have modest power requirements and have design lives comparable to those of the repeater/cable, i.e. 20 to 25 years.
- A trial is undertaken with a cable and repeaters outfitted with the three to four basic sensors to test performance, reliability and ability to withstand cable laying/recovery operations.
- Pending successful trials, sensor-bearing systems could be deployed as redundant cables are replaced and as new routes are developed. The inference is that establishment of a global monitoring network is a long-term program.

Once completed and assuming the present world-wide distribution of cables does not change significantly, deep-ocean cable monitoring would be biased towards the Northern Hemisphere. From a scientific viewpoint, this constrains a full monitoring and understanding of climate-driven ocean change because a major driver of that change is Antarctica and the Southern Ocean circulation.[19] In that light, the telecommunications cable-based data should be integrated with information gathered from other observation sites—undoubtedly a major endeavor but one that is necessary to meet the challenge of identifying the forces that control 71 per cent of Earth's surface.

II. Cabled Ocean Observatories

As of 2012, a survey of oceanographic websites revealed at least 190 coastal and deep-ocean observatories world-wide. This figure is almost four times that of 2005, thereby underlining a marked shift towards monitoring the marine environment. Such a shift is partly a response to an expanding human presence offshore associated with exploitation of living and non-living resources, greater shipping activity, disposal of wastes and increased regulation of those activities via marine protected areas and spatial planning.[20] It also reflects a greater marine research effort relating to the ocean's intimate connection with climate change especially regarding the oceanic accommodation of greenhouse gases and heat from the

---

[19] N. Bindoff *et al.*, "Position Analysis: Climate Change and the Southern Ocean" (2011) Antarctic Climate & Ecosystems, Cooperative Research Centre at 27; see also L. Carter *et al.* "Circulation and Water Masses of the Southern Ocean: A Review" in F. Florindo and M. Siegert, eds, *Antarctic Climate Evolution* (Elsevier, 2008) at 606.

[20] B.S. Halpern *et al.*, "A Global Map of Human Impact on Marine Ecosystems" (2008) 319 *Science* 948–952; see also S. van den Hove and V. Moreau "Deep Sea Biodiversity and Ecosystems: A Scoping Report on their Socio-economy, Management and Governance" (2007) Biodiversity Series 28, UNEP Regional Seas Report and Studies at 84, available online at http://www.unep-wcmc.org/medialibrary/2010/09/10/96db7c1f/Deep_Sea_Biodiversity_Ecosystems.pdf (last accessed 1 June 2013).

atmosphere.[21] Finally, there is a need for improved detection and monitoring of natural hazards as evinced recently by devastating tsunamis (Indonesia, 2004; Chile, 2010; Japan, 2011), storm surges (Hurricane Katrina, 2005; Cyclone Nargis, 2008; Hurricane Sandy, 2012) and floods (Typhoon Morakot, 2009).[22]

Observatories come in a wide range of sizes and configurations. At one end are simple oceanographic moorings: a mooring being essentially a bottom weight and subsurface buoy linked together by a line that supports instruments capable of recording information on internal data loggers or transmitting information back to shore via satellite as in the case of DART buoys. A common mooring consists of current meters and Conductivity/Temperature/Depth (CTD) sensors that respectively monitor current velocity and salinity/temperature/pressure over periods of months to a few years. At the other end of the observatory scale are the large, sophisticated cabled systems that undertake multiple experiments as well as measuring physical, biological, and chemical changes in the waters and seabed for up to 25 years, for example NEPTUNE Canada.[23] These complex systems are based on fiber optic/power cables that provide power for instruments and communications for the transfer of data to shore in real time. One of the pioneering cabled observatories is the Monterey Accelerated Research System (MARS) situated off California in 891 m of water.[24] It began full operation in November 2008 following a lengthy period of equipment development and resolution of environmental issues. A key feature of MARS is the main monitoring site, which centers on a submarine hub or *node* (see Figure 14.2). Encased in a trawler-proof frame, a *node* operates on a plug-in-and-play concept whereby several pieces of equipment can be linked to the node, which provides communications and power via a 52 km-long cable that terminates at shore-based laboratories. There, oceanographic data are processed, archived and made accessible for researchers and the public.

Also based on the node concept is the North-East Pacific Time-series Undersea Network Experiments, or NEPTUNE observatory, located off Vancouver Island, Canada (see Figure 14.3). NEPTUNE came into service in December 2009 and is currently the largest operational observatory.[25] Its backbone is an 812 km-long, fiber optic/power cable that interconnects five nodes occupying sites from the continental shelf at 20–100 m water depth to the abyssal plain in 2660 m.

---

21 Bindoff *et al.*, *supra* note 19 at 27.

22 Munich R.E., NatCatService, Great Natural Disasters since 1950. Available online at http://www.munichre.com/en/reinsurance/business/non-life/georisks/natcatservice/default.aspx (last accessed 1 June 2013).

23 See *NEPTUNE Canada: Invitation to Science*, (University of Victoria, Victoria British Colombia 2012) at 61.

24 See MARS, The Monterey Accelerated Research System, Monterey Bay Aquarium Research Institute, www.mbari.org/mars (last accessed 1 June 2013).

25 NEPTUNE Canada www.neptunecanada.com (last accessed 1 June 2013).

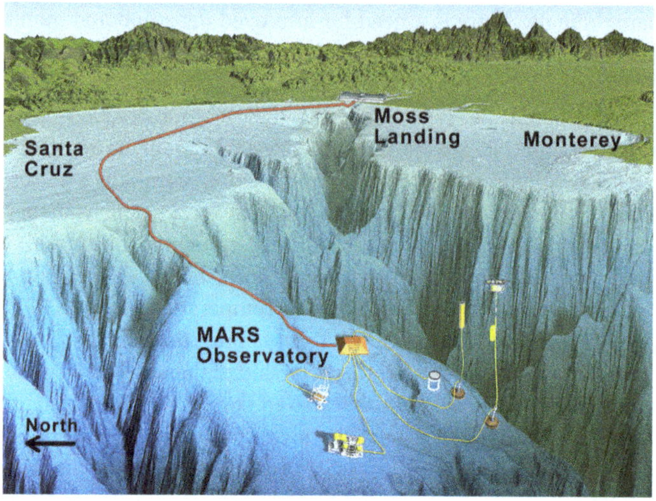

Figure 14.2 Artist's rendition of the MARS Smooth Ridge site, Monterey, CA, at 891 m water depth with the main node (orange) supporting five instrument packages. These are tethered by communication/power lines that ultimately link with the main fiber optic/ power cable (red) connecting with the Monterey Bay Aquarium Research Institute (MBARI). (Image courtesy of MBARI)

In that configuration, NEPTUNE covers a suite of different submarine settings where site-specific research can be undertaken. On the continental slope (node ODP 889, 1258 m depth), for instance, investigations are underway on: (i) gas hydrates—deposits of ice and methane; (ii) the stability of the seabed during earthquakes and tsunami and (iii) the movement of fluids and gases though seabed sediments and underlying oceanic crust. Likewise, the submarine Barkley Canyon node (400–1000 m depth) is focused on: (i) the movement of water and sediment along the canyon; (ii) canyon ecosystems and (iii) gas hydrates. Such information is highly relevant to the cable industry by virtue of the hazards posed by landslides and turbidity currents generated by unstable sediments on slopes and in canyons, and by the potential mining of hydrates as an energy source. Although operational for just three years, research results are becoming available via scientific papers and a comprehensive website.[26] Among the discoveries are: (i) the enhanced release of methane by the action of bottom currents and internal waves on gas hydrate deposits;[27] (ii) the deep-water detection of the 2010

[26] *Ibid.*

[27] L. Thomsen *et al.*, "Ocean Circulation Promotes Methane Release from Gas Hydrate Outcrops at the NEPTUNE Canada Barkley Canyon Node" (2012) 39 *Geophysical Research Letters*.

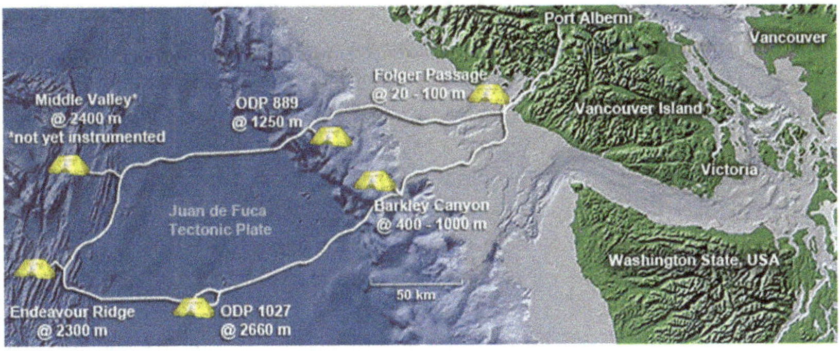

Figure 14.3 Operational nodes from NEPTUNE Canada. Five operational nodes, plus a sixth yet to be outfitted with instruments, comprise NEPTUNE Canada whose backbone is an 812 km-long cable linking the nodes with Port Alberni station. (Image courtesy of NEPTUNE.)

Chilean tsunami and its refraction approaching the Vancouver Island margin[28] and (iii) the dominant role of cool, diffuse fluid emissions from the seabed as the prime cause of subterranean heat loss in contrast to the more dramatic but less influential hot 'black smokers'.[29]

Close to completion is the Regional Scale Node (RSN) observatory complex off the northwestern United States. The complex comprises seven nodes, deployed in September 2012, and interconnected by 900 km-long fiber optic/power cable originating from a control station in Oregon.[30] Like its Canadian counterpart, NEPTUNE Canada, research will focus on: (i) the physical and biological processes affecting the continental shelf and slope, in particular the impact of climate change on ocean currents and biological communities; (ii) seabed deposits of gas hydrates and (iii) the processes associated with active tectonic plates. In addition, the RSN has deployed nodes in the vicinity of an active submarine volcano, Axial Seamount, to continue and expand upon an earlier monitoring program that has successfully forecast subsea eruptions.

RSN is part of the Ocean Observatories Initiative (OOI) that includes five other stations strategically located in the North Pacific, North and South Atlantic and the Southern Ocean. In contrast to the RSN, these other sites are occupied by more traditional moorings. One site, located in the Irminger Sea at 60°N off Greenland,

---

28 A.B. Rabinovich, "The 2010 Chilean Tsunami off the West Coast of Canada and the Northwest Coast of the United States" (2012) *Pure and Applied Geophysics*, available online at http://link.springer.com/article/10.1007%2Fs00024-012-0541-1#page-1 (last accessed 1 June 2013).

29 K. Bemis *et al.*, "Diffuse Flow On and Around Hydrothermal Vents at Mid-Ocean Ridges" (2012) 25(1) *Oceanography.*

30 Ocean Observatories Initiative (OOI) see www.oceanobservatories.org (last accessed 1 June 2013).

for example has four moorings, each supporting a host of instruments. Data are transmitted to shore by satellite link to a wind/solar-powered surface buoy. Underwater gliders, capable of making environmental measurements, shuttle between the moorings. The Irminger site was chosen because of the North Atlantic's influence on global climate and oceanic circulation: (i) North Atlantic ocean currents provide heat for Europe and the eastern United States as well as contributing to an overturning circulation previously known as the *Ocean Conveyor*; (ii) sinking surface ocean waters help remove carbon dioxide from the atmosphere and (iii) storminess in the region appears to be increasing thus affecting coastal communities and infrastructure.

Through integration of the research and data from the cabled and moored observatories, as envisaged by OOI, NEPTUNE Canada, MARS and others, scientists are making major advances in understanding how the Earth system of ice, ocean, atmosphere and land 'ticks'.

III. Regulating the Use of Submarine Cables for Marine Scientific Research

International Regulation

The use of submarine cables for marine scientific research is subject to two sets of rules within the international law of the sea; those concerning marine scientific research and those concerning submarine cables. These rules have been laid down in the United Nations Convention on the Law of the Sea (UNCLOS), which was adopted in 1982 and entered into force in 1994. It is currently in force for 165 States and the European Union. This means that it has not yet attained universal participation; some important coastal States, including the United States are not (yet) a party to the Convention. Those States are, however, bound by the rules of the customary international law of the sea. It is generally agreed that the provisions of UNCLOS relevant to the use of submarine cables and to marine scientific research essentially reflect the current rules of customary international law on those activities. This Chapter will therefore refer to the provisions of UNCLOS.

From the preceding sections it could be deduced that submarine cables can be used for both the actual collection of oceanographic data (by the incorporation of environmental sensors in the cable repeater) and/or the transport of such data collected at sea by other instruments or structures. Furthermore, the cables can be used exclusively for these purposes, or they can potentially be used for dual purposes, such as regular telecommunications data transfer and oceanographic data collection. These distinctions are relevant for determining the applicable international legal rules.

The legal regime applicable to the use of submarine cables for the collection and/or transport of oceanographic data varies according to the maritime jurisdictional zone in which the cable is (to be) located. These zones are: (i) areas under the sovereignty of a coastal State; (ii) areas under limited, functional jurisdiction of a coastal State and (iii) areas beyond the jurisdiction of coastal States.[31]

The areas under sovereignty of a coastal State include the maritime internal waters; the archipelagic waters and the territorial sea, which extends to a maximum of 12 nautical miles from the coast. The areas under functional jurisdiction are the exclusive economic zone (EEZ), which extends to a maximum of 200 nautical miles from the coast, and the continental shelf (the seabed and subsoil), which may extend well beyond 200 nautical miles in certain circumstances.[32] The areas beyond national jurisdiction include the high seas (the water column beyond the EEZ) and the international seabed area beyond the continental shelf.

Regime of Marine Scientific Research

UNCLOS does not contain a definition of the term 'marine scientific research' (MSR). During the negotiations for UNCLOS various proposals were made for a definition of MSR but a consensus appears to have been reached that a definition of the term was not necessary as the substantive provisions clearly established the meaning intended.[33]

Although UNCLOS does not contain a definition of MSR it can be implied from the Convention's provisions that this term covers any collection of oceanographic data for research carried out with the intention of open publication.[34] Other marine data collecting activities, such as resource exploration and hydrographic surveying, are subject to different legal regimes.[35]

All MSR in areas under coastal State jurisdiction requires consent from the coastal State. MSR in the territorial sea may only be conducted with the express consent of and under conditions set forth by the coastal State.[36]

For the EEZ and continental shelf UNCLOS distinguishes between two categories of MSR: research activities which come within the discretionary competence of the coastal State to grant or withhold consent; and all other research, to which the coastal State in normal circumstances should grant consent. The former

31 See Chapter 3 for an extensive overview of maritime zones under UNCLOS.

32 UNCLOS Art 76. It should be noted that the continental shelf can be considered in two senses (i) in the geological sense i.e. the plain that extends from shore to around 130 m water depth and (ii) the legal continental shelf which is the natural prolongation of the seabed that can extend beyond the 200 nm limit. This Chapter uses the UNCLOS legal definition of the continental shelf, as provided in Art 76.

33 A.H.A. Soons, *Marine Scientific Research and the Law of the Sea* (Kluwer Law and Taxation Publishers, 1982) at 124.

34 *Ibid.* at 384.

35 *Ibid.* at 125. See also J.A. Roach and R.W. Smith, *Excessive Maritime Claims. Third edition* (Martinus Nijhoff Publishers, 2012), Chapter Fifteen. Marine Data Collection.

36 UNCLOS Art 245.

category comprises, *inter alia*, research which is of direct significance for the exploration and exploitation of natural resources and research which involves the construction, operation or use of installations and structures.[37]

Applications for consent must be made through official channels at least six months before the actual starting date of the field work for the research.[38] The coastal State must respond within four months; if it does not, implied consent may be presumed.[39] When applying for consent, detailed information on the project and its means of execution must be supplied, and when consent has been given the researchers must comply with certain conditions spelled out in detail (including access to all data and research results, and assistance to the coastal State).[40] Although the rules on MSR are primarily based on the conduct of MSR from vessels, the deployment and use of any type of scientific research installations or equipment is subject to the same rules.[41]

MSR in areas beyond national jurisdiction (on the high seas and in the international seabed area) is free, which means that it is only subject to the jurisdiction of the flag State or national State of the researchers.[42]

Recently, suggestions have been made that the routine collection of oceanographic data which are immediately made generally available (sometimes referred to as 'operational oceanography') is not, or should not be, subject to the legal regime of MSR, but this is still controversial.[43] What is beyond doubt, however, is that the routine collection of marine meteorological data by voluntary observing ships under the WMO's World Weather Watch program is not covered by the regime for MSR.[44]

Regime of Submarine Cables for Marine Scientific Research

All submarine cables, for any purposes, located in the areas under coastal State sovereignty are subject to complete control of the coastal State. The coastal State is thus entitled under international law to enact any national legislation it deems fit in relation to the laying, use and maintenance of submarine cables used for the collection or transport of oceanographic data located in its territorial sea (and internal and archipelagic waters).

---

37 UNCLOS Art 246.
38 UNCLOS Arts 250 and 248.
39 UNCLOS Art 252.
40 UNCLOS Arts 248 and 249.
41 UNCLOS Arts 258–262.
42 UNCLOS Arts 256 and 257.
43 A.H.A. Soons, "The Legal Regime of Marine Scientific Research: Current Issues" in M. Nordquist *et al.*, eds, *Law, Science and Ocean Management* (Brill, 2007) at 139–166. Roach and Smith *supra* note 35 at 437–448.
44 Roach and Smith *ibid.* at 437–438.

Submarine cables on the continental shelf, however, are in principle subject to a different regime: Article 79 of UNCLOS provides for the freedom to lay and maintain submarine cables. This freedom, however, in fact only applies to cables merely transiting the continental shelf of the coastal State: cables that land in the coastal State, and therefore require the coastal State's permission to enter the territorial sea, can be subject to conditions the coastal State may impose for allowing the cable to enter its territory. These conditions may thus cover the entire segment of the cable located on the continental shelf of the coastal State involved (or even area beyond).[45] Cables merely transiting this continental shelf may only be subject to "reasonable measures" taken by the coastal State to safeguard its rights over the exploration of the continental shelf and the exploitation of its natural resources.[46] Also, when laying a transit submarine cable due regard must be given to cables and pipelines already in position; in particular, possibilities of repairing existing cables or pipelines may not be prejudiced.[47] Some coastal States claim the right to require permission for the laying and repairing of transit submarine cables on their continental shelf (including the prior surveying of the seabed before deciding on the course of the cable), but this is controversial and disputed by other States.[48]

The freedom to lay and maintain submarine cables transiting the continental shelf of a coastal State applies also to cables ***transporting*** oceanographic data collected elsewhere in the marine environment. However, transit cables ***actually collecting*** oceanographic data on the continental shelf of the coastal State will also be subject to the legal regime for marine scientific research.[49] As a result, the consent of the coastal State will have to be obtained prior to their laying and operation. It can even be maintained that their laying comes within the scope of the coastal State's discretionary power to grant or withhold consent, since such cables could be regarded as 'structures' as referred to in Article 246(5) of UNCLOS because of their semi-permanent character. These conclusions apply *a fortiori* to cables landing in the coastal State or entering its territorial sea.

---

45 There is no consensus regarding the scope of the right of a coastal State to establish conditions for a cable that is located on its continental shelf and which also enters into its territorial sea. One view holds that such conditions may go beyond those related to the "reasonable measures" that a coastal State may take for the exploration of the continental shelf and the exploitation of its natural resources, provided that they do not amount to an abuse of right by the coastal State (UNCLOS, Art 300), or otherwise violate the international obligations of the coastal State. For an alternative view, please refer to Chapter 3 and Chapter 5 which suggest that Art 79(4) does not allow a coastal State to impose additional conditions on cables that fall on both its continental shelf and territorial sea.

46 UNCLOS Art 79(2).

47 UNCLOS Art 79(5).

48 Refer Chapters 3, 4 and 6.

49 UNCLOS Part XIII.

This coastal State regulatory power would also apply to submarine cables used for dual purposes: telecommunications cables which simultaneously serve to collect oceanographic data transiting its continental shelf. This creates a significant cost and delay risk to such dual purpose cables when MSR is combined with telecommunication uses.

The rules applicable to submarine cables used for the collection and/or transport of oceanographic data beyond areas under national jurisdiction are clear: they are subject to the freedom to lay submarine cables,[50] as well as the freedom of marine scientific research.[51] This means that in principle only the national State of the entity laying and/or using the cable may regulate the activity.

National Regulation

When submarine cables used for the collection of oceanographic data are subject to the jurisdiction of coastal States, in accordance with the international legal rules identified above, they will be subject to national legislation (and, within the territorial sea close to shore, possibly also at the local level). Obviously, these national (and local) regulations can vary considerably according to the coastal State involved. Some of these national regulations may, in turn, be based on international conventions concerning the protection of the marine environment of the conservation of nature to which the coastal State is a party.

Some examples from the United States can be mentioned here. In the case of ATOC, the program was strongly influenced by the science community and the public regarding issues concerning the potential effects of sound transmission on marine mammals. Controversy was fuelled by often counter scientific and technical arguments that were difficult to reconcile because the knowledge of the day was uncertain about any impacts of artificial noise on marine mammals.[52] As a result, ATOC commissioned a US$6 million study, which after six years came to the conclusion that there was no significant biological impact.[53] Nevertheless, environmental concerns took their toll with the program ending in 2006 after a decade of operations.

Installation of the MARS observatory within the Monterey Bay National Marine Sanctuary required standard permitting procedures that were initially regarded as straight forward. However, the original plan to land the fiber optic

---

50 UNCLOS Art 87.

51 UNCLOS Art 256.

52 L.S. Weilgart and H. Whitehead, "Human-generated Undersea Noise and the Marine Environment: Objections to ATOC and LFA Sonar and Suggestions for Future Protocols", Paper presented at the Annual Meeting of the Society for Conservation Biology, Victoria, Canada, June 1997.

53 Acoustic Thermometry of Ocean Climate (ATOC) http://atoc.ucsd.edu/ and North Pacific Acoustic Laboratory (NPAL) http://npal.ucsd.edu/mammals/index.htm (last accessed 6 June 2013).

backbone cable via a disused fuel pipeline, was vetoed by the US Coast Guard because the pipeline was close to the water intakes of a power station, which could be construed as a terrorist target, bearing in mind that the installation of MARS was planned just after the 11 September 2001 terrorist attack on New York.[54] As a result a new cable landing was required. Furthermore, problems associated with a cable installation in the Olympic Coast National Marine Sanctuary led NOAA, which regulates the sanctuaries, to change requirements for laying cables. This led to a MBARI permitting process that lasted two years, cost USD1M and involved 16 permits from 16 agencies at all levels of government.

In contrast, the NEPTUNE Canada observatory was not subject to any legal requirement for an environmental impact assessment. Nevertheless, NEPTUNE commissioned an independent evaluation, in addition to consulting widely with environmental regulators, indigenous peoples, fishers and other stakeholders. Both consultation and environmental assessment initiatives made all parties aware of the benefits and potential impacts of the project and, in the case of the latter, led to project design changes that lessened any negative effects.

The increased human presence offshore, including marine research, has led governments to consider marine protection and special planning. In the words of the US Ocean Policy Task Force, such a move is to provide "... a flexible framework for coastal marine planning to address conservation, economic activity, user conflict and sustainable use of the oceans ...".[55] In New Zealand, regulations are being developed that would affect offshore activities including aspects of scientific research within its EEZ and extended legal continental shelf.[56] Should these measures and those under consideration by other States come to fruition, marine scientific research is likely to face more regulation.

---

54 Monterey Accelerated Research System (MARS), Monterey Bay Aquarium Research Institute, http://www.mbari.org/mars/ (last accessed 1 April 2013).

55 The White House Council on Environmental Quality, "Final Recommendations of the Interagency Ocean Policy Task Force, July 19, 2010" http://www.whitehouse.gov/administration/eop/ceq/initiatives/oceans, the Final Recommendations are available for download at http://www.whitehouse.gov/files/documents/OPTF_FinalRecs.pdf (last accessed 6 June 2013).

56 Ministry for the Environment, "Managing Our Oceans: A Discussion Document on the Regulations Proposed under the Exclusive Economic Zone and Continental Shelf (Environmental Effects) Bill" at 112, see http://www.mfe.govt.nz/issues/oceans/eez-regulations-consultation.html (last accessed 6 June 2013).

CHAPTER FIFTEEN

MILITARY CABLES

J. Ashley Roach

Introduction

Military cables are submarine cables used for military purposes or are military owned and/or leased. The purpose of this Chapter is to demonstrate that military cables, regardless of ownership or use, are subject to the same regime under international law that governs submarine cables. They are entitled to the same protections and are subject to the same rights and obligations provided for in the 1884 Cable Convention, the 1958 Geneva Conventions and UNCLOS. Part I of the Chapter describes the different military uses for submarine cables and Part II examines how the international conventions that govern submarine cables address military cables.[1] This Chapter does not address cyber security or information operations, as these subjects are beyond the scope of this Handbook.[2]

I. Military Uses of Submarine Cables

Historically, the military has used submarine cables for a variety of purposes, including telecommunications, acoustic monitoring and bilateral communications. These uses will be elaborated upon below.

---

[1] Portions of this Chapter are adapted from J.A. Roach and R.W. Smith, *Excessive Maritime Claims* (3rd ed., Martinus Nijhoff, 2012).

[2] For a discussion of information operations see M.N. Schmidt and B.T. O'Donnell, eds., *Computer Network Attack and International Law*, International Law Studies, vol 76 (US Naval War College Press, 2002), available online at http://www.usnwc.edu/Research---Gaming/International-Law/New-International-Law-Studies-(Blue-Book)-Series/International-Law-Blue-Book-Articles.aspx?Volume=76 (last visited 20 May 2013).

Telecommunications

Submarine cables have long been used for telecommunications purposes between military bases. For example, in 1970 a submarine telecommunications wire cable ran from the Soviet Union's missile submarine base at Petropavlovsk, under the Sea of Okhotsk, joining land cables going to Soviet Pacific Fleet Headquarters near Vladivostok and on to Moscow.[3] Over the years, military communications have increasingly used commercial submarine fiber optic networks.[4]

Listening—Hydrophones

Submarine cables have also been used for intelligence-gathering through acoustic monitoring. During the Cold War, the United States placed a number of hydrophones on the seabed off the Atlantic and Pacific coasts, and in the North Atlantic, connected to shore based facilities by submarine wire cables. This system, called Sound Surveillance System (SOSUS), became a key, long-range early-warning asset for protecting the United States against the threat of Soviet ballistic missile submarines. It also provided vital cueing information for tactical, deep-ocean, anti-submarine warfare.[5] While many of the arrays are no longer active, some have been put to other uses, such as analysis of low-frequency vocalizations from marine mammals living in the open ocean.[6]

---

3 S. Sontag and C. Drew, *Blind Man's Bluff: The Untold Story of American Submarine Espionage* (Public Affairs, 1998) at 158, 170–172.

4 Lacroix *et al.*, *A Concept of Operations for a New Deep-Diving Submarine, Appendix 1 Submarine Cable Infrastructure* (Rand, 2002) at 141, available online at http://www.rand.org/content/dam/rand/pubs/monograph_reports/MR1395/MR1395.appi.pdf (last accessed 20 May 2013); C. Rosenberg, "Navy plans fiber-optic link to Guantanamo Bay" *Washington Post/Miami Herald* (Washington, 5 July 2012), available online at http://pqasb.pqarchiver.com/washingtonpost/access/2702772501.html?FMT=ABS&FMTS=ABS:FT&date=Jul+5%2C+2012&author=Carol+Rosenberg&pub=The+Washington+Post&edition=&startpage=A.2&desc=Navy+plans+fiber-optic+link+to+Guantanamo+Bay (for purchase) (last accessed 10 November 2012).

5 E.C. Whitman, "SOSUS: The 'Secret Weapon' of Undersea Surveillance" (Winter 2005) 7 *Undersea Warfare: The Official Magazine of the U.S. Submarine Force*, available online at http://www.navy.mil/navydata/cno/n87/usw/issue_25/sosus.htm and http://www.navy.mil/navydata/cno/n87/usw/issue_25/sosus2.htm (last accessed 20 May 2013). *See also* W.C. Reed, *Red November: Inside the Secret U.S.-Soviet Submarine War* (HarperCollins, 2010) at 16–18, 253–254.

6 "Sound Surveillance System (SOSUS)" GlobalSecurity.org (undated), available online at http://www.globalsecurity.org/intell/systems/sosus.htm (last accessed 20 May 2013). For additional information see National Oceanic and Atmospheric Administration, Vents Program website, http://www.pmel.noaa.gov/vents/acoustics/sosus.html (last accessed 20 May 2013). For other new cabled seafloor ocean observation systems, see the website of NEPTUNE Canada, http://www.neptunecanada.com (last accessed 20 May 2013).

Bilateral Direct Communications (Hotline) Links

In 1949 the United States and the former Soviet Union entered into an agreement regarding the organization of commercial radio multiplex teletypewriter communication channels via a radio relay station at Tangier, to replace the existing military teletypewriter channels.[7]

The latter hotline links described below are reserved for the use of Heads of Government, although they are often operated by military personnel.

In 1963 the two countries signed a Memorandum of Understanding (MOU) regarding the establishment of a direct communications link between Washington, DC and Moscow. The direct communications link consisted of one full-time duplex wire telegraph circuit, routed Washington-London-Copenhagen-Stockholm-Helsinki-Moscow and used for the transmission of messages between Heads of Government. It also consisted of a full-time duplex radio telegraph circuit and the necessary telegraph-teleprinter equipment. The telegraph circuits of the link and of the terminal equipment were to be in conformance with the recommendations of the International Telegraph and Telephone Consultative Committee and the International Radio Consultative Committee.[8] With the advent of secure satellite communications, this 'hotlink' MOU was modified in 1971 to provide for the use of satellite circuits.[9] The direct communication link was further modified in 1984 to add facsimile transmission capability to the 'Hot Line'.[10]

---

7 Agreement on the Organization of Commercial Radio Teletype Communication Channels Between the United States of America and the Union of Soviet Socialist Republics, signed at Moscow 24 May 1946, 60 Stat 1606, TIAS 1527, 11 Bevans 1291, 4 UNTS 201 (entered into force 24 May 1946), available online at http://treaties.un.org/doc/Publication/UNTS/Volume%204/v4.pdf (last accessed 20 May 2013).

8 Memorandum of Understanding (with Annex) regarding the establishment of a direct communications link, signed at Geneva 20 June 1963, 14 UST 825, TIAS 5362, 472 UNTS 164 (entered into force 20 June 1963), available online at http://treaties.un.org/doc/Publication/UNTS/Volume%20472/volume-472-I-6839-English.pdf (last accessed 18 March 2013). *See also* http://www.state.gov/t/isn/4785.htm for background and the text of the MOU and Annex (last accessed 20 May 2013).

9 Agreement supplementing and modifying the above-mentioned [1963] Agreement on measures to improve the USA-USSR direct communications link (with annex), signed at Washington 30 September 1971, 22 UST 1598, TIAS 7187, 806 UNTS 402 (entered into force 30 September 1971), available online at http://treaties.un.org/doc/Publication/UNTS/Volume%20806/volume-806-I-6839-English.pdf (last accessed 20 May 2013). *See also* http://www.state.gov/t/isn/4787.htm for background and the text of the agreement and annex (last accessed 20 May 2013).

10 Agreement Between the United States of America and the Union of Soviet Socialist Republics to Expand the U.S.-USSR Direct Communications Link, signed at Washington 17 July 1984, TIAS 11428 (entered into force 17 July 1984), available online at http://www.state.gov/t/isn/4786.htm for background and the text of the agreement (last accessed 20 May 2013). Although registered with the UN, the text was not published in UNTS. 2193 UNTS 51.

Figure 15.1 The USNS *Zeus*, a cable laying/repair ship operated by the United States Navy Military Sealift Command; it transports, deploys, retrieves and repairs undersea cables. Built specifically for the Navy, the USNS *Zeus* can lay up to 1000 miles of cable in depths up to 9000 feet during a single deployment before having to restock its cable supply. (Photography courtesy of United States Navy Military Sealift Command)

The communications link was utilized on a number of occasions, such as during the Arab-Israeli wars of 1967 and 1973; however the cable circuits are now no longer in use.[11]

In 1967 the Soviet Union and the United Kingdom entered into an agreement for the establishment of a direct communication link between the residence of the Prime Minister of the United Kingdom in London and the Kremlin. The terms of the agreement were substantially similar to the 1963 US-USSR agreement described above, except that the link was routed UK-Holland-Denmark-

[11] The agreement was further amended by exchange of diplomatic notes in Washington, DC, on 24 June 1988, providing, *inter alia*, for ceasing the cable transmission circuits employed for the existing teletype system after an improved disk cypher text message was operational. M.M. Whiteman, *Cumulative Digest of United States Practice in International Law 1981–1988* (US Government Printing Office, 1995) vol II, at 2950 (Cumulative Digest). The Nuclear Risk Reduction Centers rely on satellite communications. Protocol II Article 1 of the Agreement Between the Union of Soviet Socialist Republics and the United States of America on the establishment of nuclear risk reduction centers (with protocols), signed at Washington 15 September 1987, 1530 UNTS 386 (entered into force 15 September 1987), available online at http://treaties.un.org/doc/Publication/UNTS/Volume%201530/volume-1530-I-26557-English.pdf (last accessed 20 May 2013). For further information on the United States Nuclear Risk Reduction Center *see* http://www.state.gov/documents/organization/199776.pdf (last accessed 20 May 2013).

Poland-USSR.[12] In 1987 the Soviet Union and the United Kingdom entered into an agreement to improve the direct communications link between the residence of the Prime Minister of the United Kingdom in London and the Kremlin. The terms of this agreement were substantially the same as the 1971 US-USSR agreement described above.[13]

II. International Conventions Governing Military Uses of, and/or Military-Owned/Leased, Submarine Cables

Neither the 1884 Cable Convention,[14] the 1958 Geneva Conventions on the Continental Shelf[15] and on the High Seas,[16] nor the 1982 UN Convention on the Law of the Sea[17] (UNCLOS) address military uses of submarine cables. Rather, these instruments address the rights and duties of States to lay, maintain, repair and protect *all* submarine cables (as discussed in Chapter 3). There is nothing to suggest that military cables should be afforded different treatment. The rights and obligations of the owners and lessors of military cables, and of the users of submarine cables for military purposes, are the same as those of private natural or juridical entities and other governmental entities. Similarly, the obligations for protection of submarine cables and penalties for damage or injury, as set out in Articles 113, 114 and 115 of UNCLOS, would also apply to military cables.

---

12 Agreement concerning the establishment of a direct communication link between the Residence of the Prime Minister of the United Kingdom in London and the Kremlin, signed in London 25 August 1967, 632 UNTS 49 (entered into force 25 August 1967), available online at http://treaties.un.org/doc/Publication/UNTS/Volume%20632/volume-632-I-9007-English.pdf (last accessed 28 January 2013).

13 Agreement on the improvement of the direct communication link between the Residence of the Prime Minister of the United Kingdom in London and the Kremlin, signed in Moscow 31 March 1987, 1655 UNTS 393 (entered into force 31 March 1987), available online at http://treaties.un.org/doc/Publication/UNTS/Volume%201655/v1655.pdf (last accessed 20 May 2013).

14 Convention on the Protection of Submarine Telegraph Cables, adopted 14 March 1884, 24 Stat 989, TS No 380, as amended 25 Stat 1414, TS Nos. 380-1 and 380-2, 380-3, 1 Bevans 89, 112, 114 (entered into force 1 May 1888), full text in Appendix 3, available online at http://cil.nus.edu.sg/wp/wp-content/uploads/2009/10/Convention_on_Protection_of_Cables_1884.pdf (last accessed 20 May 2013).

15 Convention on the Continental Shelf, adopted 29 April 1958, 499 UNTS 311 (entered into force 10 June 1964), available online at http://untreaty.un.org/ilc/texts/instruments/english/conventions/8_1_1958_continental_shelf.pdf (last accessed 20 May 2013).

16 Convention on the High Seas, adopted 29 April 1958, 450 UNTS 82 (entered into force 30 September 1962), available online at http://untreaty.un.org/ilc/texts/instruments/english/conventions/8_1_1958_high_seas.pdf (last accessed 20 May 2013).

17 United Nations Convention on the Law of the Sea, adopted 10 December 1982, 1833 UNTS 3 (entered into force 16 November 1994), available online at http://www.un.org/Depts/los/convention_agreements/texts/unclos/unclos_e.pdf (last accessed 20 May 2013).

Arguably, one critical issue that may arise is whether the use of military cables is consistent with the legal regime governing the Exclusive Economic Zone (EEZ) and continental shelf. This is part of the larger debate on whether 'military activities' are permitted in these maritime spaces.[18] Given that the right to lay, repair and maintain submarine cables is a freedom afforded to all States in the EEZ and continental shelf,[19] the question remains whether there is a similar freedom to lay military cables or whether the laying of such cables would be subject to the consent of the coastal State. The answer to this will, to a large extent, depend on whether military activities in the EEZ and/or the continental shelf are permitted.

It is worth noting that the treatise on the law of the sea written by Churchill and Lowe seems to be the only modern treatise to note disagreements on the compatibility of the laying of SOSUS cables on the continental shelf with the various provisions of the law of the sea. Those authors did not take a position on that issue, nor did they mention other military uses of submarine cables.[20] The Virginia Commentary on UNCLOS does not address military uses of submarine cables.[21] In addressing military uses of the oceans, other treatises on the law of the sea make no mention of military uses of submarine cables.[22]

---

18 For a good summary of the debate between maritime powers such as the United States and coastal States such as China; see R. Pedrozo, "Preserving Navigational Rights and Freedoms: The Right to Conduct Military Activities in China's Exclusive Economic Zone" (2010) 9 *Chinese Journal of International Law* at 23 and Zhang Haiwen, "Is it Safeguarding the Freedom of Navigation or Maritime Hegemony of the United States? Comments on Raul (Pete) Pedrozo's Article on Military Activities in the EEZ" (2010) 9 *Chinese Journal of International Law* at 43.

19 UNCLOS, Art 58(1) and Art 79.

20 R.R. Churchill and A.V. Lowe, *The Law of the Sea* (3rd ed., Manchester University Press, 1999) at 427–428 (summarizing various views on the compatibility of SOSUS cables laid on the continental shelf with various provisions of UNCLOS). *See also* T. Mensah, "The Rights of Submarine Cable Owners and Coastal States Under the United Nations Convention on the Law of the Sea" (2000) 114 *Maritime Studies* 10. The Annotated Supplement to the Commander's Handbook on the Law of Naval Operations concludes that "SOSUS arrays can be lawfully laid on other nations' continental shelves beyond the territorial sea without notice or approval." A.R. Thomas and J.C. Duncan, eds., *Annotated Supplement to The Commander's Handbook on the Law of Naval Operations*, International Law Studies, vol 73 (US Naval War College Press, 1999) at 131 note 64, available online at http://www.usnwc.edu/Research---Gaming/International-Law/New-International-Law-Studies-(Blue-Book)-Series/International-Law-Blue-Book-Articles.aspx?Volume=73 (last accessed 20 May 2013).

21 S. Nandan and S. Rosenne, eds., *United Nations Convention on the Law of the Sea 1982: A Commentary*, (Martinus Nijhoff, 1993) Volume II, 564 para 58.10(b), 908–917 (EEZ, continental shelf Art 79); and Volume III (1995) (discussing high seas Arts 87, 112–115) (Va Commentary).

22 D.R. Rothwell and T. Stephens, *The International Law of the Sea* (Hart Publishing, 2010) at chapter 12; D.P. O'Connell, *The International Law of the Sea* (I.A. Shearer ed.,

One of the common arguments raised in support of the position that military activities, including the use of military cables, are prohibited in the EEZ/continental shelf of a coastal State is that these activities are contrary to the 'peaceful uses of the sea'. There are several potentially relevant provisions in UNCLOS that address 'peaceful uses of the seas' or 'peaceful purposes': Articles 88, 141, 143(1), 147(2)(d), 155(2), 240(a), 242(1), 246(3) and 301. The meaning of those terms can be understood by tracing the negotiating history and analysis of the Convention.

During the general debate on the 'peaceful uses of ocean space: zones of peace and security' at the fourth session, 1976, of Third Conference on the Law of the Sea the representative of Ecuador noted that the basis for understanding the term 'peaceful uses' had been laid down in 1970 in a General Assembly resolution[23] and further refined since then. He stated:

> ... [T]he use of the ocean space for exclusively peaceful purposes must mean complete demilitarization and the exclusion from it of all military activities.[24]

In response, the representative of the United States stated:

> The term "peaceful purposes" did not, of course, preclude military activities generally. The United States had consistently held that the conduct of military activities for peaceful purposes was in full accord with the Charter of the United Nations and with the principles of international law. Any specific limitation on military activities would require the negotiation of a detailed arms control agreement. The Conference was not charged with such a purpose and was not prepared for such negotiation. Any attempt to turn the Conference's attention to such a complex task could quickly bring to an end current efforts to negotiate a law of the sea convention.[25]

On 7 December 1982, at the plenary meeting of the eleventh session of UNCLOS negotiations, the representative for Brazil stated:

> 26.... [T]he Convention on the Law of the Sea is much less explicit concerning the security interests of the coastal State in the area between 12 and 200 miles. It was impossible to overcome the intransigence of the major naval Powers. As a result of the basic rule of consensus adopted by this Conference, gaps and ambiguities remain in the text of the Convention. However, these problems can be solved by resorting

---

Clarendon Press, 1982); M.S. McDougal and W.T. Burke, *The Public Order of the Oceans* (Yale University Press, 1962) at 704–706, 733–734, 778–782.

23 Declaration of Principles Governing the Sea-Bed and the Ocean Floor, and the Subsoil Thereof, beyond the Limits of National Jurisdiction, adopted by General Assembly Resolution 2749 (XXV), available online at http://www.un.org/ga/search/view_doc.asp?symbol=A/RES/2749%28XXV%29&Lang=E&Area=RESOLUTION.

24 Volume 5, *Official Records of the Third United Nations Conference on the Law of the Sea*, Fourth Session, 67th Plenary Meeting (1976), Document: A/CONF.62/SR.67, para 2 at 56, available online at http://untreaty.un.org/cod/diplomaticconferences/lawofthe-sea-1982/docs/vol_V/a_conf-62_sr-67.pdf (last accessed 20 May 2013).

25 *Ibid.*, para 81 at 62.

> to the option defined in article 310 of the Convention, which allows formal declarations at the time of signature, ratification or adherence, "with a view, *inter alia*, to the harmonization of [national] laws and regulations with the provisions of this Convention".
>
> ....
>
> 28. In the first place, it is our understanding that the provisions of article 301, which prohibit the threat or use of force on the sea against the territorial integrity or independence of any State, apply particularly to the maritime areas under the sovereignty or jurisdiction of the coastal State. In other words, we understand that the navigation facilities accorded third world countries within the exclusive economic zone cannot in any way be utilized for activities that imply the threat or use of force against the coastal State. More specifically, it is Brazil's understanding that the provisions of the Convention do not authorize other States to carry out military exercises or manoeuvres within the exclusive economic zone, particularly when these activities involve the use of weapons or explosives, without the prior knowledge and consent of the coastal State.[26]

In exercising its right of reply, the United States stated:

> Military operations, exercises and activities have always been regarded as internationally lawful uses of the sea. The right to conduct such activities will continue to be enjoyed by all States in the exclusive economic zone. This is the import of article 58 of the Convention. Moreover, Parts XII and XIII of the Convention have no bearing on such activities.[27]

A 1985 report of the UN Secretary-General concluded that:

> ... military activities which are consistent with the principles of international law embodied in the Charter of the United Nations, in particular with Article 2, paragraph 4, and Article 51, are not prohibited by the Convention on the Law of the Sea.[28]

The correct situation is thus summarized in a US Commentary accompanying the transmittal of UNCLOS to the US Senate for consideration in 1994, which noted:

---

26 *Official Records of the Third United Nations Conference on the Law of the Sea*, Volume XVII, 187th Plenary Meeting (1982), paras 26 and 28 at 40, available at http://untreaty.un.org/cod/diplomaticconferences/lawofthesea-1982/docs/vol_XVII/a_conf-62_sr-187.pdf (last accessed 20 May 2013).

27 *Official Records of the Third United Nations Conference on the Law of the Sea*, Volume XVII, UN Doc. A/CONF.62/WS/37 and Add. 1 and 2, Note by the Secretariat, at 244, available at http://untreaty.un.org/cod/diplomaticconferences/lawofthesea-1982/docs/vol_XVII/a_conf-62_ws_37%20and%20add-1%20and%202.pdf (last accessed 20 May 2013).

28 *General and Complete Disarmament—Study on the Naval Arms Race*, Report of the Secretary-General, UN doc. A/40/535, para 188, available at http://www.un.org/disarmament/HomePage/ODAPublications/DisarmamentStudySeries/PDF/SS-16.pdf (last accessed 20 May 2013), quoted in *Va Commentary* Volume III, para 88.7(c) at 91, *supra* note 21. See also B.H. Oxman, "The Regime of Warships Under the United Nations Convention on the Law of the Sea" (1984) 24 *Virginia Journal of International Law* 809–863, at 829–832.

> Article 301 reaffirms that all States Parties, whether coastal or flag States, in exercising their rights and performing their duties under the Convention with respect to all parts of the sea, must comply with their duty under article 2(4) of the United Nations Charter to refrain from the threat or use of force against the territorial integrity or political independence of any States.
>
> Other provisions of the Convention echo this requirement. Article 88 reserves the high seas for peaceful purposes, while articles 141 and 155(2) reserves the Area for peaceful purposes. Under articles 143(1), 147(2)(d), 240(a), 242(1) and 246(3), marine scientific research is required to be conducted for peaceful purposes.
>
> None of these provisions creates new rights or obligations, imposes restraints upon military operations, or impairs the inherent right of self-defense, enshrined in article 51 of the United Nations Charter. More generally, military activities which are consistent with the principles of international law are not prohibited by these, or any other, provisions of the Convention.[29]

A 2007 executive report of the US Senate Foreign Relations Committee on UNCLOS stated that:

> The first understanding states that nothing in the Convention impairs the inherent right of self-defense or rights during armed conflict, including Convention provisions that refer to "peaceful uses" or "peaceful purposes." This understanding, which is a statement of fact, underscores the importance the United States attaches to its right under international law to take appropriate actions in self-defense or in times of armed conflict, including, where necessary, the use of force.[30]

The Committee on Foreign Relations recommended that upon accession to UNCLOS the US append an interpretative declaration, which would set out the US understanding of matters contained in the Convention. With regard to the term 'peaceful purposes' the committee proposed the following understanding:

> The United States understands that nothing in the Convention, including any provisions referring to "peaceful uses" or "peaceful purposes," impairs the inherent right of individual or collective self-defense or rights during armed conflict.[31]

In depositing its instrument of ratification of the Convention in 1996 the Netherlands stated:

---

29 Letter of Submittal accompanying Transmittal from the President of the United States to the United States Senate, 7 October 1994, available online at http://www.jag.navy.mil/organization/documents/Senate_Transmittal.pdf. For further discussion of the term 'peaceful purposes' during the UNCLOS negotiations, see also *Va Commentary*, Volume III, *supra* note 21, at 90–92; Volume IV *ibid.* 461; Volume V *ibid.* 153–155; and Volume VI *ibid.* 149, 171.

30 Executive Report of the US Committee on Foreign Relations 110–9, 110th Congress, 1st Session, Convention on the Law of the Sea 19 December 2007, at 11, available through a link at http://www.foreign.senate.gov/treaties/details/103-39 (last accessed 14 November 2012).

31 *Ibid.* at 19.

> Article 301 must be interpreted, in accordance with the Charter of the United Nations, as applying to the territory and territorial sea of a coastal State.[32]

Accordingly, based on the above, there is no prohibition against military activities generally in the EEZ/continental shelf of a coastal State and consequently, the use of submarine cables for military purposes is consistent with international law as reflected in UNCLOS.

III. Analogous Communications Treaties

Apart from the rights and obligations provided for in UNCLOS, there are also other international conventions which may regulate the use of military cables. For example, the Constitution of the International Telecommunication Union (ITU) provides in Article 48:

> **Installations for National Defence Services**
>
> 1. Member States retain their entire freedom with regard to military radio installations.
> 2. Nevertheless, these installations must, so far as possible, observe statutory provisions relative to giving assistance in case of distress and to the measures to be taken to prevent harmful interference, and the provisions of the Administrative Regulations concerning the types of emission and the frequencies to be used, according to the nature of the service performed by such installations.
> 3. Moreover, when these installations take part in the service of public correspondence or other services governed by the Administrative Regulations, they must, in general, comply with the regulatory provisions for the conduct of such services.[33]

Satellite Communications

'Radio installations' now include satellite communications systems, to which the same caveat applies.[34] Militaries now use military and commercial satellites for many purposes, including communications.[35]

---

[32] United Nations, *Multilateral Treaties Deposited with the Secretary-General: Status*, available online at http://treaties.un.org/pages/Participation/Status.aspx (last accessed 13 November 2012).

[33] Available online at http://www.itu.int/net/about/basic-texts/constitution/chaptervii.aspx (last accessed 20 May 2013). Earlier versions of this reservation may be found in B. Chen, *Studies in International Space Law* (Oxford, 1997) at 97.

[34] Radio Regulations, Chapter I (2008 ed), available online at http://www.itu.int/pub/R-REG-RR-2008 (last accessed 20 May 2013).

[35] See for example, UK Parliamentary Office of Science and Technology, "Military Uses of Space," (POSTnote No 273, December 2006), available online at http://www.parliament.uk/business/publications/research/briefing-papers/POST-PN-273 (last accessed 20 May 2013).

It may be noted that the 1971 Agreement establishing INTELSAT precluded the use of the INTELSAT space segment, i.e. the space segment owned by INTELSAT, for military purposes.[36] The 1982 Treaty establishing EUTELSAT contained a similar prohibition.[37] Nevertheless, and perhaps as a result, commercial satellites serving military customers have flourished.[38]

Conclusion

Submarine cables used for military purposes or owned/leased by the military are as vital to national defence and security as submarine fiber optic cables are for telecommunications. Accordingly, there is every reason to afford military cables the same rights and protections that are given to submarine cables under UNCLOS. As discussed above, the argument that the use of military cables is a 'military activity' that can be prohibited by the coastal State is not consistent with UNCLOS and, hence, should not be used by coastal States as a justification to regulate such cables.

36 Articles III(d) and (e)(iii), Agreement Relating to the International Telecommunications Satellite Organization (INTELSAT), opened for signature 20 August 1971, 23 UST 3813, TIAS 7532, 1220 UNTS 21 (entered into force 12 February 1973), available online at http://treaties.un.org/doc/Publication/UNTS/Volume%201220/volume-1220-I-19677-English.pdf (last accessed 13 November 2012).

37 Article III(f)(iii), Convention Establishing the European Telecommunications Satellite Organization (EUTELSAT), opened for signature 15 July 1982, 1519 UNTS 175 (entered into force 1 September 1985), available online at http://treaties.un.org/doc/Publication/UNTS/Volume%201519/volume-1519-I-26342-English.pdf (last accessed 13 November 2012).

38 See the website of Intelsat General Co, http://www.intelsatgeneral.com/about-us (last accessed 13 March 2013).

CHAPTER SIXTEEN

SUBMARINE CABLES AND OFFSHORE ENERGY

Wayne F. Nielsen and Tara Davenport

Introduction

Unlike submarine fiber optic cables for telecommunications, submarine cable systems for communications and power for offshore oil and gas inter-platform connectivity and for alternative energy in the form of wind farms are relatively new segments to the industry. However, due to the ever-increasing demand for energy, both renewable and non-renewable, there is a great interest in the potential for these types of submarine cables. This Chapter provides an overview of submarine cables used for oil and gas infrastructure and submarine cables used for alternative energy. The issues that will be discussed include the development of these special purpose energy cables, the drivers for the cables, the industry, the legal regime governing these cables and the law and policy challenges that the cables pose.

I. Submarine Cables for Offshore Oil and Gas Infrastructure

The Development of Oil and Gas Submarine Cables

It is estimated that there are more than 6500 offshore oil and gas installations around the world spread across 53 countries in geographically diverse areas.[1] Communications between onshore facilities and offshore oil and gas facilities have historically been a challenge for the oil and gas industry. Offshore oil fields are being located increasingly further from shore and are, as a result, not well

[1] See Energy Systems Research Unit, University of Strathclyde Engineering, available online http://www.esru.strath.ac.uk/EandE/Web_sites/98-9/offshore/platintr.htm (last accessed 1 June 2013).

connected to the telecommunications infrastructures available on land.[2] Further challenges include the fact that there is a limited amount of space for equipment, the potential movement of the structure in bad weather, lack of power and a corrosive environment.[3]

Thirty years ago, communications between offshore facilities and onshore locations was restricted to a two-way radio and daily reports and oilfield workers stationed offshore were virtually cut off from the rest of the world.[4] However, developments in communications technology have meant that the way the offshore industry works has been transformed by improved communications systems.[5] In this regard, there are three technologies utilized by oil and gas companies to facilitate communications, namely, radio frequency such as microwave (including broadband *WiMAX* (Worldwide Interoperability for Microwave Access)), satellite and submarine fiber optic cables.

The use of radio frequency for communications has been widely utilized in the oil and gas industry[6] and microwave telecommunications technology can offer communications at shorter distances, such as 20–25 miles offshore.[7] If communications need to travel over a longer distance, satellite is the next feasible option, and satellite communications require a "very small aperture terminal at the offshore site; a broadband satellite connection in space; and a teleport onshore."[8] However, the development of oil fields in deeper waters and the installation of bigger platforms have necessitated communication technology that could travel over longer distances. This has provided a great impetus to the development of submarine fiber optic cables to connect offshore structures.[9]

Early developments took place in the late 1970s when Offshore Telephone deployed a metallic cable network in the Gulf of Mexico and in the mid-1990s when Petrocom deployed the FiberWeb inter-platform system. Both systems failed miserably and were later abandoned.[10] The first 'successful' submarine cable for offshore oil and gas telecommunications to a platform was installed in the

---

2 G. Arnos, "Bandwidth in the Oil Patch: Requirements and Challenges" (September 2005) Issue 22 *Submarine Telecoms Forum Magazine* 17–20, at 17.

3 G. Berlocher, "Subsea Fiber in the Energy Industry" (September 2009) Issue 46 *Submarine Telecoms Forum Magazine* 11–13, at 11.

4 "How do Offshore Communications Work?" RIGZONE Insight, available online at http://www.rigzone.com/training/insight.asp?i_id=337 (last accessed at 1 June 2013).

5 *Ibid.*

6 Radio Frequency technology includes microwave technology which has been frequently used as a "backbone" for many offshore networks, see Berlocher, *supra* note 3 at 11.

7 *Ibid.*

8 "How do Offshore Communications Work?" RIGZONE Insight, *supra* note 4.

9 R. Munier and K. Haaland, "BP GoM: Next Generation Offshore Fiber" Ocean News and Technology (date not indicated) available online at http://www.subcom.com/pdfs/articles/TT-BP-gom.pdf (last accessed 1 June 2013).

10 See Berlocher, *supra* note 3, at 12.

early 1990s in the North Sea for BP. The offshore fiber cable market segment began to grow in the late 1990s and early 2000s.[11] In 1998, PetroBras developed a 451 km offshore platform cable system in the Campos Basin. In 2001, BP developed the Central North Sea Fiber Optic Cable, a 300 km submarine cable and microwave system serving 16 platforms, being the first fiber optic cable that linked the Scottish mainland with offshore platforms. In 2002, Saudi Aramco developed the 250 km Offshore Fiber Optic Cable System. During this period, the number of platforms connected by fiber only grew by a few, presumably while owners were still feeling out the utility of fiber.

2005 proved to be a critical year. It was in this year that Hurricanes Dennis, Katrina and Rita seriously affected the platforms in the Gulf of Mexico. The hurricanes impacted platforms and caused serious communications failures, resulting in millions of dollars in lost production. Even platforms not affected by the hurricanes and which were dependent on relays to affected nearshore platforms lost connectivity.[12] For BP, this was a catalyst to invest approximately US$80 million in the BP Gulf of Mexico Fiber Optic Network (BP GoM FON) to provide continuous connectivity to the company's deepwater platforms.[13]

As the technology was tested and became apparently more reliable, new systems added a 25 per cent increase to lit platforms worldwide. Offshore oil and gas systems use submarine fiber optic cables to link onshore oil and gas facilities to a variety of assets, including conventional fixed platforms, compliant towers, vertically moored tension leg and mini-tension leg platforms, spars, semi-submersibles, floating production, storage, sub-sea completion and tie-backs, and offloading facilities. Oil and gas operators implement digital systems for their facilities, which include real-time monitoring and sensors, collaboration, video surveillance, work management systems and other applications that require continuous connectivity.

The low latency and reliability of fiber has proven to be an asset to platform owners, as reflected by the steady increase in total kilometers of fiber added per year.[14] The demand for fiber doubled in 2008, almost 10 years after fiber started to become widely accepted.[15] In 2001, the total length of all fiber cable in use

---

11 S. Jarvis, "Important and Necessary: The Rising Requirement of Oil" (September 2012) Issue 65 *Submarine Telecoms Forum Magazine* 7–14, at 7.

12 Munier and Haaland, *supra* note 9 at 45.

13 D. Paganie, "BP Leverages Technology to Optimize Deepwater Performance" Offshore Magazine (date not indicated) available online at http://www.offshore-mag.com/articles/print/volume-67/issue-8/e-technology/bp-leverages-technology-to-optimize-deepwater-performance.html (last accessed 1 June 2013).

14 Jarvis, *supra* note 11 at 7.

15 *Ibid.*

was less than 1800 km, by 2020 that number is anticipated to increase over 550 per cent.[16]

Despite the recent push for more renewable energy sources, the need for oil has increased, thereby increasing the demand for submarine fiber optic cables for communications.[17] The total of fiber-connected platforms is a good indicator of the market potential for submarine fiber growth in this niche market. Looking ahead, the number of fiber-connected platforms will double by 2020, showing a huge uptick in fiber demand in 2012 and 2017. As the applications for fiber expand, the total number of systems and the required fiber is expected to grow exponentially. According to one report:

> As new systems are deployed, it's important to note the regions of growth and where platform owners are focusing their assets. In the past, owners have focused their attention on traditional oil and natural gas fields in the North Sea and Gulf of Mexico; seeing where their new systems are being planned, it is apparent that owners are now expanding their fiber operations into other developing regions. Most predominately, new systems are being planned for South Asia and Australia, showing that more non-traditional owners are embracing the submarine fiber solution.[18]

Similarly, "planning for fiber communications in West Africa's offshore industry has moved past the concept stage."[19] Assuming that oil and gas companies invest only in offshore telecommunications for new or relatively new projects, an estimated 100 projects have the potential for submarine system development in the ten-year timeframe of 2006 to 2016.[20] Given that the average offshore submarine telecommunications system is valued at approximately $50M, a sizable market in the future exists.

Drivers of Oil and Gas Submarine Cables

While the type of communication technology required will generally depend on the particular characteristics of the oil field, including its distance from shore and the water depths,[21] there are several drivers propelling the growth of submarine fiber cables in oil and gas installations. Submarine fiber offers a viable alternative to both radio/microwave and satellite. While microwave is capable of delivering bandwidth, the primary drawback is its limitation of 40 kilometers between sites. Many assets are hundreds of kilometers from the beach, and those in deep water are floating and movement makes their alignment for telecommunications

---

16 *Ibid.* at 8.

17 Berlocher, *supra* note 3, at 12.

18 Jarvis *supra* note 11 at 7.

19 S. Lentz, "Fibre Installations: Build a Subsea Connection Point" (September 2011) Issue 59, *Submarine Telecoms Forum Magazine* 19–22, at 21.

20 M. Cleveland, "The State of the Offshore O&G Market" (September 2011) Issue 59 *Submarine Telecoms Forum Magazine* 6–10, at 9.

21 Berlocher, *supra* note 3 at 13.

difficult to maintain. Finding and maintaining suitable repeater stations for the life of the offshore asset is nearly impossible. Similarly, bandwidth provided by satellite is expensive and is vulnerable to loss of signal in bad weather.[22]

Submarine fiber offers distinct advantages. First, it provides low latency and virtually limitless amounts of bandwidth. Fiber can connect distant points from the beach without relying on other structures.[23] Second, it reduces personnel on board and enhances control of offshore systems by onshore facilities thereby reducing operating costs and personnel risks.[24] The general offshore business conditions are evolving. Work on a platform can be dangerous, and health, safety and environmental concerns are a growing issue.[25] The speed of business has escalated, requiring collaboration and decision-making on a quicker timeline.[26] In addition, a younger workforce is less inclined to work offshore for several weeks at a time, making it a challenge for energy companies to locate skilled labor.[27] Berlocher notes that:

> All of these issues are being addressed by new initiatives, such as Field of the Future® (BP) iField® (Chevron), and Smart Fields® (Shell). Using a different paradigm to manage new fields, these companies have created centers of excellence onshore and make use of telecommunication and computing power to improve the management of offshore production, and to reduce risks. BP's Thunderhorse platform is a good example. Personnel on the platform collaborate with a dedicated team in West Houston, making decisions together, with multiple stakeholders involved.[28]

BP has reportedly said that this connectivity "is the grease that drives productivity at BP" and in 2011, observed that an additional 80,000 barrels a day of extra production is a result of using this technology coupled with a saving of more than $100M of capital expenditure.[29]

The third major advantage is the fact that fiber cables are relatively unaffected by violent storms as they are installed on the seafloor.[30] They are able to survive storms "leading to faster re-manning and reduced downtime after abandonment due to weather."[31]

---

22 Berlocher, *ibid.* at 11.

23 *Ibid.*

24 G. Arnos, "Design Challenges for Undersea Systems Serving Offshore Production Platforms" (September 2007) Issue 34 *Submarine Telecoms Forum: Offshore Oil and Gas Telecoms Issue.*

25 Berlocher, *supra* note 3, at 12.

26 *Ibid.*

27 *Ibid.*

28 *Ibid.*

29 D. Latin, "BP—Using Connectivity to Drive Productivity" (September 2009) Issue 46 *Submarine Telecoms Forum Magazine* 22–23, at 22.

30 Berlocher, *supra* note 3 at 12.

31 Lentz, *supra* note 19 at 21.

The advantages of fiber communications are well illustrated by the 1216 km BP GoM FON developed to connect eight platforms in the deep waters of Gulf of Mexico discussed above. Unlike past submarine networks in this basin, BP's fiber was plowed to a water depth of 1000 meters, thereby minimizing the chance of a disruption caused by anchorage or fishing trawls.[32] Should the system be broken, data could flow in either direction back to the beach.[33] Since the system utilizes submarine repeaters, there was no reliance on other platforms to regenerate fiber optic signals.[34] The system allows platforms to be monitored remotely when storms necessitate the evacuation of crew; the reduced manning requirements also cut down on helicopter flights because centralized management are located in shore command centers; and it is more reliable than both RF and satellite, which go offline in high winds and storms. For example, if one platform is affected by a hurricane, or if a drillship drags an anchor across one of the fiber cables and breaks it, communication would continue at the other platforms.[35] As stated by Otto and Nielsen:

> Fiber is clearly the lead technology as a result of its ability to reach nearly unlimited distances using submarine repeaters, near unlimited bandwidth, and tolerance to poor weather conditions, including rain fade and wind damage.[36]

Installation of Submarine Cables for Oil and Gas Installations

The installation of submarine fiber optic cables faces unique challenges not faced in the installation of telecommunications cables, stemming from the issues relating to the final connection to the platform.[37]

For platforms that have already been built, long distance communications cables can be connected to existing or newly installed risers. For example, for BP's GoM FON, TE Subcom was commissioned to build and install the fiber optic loop and they, in turn, sub-contracted the manufacture and supply of the system to Ocean Design Inc, which specializes in the supply of subsea, wet-mate, fiber-optic connection systems to the offshore oil and gas industry.[38] The system was buried from shore out to 800 meters and was connected to a newly installed fiber optic cable riser. TE SubCom's cableship was able to operate inside the 500 meter

---

32 *Ibid.*

33 *Ibid.*

34 *Ibid.*

35 Paganie, *supra* note 13.

36 G. Otto and W. Nielsen, "Drivers and Technologies for Next Generation Digital Connectivity in Offshore Oil and Gas Production Facilities" (September 2007) Issue 34 *Submarine Telecoms Forum* at 14.

37 Lentz, *supra* note 19 at 19.

38 "ODI Awarded Contract by Tyco Telecommunications for Subsea Fiber Optic Connection System for Gulf of Mexico Project" News Report, Teledyne Technologies available at http://www.teledyne.com/news/odi110706.asp (last accessed 7 June 2013).

safety zone for the platform and used its remotely operated vehicle (ROV) to make the wet-mate connections on the seafloor.[39]

Alternatively, in many cases, a long distance communications cable can be connected to a subsea umbilical which means that the need for new cable risers and cableship operations close to the platform are avoided.[40]

In a new offshore installation, communications fibers are installed and terminate a few kilometres from the platform. A cable termination module with subsea connectors is installed on the platform and the cable-laying vessel will use an ROV to connect the fibers to the platform, also ensuring that work is performed at a safe distance without interfering with platform operations.[41]

It is clear from the above, that the installation of fiber optic cables for the oil and gas industry requires more upfront engineering, higher capital requirements, longer lead-times, and lacks the mobility of other communications technologies. The cost in particular can be prohibitive, amounting to millions of dollars. These costs include "mobilization, shore stations, landings, transmissions electronics, project management costs."[42] However, the expenses ultimately saved in terms of reduced personnel and enhanced productivity may more than make up for the increased costs in the beginning of the project.

An Overview of Industry

The Typical Oil and Gas Company

Government-owned firms or entities control the vast majority of the world's oil reserves. According to the Wall Street Journal, multinational oil companies produce just 10 per cent of the world's oil and gas, while state-owned companies control more than 75 per cent of all crude oil production:

> [T]he 13 largest energy companies on Earth, measured by the reserves they control, are now owned and operated by governments. Saudi Aramco, Gazprom (Russia), China National Petroleum Corp., National Iranian Oil Co., Petróleos de Venezuela, Petrobras (Brazil) and Petronas (Malaysia) are all larger than ExxonMobil, the largest of the multinationals.[43]

Whether government-owned or multinational, the 'typical' oil company is driven solely by the need to accomplish extraction and production of a hydrocarbon commodity at the most efficient rate practicable. Telecommunication cables are

---

39 Munier and Haaland, *supra* note 9, at 45.

40 Lentz, *supra* note 19 at 19.

41 *Ibid.*

42 *Ibid.* at 20.

43 See I. Bremmer, "*The Long Shadow of the Visible Hand: Government-owned firms control most of the world's oil reserves. Why the power of the state is back*" Wall Street Journal, 22 May 2010, available online at http://online.wsj.com/article/SB100014240527487048 52004575258541875590852.html (last accessed 26 February 2013).

generally viewed as a means to that end and not an end in themselves. Unlike the engineering function of a telecommunications provider, the telecommunication arm of an oil company is generally a support function, subservient to production; extraction is the primary role of an oil and gas company. However, according to Otto and Nielsen:

> The implementation of digital connectivity for offshore production facilities using fiber optics is increasingly becoming a strategic initiative within the oil industry. A successful implementation is defined by extended reliability and sustained access to the high capacity digital infrastructure.[44]

As telecommunications become more strategic, so too will their utility and usefulness be realized by oil and gas companies.

Project Ownership Models

There is a need to give careful consideration to the ownership and operating model for a submarine cable that serves the oil and gas sector. While there are a multitude of models and variations to any option provided, Otto and Nielsen cite the following owner and consortium options available:

- Dedicated system that is owned and operated by the dominant end-user company, i.e. the oil and gas company. This is typically employed when a user requires accessing fiber in a timely and long term manner, which is critical to their business and more ideal alternative models are not readily available.
- Level of guaranteed ownership and access rights. This is useful when there are multiple company consumers desiring fiber connectivity, but there is limited time to build a multi-company agreement or consortium.[45]

The guaranteed ownership/access option is useful when future assets locations are questionable, or there is a desire for a level of guaranteed ownership and access rights. Each of the dominant companies possess a cable with a shore landing connected to their facilities. There needs to be either initial agreement on the supplier, or a high level of interoperability to tie the systems together in order for this to be successful.[46]

The owner of any direct company ownership model can sell extra connections and excess to third parties directly or through external third parties.[47] Ownership models can be analyzed on a case-by-case basis; they can then evolve, as economics and risk management allow, into joint ownership consortiums, managed

---

44 Otto and Nielsen, *supra* note 36, at 12.
45 *Ibid.* at 15.
46 *Ibid.*
47 *Ibid.*

services or other forms.[48] Any model other than 100 per cent ownership by the operating company poses long term risks.

The International Legal Regime Governing Submarine Cables Used for Oil and Gas Infrastructure

Oil and Gas Resources

The majority of the oil and gas resources found in the seabed of the oceans are under the control and jurisdiction of the coastal State. First, the territorial sea regime gives the coastal State sovereignty over the seabed and subsoil up to a distance of 12 nautical miles (nm).[49] Second, under the exclusive economic zone (EEZ) regime, the coastal State is given sovereign rights for the purpose of exploring and exploiting, conserving and managing the natural non-living resources of the seabed and its subsoil.[50] The continental shelf regime also gives the coastal State sovereign rights over natural non-living resources[51] up to a distance of 200 nm, or to the outer edge of its continental margin if it meets certain geophysical criteria.[52] It is estimated that approximately 87 per cent of the world's known offshore hydrocarbon fields is under coastal State jurisdiction as a result of the EEZ and continental shelf regime.[53]

A critical aspect of the coastal State's sovereign rights over its oil and gas resources is the jurisdiction given to it over artificial islands, installations, and structures that would be utilized for oil and gas exploration and exploitation.[54] Accordingly, in both the EEZ[55] and continental shelf,[56] the coastal State has the exclusive right to construct and to authorize and regulate the construction, operation and use of artificial islands, installations and structures for the purposes provided for in Article 56 of UNCLOS[57] and other economic purposes; as well as installations and structures that may interfere with the exercise of the rights of the coastal State in the zone.[58] Accordingly, all oil and gas platforms and related

---

48 *Ibid.*
49 UNCLOS Art 2.
50 UNCLOS Art 56(1).
51 UNCLOS Art 77.
52 UNCLOS Art 76(1) and 4.
53 R. Churchill and V. Lowe, *The Law of the Sea* (3rd ed. Manchester, 1999) at 162.
54 UNCLOS Art 56(1)(b).
55 UNCLOS Art 60.
56 UNCLOS Art 80.
57 Article 56(1) provides that the coastal State has "sovereign rights for the purpose of exploring and exploiting, conserving and managing the natural resources, whether living or non-living, of the waters superjacent to the seabed and of the seabed and its subsoil, and with regard to other activities for the economic exploitation and exploration of the zone, such as the production of energy from the water, currents and winds."
58 UNCLOS Art 60(1).

infrastructure in the EEZ or continental shelf of a coastal State are under the jurisdiction of the coastal State and are subject to its law and regulations.

Submarine Cables used for Oil and Gas Installations

Submarine cables used for oil and gas installations are serving infrastructure which are under the jurisdiction of the coastal State and, consequently, such submarine cables are also within the jurisdiction of the coastal State. Within the territorial sea, the coastal State has complete jurisdiction over submarine fiber optic cables used to serve oil and gas installations pursuant to its sovereignty over the territorial sea.[59]

For submarine fiber optic cables used to serve oil and gas installations within the EEZ and continental shelf, UNCLOS expressly recognizes that the coastal State has jurisdiction over cables constructed or used in connection with the exploration of its continental shelf or exploitation of its resources or the operations of artificial islands, installations and structures under its jurisdiction.[60] This is in contrast to the general freedom to lay, repair and maintain submarine cables in these maritime zones.[61]

It also appears that the obligations of States Parties to UNCLOS to adopt legislation to protect submarine cables in the EEZ set out in Articles 113–115 also applies to submarine fiber optic cables used for oil and gas installations.[62] Article 113, for example, applies to both submarine cables used for telegraphic or telephonic communications as well high-voltage power cables. Submarine fiber optic cables used for oil and gas installations can be considered to be cables used for "telephonic communications". On the other hand, it could be argued that Articles 113–115 were intended to protect cables that were laid pursuant to the freedom to lay cables in the EEZ/continental shelf recognized in Articles 58 and 79 of UNCLOS. Submarine cables used for oil and gas installations are not installed pursuant to the freedom to lay cables under the EEZ/continental shelf and are under the jurisdiction of the coastal State. On this interpretation, Articles 113–115 of UNCLOS would not apply. But the proven practical utility of Articles 113–115 and the common sense protections they provide to mariners and cable owners alike make this position difficult to sustain from a public policy perspective. Certainly these cables and pipelines should be charted (see Figure 11.2). However, as will be explained below, this uncertainty may cause issues in the protection of submarine cables used for oil and gas installations.

---

59 UNCLOS Art 2.

60 UNCLOS Art 79(4).

61 UNCLOS expressly recognizes the freedom to lay, repair and maintain cables in the EEZ (Art 58) and on the continental shelf (Art 79). This is dealt with in Chapter 3 on The Overview of the International Legal Regime Governing Submarine Cables and Chapter 5 on the Manufacture and Laying of Submarine Cables.

62 The Protection of Submarine Cables from Competing Uses is dealt with in Chapter 11.

Law and Policy Challenges for Submarine Cables used in Oil and Gas Installations

The main challenge for the cable industry, the oil and gas industry and governments is the need to balance competing uses of the seabed and water column. First, there is the issue of balancing competing uses during the installation of submarine cables used for oil and gas facilities. These submarine cables will inevitably have to cross pipelines already in existence. Careful consideration will have to be paid to such pipelines, especially in light of the underlying principle in Article 114 that owners of a submarine cable in the EEZ or high seas shall be liable for the cost of repair of injury to another submarine cable or pipeline that occurred in the course of laying and repair operations. To address this, during installation, companies carrying out installation of oil and gas submarine cables will lay "concrete mattresses" on top of the cable to ensure its protection.[63] (Figure 13.6.) The International Cable Protection Committee (ICPC) has also issued "Telecommunications Cable and Oil Pipeline /Power Cables Crossing Criteria."[64] While not a legal requirement under international law, it is highly recommended that the relevant parties consult in the initial stages of planning and negotiate an agreement or arrangement to cover any pipeline/cable crossing which covers a range of issues, including *inter alia*, notifications, accurate position information, scheduling, an exchange of observers from both the crossed and crossing systems, and the agreed engineering process and safeguards for the crossing. In some cases, the agreement may be voluntarily extended to the rights and liabilities of both parties, the exclusion of consequential losses, definition of the specific area where the crossing will occur, future maintenance of pipelines and cables and expiry of the agreement.[65]

The second issue is the potential for conflict between submarine cables used for telecommunications (the laying and repair of which is a freedom of the sea) and oil and gas facilities, which are under the jurisdiction of the coastal State. As noted in a 2012 Submarine Telecoms Industry Report:

> Like Australia, many nation states have, or are in the process of, dividing their EEZ into leasable blocks for Oil & Gas exploration and later production. The challenge in drafting the legislation for the necessary leases and licenses will be to protect the rights of the lessee while maintaining the requirements of UNCLOS to allow

---

63 See P. Boulanger, "Developing submarine cable system for oil and gas applications" Alcatel-Lucent Corporate Blog, 19 December 2012, available online at http://www2.alcatel-lucent.com/blogs/corporate/2012/12/developing-submarine-cable-system-for-oil-and-gas-applications/ (last accessed 1 June 2013).

64 ICPC Recommendation No. 9A, Issued on 26 January 2007.

65 The Crown Estate, "Submarine Cables and Offshore Renewable Energy Installations: Proximity Study" (The Crown Estate, 2012) available online at http://www.thecrownestate.co.uk/media/313713/submarine_cables_and_offshore_renewable_energy_installations_proximity_study.pdf (last accessed 1 June 2013).

> submarine cables free access to cross these blocks and permit maintenance and repair of cables within the blocks, when required.[66]

The challenge referred to in the above passage is often complicated by the fact that the ministries for telecommunications and natural resource development are different and lack jurisdiction over each other's activities with no established means of accommodating mutual interests. This leads to situations where oil and gas leaseholders on occasion take the incorrect position that they control all activities in their EEZ concession and can deny access to cables or demand exorbitant penalties, privileges and security arrangements as a condition of permission to allow a cable to transit the concession. Such energy industry over-reach is a major problem in the Arabian Gulf. If the coastal State lacks this power under UNCLOS, any concession holder certainly lacks such power. There is a need for each party to show due regard (as required by UNCLOS)[67] to the other party's interests, but ultimately, reasonableness and logical safety considerations should be the foundation of any crossing agreement or arrangement.

The third issue is the protection of submarine cables used for oil and gas facilities. Like cables used for general telecommunications purposes, the biggest risk to the system is from "external aggression such as fishing, pipeline crossings, anchors and other materials on the seabed being laid or dragged across the cable."[68] While Articles 113–115 oblige States to adopt legislation to protect submarine cables[69] as explained in the previous section, it is not entirely clear whether these obligations apply to submarine cables used for oil and gas facilities. Many States may not be sure how to protect such infrastructure. For example, the Australian Government has enacted Schedule 3A to the 1997 Telecommunications Act which is intended to address the protection of international and domestic submarine telecommunications cables. At the time of its drafting, legislators did not anticipate the increasing use of submarine cables for oil and gas installations.[70] It is therefore not clear whether this legislation would also apply to the protection of cables used for oil and gas facilities. While the Australian Government has enacted legislation to cover offshore platforms in the 2006 Offshore Petroleum and Greenhouse Gas Storage Act, this also does not cover submarine cables used for oil and gas facilities.[71] It has been observed that:

---

66 "Submarine Cable Industry Report" (July 2012) Issue 1 *Submarine Telecoms Forum* at 30.

67 UNCLOS Arts 56(2) and 58(3).

68 See BP Gulf of Mexico Fiber Optic Network Website available online at http://www.gomfiber.com/location/default.htm (last accessed 1 June 2013).

69 Many States have not implemented their obligations under Arts 113–115 which has undermined the effectiveness of these provisions—see Chapter 11 on the Protection of Submarine Cables from Competing Uses.

70 "Submarine Cable Industry Report" (July 2012) Issue 1 *Submarine Telecoms Forum* at 30.

71 *Ibid.*

> The Australian Government has a dilemma to resolve in the next year or so as to which legislation should be applicable to these types of cables and how this will apply if the cable is owned by an Australian carrier providing a service to the Oil and Gas company or the cable is an integral part of the offshore facility owned and operated by the Oil and Gas Company.[72]

This is likely a dilemma that will be faced by many countries that utilize submarine cables for their oil and gas facilities. The solution is to ensure that domestic legislation implementing Article 113 covers all types of submarine cables including the provision of meaningful penalties for the damage to cables by willful or culpably negligent conduct. There would seem to be no reason why such legislation, which applies to telecommunication and power cables, should also not apply to telecommunications and power cables used to connect offshore energy structures.

II. Submarine Cables for Alternative Energy

Renewable energy i.e. energy that emanates from resources that are continually replenished, is the goal of the alternative energy industry. "Clean, consistent energy that will never run out and, best of all, will forever turn a profit because there must always be machines to produce it and people to produce the machines."[73] It has been estimated that the installed global renewable electricity capacity nearly doubled between 2000 and 2011, although it is still a relatively small portion of total energy supply both globally and in the United States.[74]

Offshore renewable energy includes offshore wind energy, ocean wave energy and ocean current energy, the development of which are all at different stages.[75] Submarine **power** cables[76] are used in offshore wind energy infrastructure, accordingly, an overview of offshore wind farms and how such wind farms utilize submarine power cables, is given below.

---

72 *Ibid.*

73 Jarvis, *supra* note 11 at 7.

74 See National Renewable Energy Laboratory (NREL), US Department of Energy, *2011 Renewable Energy Data Book* available online at www.nrel.gov/docs/fy13osti/54909.pdf (last accessed 7 June 2013).

75 For more information on the various sources of offshore energy, please see the Offshore Renewable Energy Guide at the Bureau of Ocean Energy Management (BOEM) available at http://www.boem.gov/Renewable-Energy-Program/Renewable-Energy-Guide/index.aspx (last accessed 7 June 2013).

76 Chapter 13 deals in more detail with submarine power cables. It should also be borne in mind that submarine power cables used for offshore wind farms may include a fiber optic cable with the power cable to monitor equipment and gather data, but the submarine cables used for offshore wind farms are almost all submarine power cables.

Overview of Offshore Wind Farms

Offshore wind power refers to the construction of wind farms in the oceans to generate electricity from the wind. Countries around the world have begun efforts to create large-scale offshore energy farms using wind turbines to exploit the strong winds that are found over the oceans.[77] Generally, higher wind speeds are available offshore compared to land, which means that offshore wind energy can supply a higher proportion of electricity in comparison.[78]

The first offshore wind farm was installed in Denmark in 1991, that being the 5 megawatts (MW) Vindeby project. Offshore wind grew sporadically through the 1990s with Sweden and the Netherlands adding capacity. Denmark then added 400 MW of offshore wind farm capacity from 2001 to 2003. It is reported that worldwide, there are 4.45 gigawatts (GW)[79] of offshore wind energy installed and an additional 30.44 GW approved.[80] Over 50 projects are operational in coastal waters of countries such as Belgium, Denmark, Germany, Italy, Japan, the United Kingdom, Norway, the Netherlands, Portugal, South Korea, Sweden and others.[81] Europe leads the world in wind farm development. According to the European Wind Energy Association:

> A total of 1,371 offshore turbines have been installed and connect to electricity grids in European waters totaling 3,812.6 megawatts (MW) spread across 53 wind farms in 10 countries. The offshore wind capacity installed by the end of 2011 will produce, in a normal wind year, 14 tera-watt hours (TWh) of electricity, enough to cover 0.4% of the EU's total consumption.[82]

Within Europe, the United Kingdom is the largest market with 2094 megawatts (MW) installed, representing in excess of 50 per cent of all installed offshore wind capacity in Europe. Denmark is second with 857 MW (23 per cent), followed by the Netherlands (247 MW, six per cent), Germany (200 MW, five per cent), Belgium (195, five per cent), Sweden (164, four per cent), Finland (26 MW in near-shore projects) and Ireland 25 MW; Norway and Portugal both have full-scale floating turbines (2.3 MW and 2 MW, respectively).[83]

---

77 See Offshore Wind Energy on BOEM Website, *supra* note 75.

78 *Ibid.*

79 1 gigawatt is equivalent to 1 billion watts of electricity. It has been said that 1 GW of wind power will supply between 225,000 to 300,000 average US homes with power annually: See Offshore Wind Energy on BOEM Website, *supra* note 75.

80 *Ibid.*

81 For a list of offshore wind power projects, both planned and installed, see The Wind Power Website available online at www.thewindpower.net/windfarms_offshore_en.php (last accessed 12 March 2013).

82 B. Hamon, "Survey of Offshore Wind Farm Project in EU and Their Connecting Grid Systems" at Global Energy Network Institute (July 2012) at 5.

83 *Ibid.* at 6.

Figure 16.1 Wind farm at sunset. (Image Courtesy of Global Marine Systems)

In the first half of 2012, 520 MW of new offshore wind capacity was installed in Europe; roughly 80 per cent was located in the Irish Sea and North Sea waters of the United Kingdom, and the rest was built by Belgium, Denmark and Germany.[84] A notable example of an effective wind farm is the Alpha Ventus Wind Farm which was completed by Germany in 2010 and which promised enough energy to power 50,000 homes (220 gigawatts).[85] Despite some initial issues with overheating, Alpha Ventus proved to be very effective.[86] A number of new wind farm projects have since been commenced in other countries,[87] and increasing interest expressed by other states. China, for example, has enacted a Renewable Energy Law that sets targets for its dominant electricity producers to source at least 15 per cent of their energy from wind by 2020 from both offshore and terrestrial assets.

The United States does not have any operational projects yet, but according to the US Bureau of Ocean Energy Management (BOEM), there are thousands of megawatts (MW) in the planning stages.[88]

Wind Farm Submarine Power Cable Systems

All the power generated by wind turbines needs to be transmitted to shore and connected to the power grid.[89] Offshore wind farm systems are linked via submarine power cables to the tower of the wind turbine where electricity

84 M.J. Roney, "Offshore Wind Development Picking Up Pace" (2012) http://www.earth-policy.org/plan_b_updates/2012/update106 (last accessed 12 March 2013).

85 Jarvis *supra* note 11 at 7.

86 *Ibid.*

87 *Ibid.*, at 8.

88 See Offshore Wind Energy on BOEM Website, *supra* note 75.

89 *Ibid.*

Figure 16.2 The Atlantic Wind Connection cable system grid is a planned 54 gigawatt system using a 563 km power cable off the coasts of New Jersey, Maryland, and Virginia that will provide both power along the US east coast and connections to new offshore wind farms. The system links wind farms using submarine cables and voltage conversion stations like that shown in this drawing. (Image courtesy of Atlantic Wind Connection)

is created. The various turbines in a wind farm system then connect by an array of submarine power cables to the step-up transformer where electricity is collected and transferred by an interconnect power cable system and from the substation to the onshore power grid. The amount of cable required depends on many factors, including "how far offshore the project is located, the spacing between turbines, the presence of obstacles that require cables to be routed in certain directions."[90]

Demand for submarine power cables is growing steadily as national governments and regional organizations expand their efforts in offshore renewable power generation, linking remote land masses and interconnecting their national grids.[91] These projects often involve offshore wind farms with submarine connections or power generation.[92] According to Pike Research, as cable technol-

90 *Ibid.*

91 Submarine Electricity Transmission HVDC and HVAC Submarine Power Cables: Supply Constraints, Demand Drivers, Technology Issues, Prominent Projects, Key Industry Players, and Global Market Forecasts, available online at http://www.pikeresearch.com/research/submarine-electricity-transmission (last accessed 7 June 2013).

92 *Ibid.*

ogy advances, more projects are proposed that require longer, deeper and higher-capacity cables:

> Even the most conservative growth models show that the industry will expand rapidly, but its analysis finds that the supply chain will not be capable of meeting the full demand in this growing market. While there are very few companies that are capable of performing each step in the installation process, from surveying to final installation, the tightest bottleneck in this already constrained supply chain is the cable manufacturers. Only a handful of companies are capable of building the complex cables that are required by this market. In order to meet the demands of proposed projects, cable manufacturers will need to increase their production significantly.[93]

During this transition, some projects will likely encounter delays due to the limited supply chain and dependence upon subsidies. Additional market players and increased capacity by existing suppliers will increase the industry's response to an ever-growing demand.[94]

Drivers and Economics of Wind Farms and Submarine Power Cables for Wind Farms

Wind energy, both offshore and onshore has emerged as an attractive solution to the world's energy challenges.[95] There are several drivers that have provided an impetus to the growth of the offshore wind energy and consequently the growth of the industry for submarine power cables for wind farms.[96] First, wind is an infinite resource in that enough wind blows across the globe to cope with the ever increasing electricity demand. Second, there is an increasing concern over remaining fossil fuel reserves, import dependence and security of supply (although new supplies of shale gas and oil make this concern dynamic). Third, concerns about environmental pollution relating to the extraction and transport of such fossil fuels has also been an important factor resulting in an increased focus on offshore wind farms as a renewable energy source. Fourth, the accessibility of shallow coastal waters has allowed the application of larger projects, as opposed to land based designs.

However, there are also certain challenges in the deployment of offshore wind farms, including the high costs of constructing such projects and supporting

93 See "Submarine Electricity Transmission" available online at http://www.prnewswire.com/news-releases/submarine-electricity-transmission-165398756.html (last accessed 7 June 2013).

94 See "Submarine Electricity Transmission" 8 September 2012, available online at http://www.evwind.es/2012/08/09/submarine-electricity-transmission-2/21386 (last accessed 7 June 2013).

95 See "Analyzing Offshore Wind Energy" available online at https://www.asdreports.com/shopexd.asp?id=35969&desc= (last accessed 7 June 2013).

96 *Ibid.*

facilities as well as demanding technical systems.[97] As a result, over 70 governments offered financial incentives which have also contributed to the growing interest in offshore wind energy.[98] For example, it has been said that the tailored financial incentives provided by European Union (EU) member States and the EU itself has significantly contributed to an additional 577 megawatts of added offshore wind capacity in 2009.[99] Similarly, in a 2012 New York Times article, Stanley Reed opined:

> Offshore wind is feasible only because the governments of Britain and other countries provide generous price supports. Britain's direct offshore wind subsidy amounted to £288 million, or $464 million, in the 2010–11 fiscal year.[100]

International Legal Regime Governing Submarine Cables used for Wind Farms

Wind Farms

For wind farms constructed in the territorial sea, the coastal State has complete sovereignty and jurisdiction pursuant to its general sovereignty over the territorial sea.[101] Within the 200 nm EEZ, UNCLOS gives the coastal State sovereign rights "with regard to other activities for the economic exploitation and exploration of the zone, such as the production of energy from the water, currents and winds,"[102] as well as jurisdiction over artificial islands, installations and structures.[103] Accordingly, similar to oil and gas installations, wind farms within the 200 nm EEZ are under the jurisdiction of the coastal State. This would also apply to wind farms on the continental shelf within 200 nm.[104]

---

97 A. Zehnder and Z. Warhaft, "University Collaboration on Wind Energy" Cornell University, 27 July 2011 at 19 available at www.sustainablefuture.cornell.edu/attachments/2011-UnivWindCollaboration.pdf. Generally, offshore wind turbine installations have higher capital costs than land-based installations because of turbine upgrades required for operation at sea and increased costs related to turbine foundations, balance-of system infrastructure, interconnection and installation (at 23).

98 See for example, R. Bell, "UK has best incentives for offshore wind farms" Telegraph, 7 November 2007 available online at http://www.telegraph.co.uk/earth/earthnews/3313411/UK-has-best-incentives-for-offshore-wind-farms.html.

99 See S. Asaad and B. des Roos, "Government Support for Offshore Wind: What can the US learn from Europe" North American Clean Energy, May/June 2010, available online at http://www.taylor-dejongh.com/wp-content/uploads/2010/07/Government-Support-for-Offshore-Wind.pdf (last accessed 23 March 2013).

100 S. Reed, "Green Power on the Edge of Practicality" New York Times, 4 October 2012, available online at http://www.nytimes.com/2012/10/04/business/energy-environment/04iht-green04.html (last accessed 23 March 2013).

101 UNCLOS Art 2.

102 UNCLOS Art 56(1).

103 UNCLOS Art 60.

104 Both the continental shelf and the EEZ provide distinct rights to the seabed up to a distance of 200 nm and in many cases, the 200 nm EEZ and 200 nm continental shelf of a coastal State will overlap: See UNCLOS, Art 56 and Art 79.

However, UNCLOS does not expressly give the coastal State sovereign rights with regard to the production of energy from the water, currents and winds in the extended continental shelf i.e. beyond 200 nm.[105] It does nevertheless, give the coastal State the exclusive right to construct and to authorize and regulate the construction, operation and use of artificial islands, installations and structures for the purposes provided for in Article 56 and other economic purposes.[106] It therefore would appear that wind farms constructed on the extended continental shelf would be under the jurisdiction of the coastal State.

Submarine Power Cables

As is the case with submarine fiber optic cables used to serve oil and gas installations, submarine power cables for wind farms are under the jurisdiction of the coastal State both within the territorial sea and in the EEZ/continental shelf.[107] Accordingly, the coastal State has every right to regulate the route and laying, repair and maintenance operations of a submarine power cable used for wind farms.

Law and Policy Challenges for Submarine Power Cables Used for Wind Farms

Impact on the Environment

Generally, the main concern relating to submarine cables and the environment is the disturbance to the seabed, especially if the cable is buried. However, submarine fiber optic cables (used for either general telecommunications purposes or for oil and gas installations) are thought to have little impact on the marine environment. They are usually 17–50 mm wide and their footprint is small.[108] On the other hand, submarine power cables can measure up to 300 mm depending on current capacity and amount of armor protection, relatively larger than the 17–50 mm diameter of a submarine fiber optic cable. Their comparative weights also differ, with submarine power cables weighing up to 140 kg per meter whereas fiber optic cables weigh between 0.7 kg/m to 4.8 kg/m.[109]

---

105 There is no equivalent recognition of the coastal State's rights "with regard to other activities for the economic exploitation and exploration of the zone, such as the production of energy from the water, currents and winds," in Part VI of UNCLOS on the continental shelf. The rights of the coastal State over the continental shelf are confined to sovereign rights over non-living natural resources of the seabed and subsoil: See UNCLOS Art 79.

106 UNCLOS Art 80.

107 UNCLOS Art 79(4).

108 See also Chapter 7 on The Relationship between Submarine Cables and the Marine Environment.

109 International Cable Protection Committee, 'About Submarine Power Cables' available online at http://www.iscpc.org/publications/About_SubPower_Cables_2011.pdf. See also Chapter 7 of this Handbook.

However, as noted in Chapter 7 on the Relationship between Submarine Cables and the Marine Environment and reiterated in the Report by the United Nations Environment Programme (UNEP) and the International Cable Protection Committee (ICPC):

> Unless a cable fault develops, the seabed may not be disturbed again within the system's design life. Furthermore, the one-off disturbance associated with cable placement is restricted mainly to a strip of seabed less than 5–8 m wide. For comparison, bottom trawl and dredge fishing operations are repetitive and more extensive (e.g. National Research Council, 2002; UNEP, 2006). A single bottom trawl can be tens of metres wide, sweep substantial areas of seabed in a single operation and is likely to be repeated over a year at the same site.[110]

There have been no studies so far to suggest that the greater width and weight of power cables causes more seabed disturbance than submarine fiber optic cables. Indeed, a study conducted on the SwePol Link submarine electrical energy transfer system between Sweden and Poland demonstrated that there had been no visible changes to the surface of the sea bottom.[111] Accordingly, the difference of impact on the seabed of a wind farm power cable versus offshore oil and gas submarine fiber optic cables is negligible.

Another possible consequence of submarine power cables to the marine environment has been identified. Submarine power cables can generate electromagnetic fields depending on the applied voltage.[112] Some regional organizations have expressed concern that some marine organisms may be sensitive to electromagnetic fields and that these fields may influence an animal's navigation, feeding, orientation and/or detection of other animals.[113] There has been a concern that "cables associated with the large number of planned offshore wind farms may, for example, disrupt the migration of sensitive anadromous fish species on their route into the rivers where they reproduce."[114]

---

110 See L. Carter *et al.*, "Submarine Cables and the Oceans—Connecting the World" Report of the United Nations Environment Program and the International Cable Protection Committee (2009) 'UNEP/ICPC Report' at 34. Available online at http://www.unep-wcmc.org/medialibrary/2010/09/10/352bd1d8/ICPC_UNEP_Cables.pdf (last accessed 22 March 2013).

111 E. Andrulewicz *et al.*, "The Environmental Effects of Installation and Functioning of the Submarine SwePol Link HVDC Transmission Line: a Case Study of the Polish Marine Area of the Baltic Sea" (2003) 49 *Journal of Sea Research* 337–345.

112 For more information on this, please refer to Chapter 7 on the Relationship between Submarine Cables and the Marine Environment and Chapter 14 on Power Cables.

113 *Ibid.*

114 OSPAR Commission, *Report on the Assessment of the Environmental Impacts of Cables*, 2009 at 14 available online at http://qsr2010.ospar.org/media/assessments/p00437_Cables.pdf (OSPAR Report).

However, studies on the potential effect of electromagnetic fields on animals have so far been inconclusive[115] and there are also several mitigation measures that can be taken to reduce the electromagnetic fields generated by power cables.[116] For a more detailed discussion on this, please refer to Chapter 7 on The Relationship between Submarine Cables and the Marine Environment and Chapter 14 on Submarine Power Cables.

Competing Uses

As with submarine cables that service offshore oil and gas installations, one of the major issues for governments, the cable industry and the renewable energy industry is competing uses. First, there is the potential conflict between offshore wind farms and telecommunications cables which are laid in the same vicinity. Because offshore wind farms are a newer technology, they tend to be built in areas where existing telecommunications cables have already been laid. The installation of offshore wind turbine towers near undersea cables increases the probability of cable damage due to the risk of seafloor scouring which is the "effect of currents eroding sediment in the areas around a structure on the sea floor."[117] This causes cables to be exposed to potential threats.[118] Further, the presence of large wind farm developments also restricts the installation of cables. This is because cable operators will not lay cables in an area near wind farms which means that telecommunications cables will be limited to narrow corridors "that dramatically limit infrastructure route diversity and increase risks of system outages from damage to multiple cables at once."[119] Cable repair ships will also not easily be able to repair cables if located near wind farms due to maneuverability and space issues.

---

115 See Chapter 7 on the Relationship between Submarine Cables and the Marine Environment. Also see OSPAR Report at 14–15.

116 See Chapter 7 on Cables and the Relationship between Submarine Cables and the Marine Environment.

117 Comments of the North American Submarine Cable Association (NASCA) Before the Bureau of Ocean Management, U.S. Department of the Interior In the Matter of Atlantic OCS Proposed Geological and Geophysical Activities, Mid-Atlantic and South Atlantic Planning Areas Draft Programmatic Environmental Impact Statement, 30 May 2012, OCS EIS/EA BOEM 2012-005 available online at http://www.wiltshiregrannis.com/siteFiles/News/F7FE3C69FE76D28CD58D601E45D5C56B.pdf at 20 (Statement of NASCA).

118 *Ibid.*

119 *Ibid.*

To address such issues, the United Kingdom's Crown Estate,[120] together with Subsea Cables UK,[121] the regional cable protection organization, and RenewableUK,[122] the professional body for the UK wind and marine renewable energy industries, have worked together to devise guidelines on the proximity between offshore wind farms and submarine cables.[123] Their collaborative activities involved not only computer simulations but an actual field test using a cable vessel maneuvering in a simulated repair near wind turbines. The Guidelines produced as a result include recommendations on the minimum separation needed between wind farms and undersea cables. If there is a distance of one nm between a wind turbine and a telecommunication cable, no proximity agreement is required. If the distance is less than one nm, then the affected parties should enter into a "due regard" dialogue with a starting point of 750 meters as a proper separation. Separation of less than 500 meters between a wind turbine and the telecommunications cable is likely to be unsafe.

The ICPC has also issued a Recommendation on the proximity of Wind Farm Developments and Submarine Cables.[124] The recommendation calls for a distance between the telecommunications cable and the wind farm structure (including whirling turbine blades) that will allow a cableship to safely conduct operations in the event that the telecommunications cable needs repair. In water depths of 40 meters, the recommended separation is 1100 meters. The safe distance, however, is not fixed. The distance can be reduced or increased to take into account the seabed, crossing angles, currents, local conditions, and other factors that a cableship master exercising prudent seamanship would consider in designing a repair plan in proximity to offshore wind turbines, tidal current generators, or other new technologies that will share the seabed with cables.

The second potential conflict exists between power cables servicing offshore wind farms and telecommunications cables. In this regard, it is useful to note the observations by the North American Submarine Cables Association:

---

120 The Crown Estate manages on behalf of the Crown, approximately 50 per cent of the foreshore, tidal riverbeds, together with most of the seabed out to 12 nm territorial limit and the rights to renewable energy developments within the UK's Renewable Energy Zone (REZ).

121 For more information on Subsea UK, please refer to their website at http://www.subseauk.com/ (last accessed 8 June 2013).

122 For more information on RenewableUK, please refer to their website at http://www.renewableuk.com/ (last accessed 8 June 2013).

123 See Subsea Cables UK Guideline No. 6 on *The Proximity of Offshore Renewable Energy Installations and Submarine Cable Infrastructure in UK Waters*, Issue 4 (August 2012) available online at http://www.thecrownestate.co.uk/media/343985/Subsea%20Cables%20UK%20Guideline%20%28SCUK%29%20No.%206.pdf (last accessed 7 June 2013).

124 ICPC Recommendation No. 13 Proximity of Wind Farm Developments and Submarine Cables (27 September 2010).

> Offshore energy systems, including wind farms and other alternative energy sources, run power transmission cables back to shore. With respect to wind farms, these often consist of multiple cables (typically three to six for larger operations) running in parallel with 50 to 100 meter separation to meet capacity requirements. Therefore, when a cable crossing situation arises, it now poses a risk of "sterilizing" a much larger section of crossed telecommunications cable than a standard telecom-to-telecom crossing. For an undersea telecommunications cable owner planning to install a new undersea telecommunications cable that will cross an energy export cable, installation costs (for negotiating multiple cable crossing agreements) and risks both to the cable and to its commercial agreements in the event of delay have dramatically increased.[125]

In this regard, reasonable crossing agreements or arrangements based on due regard as discussed above may be very valuable in mitigating this risk.

Conclusion

When the first submarine telegraph cable was laid in 1850, it would have been difficult to foresee the multiple uses that would emerge from such technology. However, as illustrated above, wind farm and offshore oil and gas submarine cable systems appear on the cusp of revolutionizing the way in which essential industries operate. There inevitably remain challenges, particularly in the balancing of competing uses of the seabed, but the collaboration between the Crown Estate, Subsea UK and RenewableUK in devising guidelines to minimize such conflicts sets an excellent precedent on how such issues can be resolved. It is critical for all relevant stakeholders to come together to communicate and cooperate and only then will the full potential of the many different uses of the oceans be realized.

125 Statement of NASCA, *supra* note 117 at 20.

PART VI

APPENDICES AND KEYWORD INDEX

APPENDIX ONE

TIMELINE OF THE SUBMARINE CABLE INDUSTRY

Prepared by Stewart Ash

1720 Stephen Gray formulates the first principles of electricity

1753 A letter to the Scots Magazine entitled 'An Expeditious Method of Conveying Intelligence' is the first known record of a proposal for an electrical telegraph

1791 Claude & Rene Chappe demonstrate their *Pendulum System*, the first mechanized system to transmit an organized code

1794 The French State Telegraph opens between Paris and Lille, based on the Chappe *semaphore* or 'optical telegraph' with the ability to transmit 196 codes

1800 Alessandro Volta demonstrates the 'Voltaic Pile' producing an electric current

1809 Dr Samuel Thomas Soemmerring presents his design for an electro-chemical telegraph to the Münchener Akademie der Wissenschafte

1811 Soemmerring and Baron Pavel L'vovitsch Schilling lay and successfully test the first electrically insulated cable across the river Isar, in Germany

1812 Schilling detonates powder mines through an electrical cable laid across the river Neva, near St Petersburg

1815 Sir Home Popham's system of semaphore is adopted by the British Admiralty

1820 Christian Oersted discovers the effect of an electric current on a magnetic needle. Andre Marie Ampere explains the significance of this discovery

1820 Augustine-Jean Fresnel develops the equations for light trapped in a flat glass plate

1821 Michael Faraday observes electro-magnetic rotation and Ampere establishes the laws of electro-magnetic action

1823 Francis Ronalds publishes a booklet entitled 'Description of an Electric Telegraph'. He offers the system to the British Admiralty but they refuse it, having already adopted the Chappe optical telegraph

1825 William Sturgeon demonstrates the first electro magnet

1831 Michael Faraday discovers electro-magnetic induction

Table (*cont.*)

1832	Schilling demonstrates what is thought to be the first working electromagnetic telegraph in his apartment in St Petersburg.
1835	Shilling and Dr Wilhelm Soemmerring present their electromagnetic telegraph to Physikalischer Verein
1837	William Forthergill Cooke and Charles Wheatstone are granted the world's first patent for an electrical telegraph system
1838	Samuel Morse and Alfred Vail demonstrate the first electromagnetic telegraph in the US. They used a telegraphy code that is now known as American Morse Code or 'railroad code'
1840	On 9 March Professor Charles Wheatstone proposes a submarine telegraph cable between England and France to the British Government
1840	Samuel Morse is granted a patent for an electrical telegraph in the US
1840	R.S. Newall is granted a patent for a soft cored steel rope. This is the base patent for future armored submarine cables
1842	Morse and Vail lay a telegraph cable across New York harbor but their demonstration of submarine telegraphy excited little or no interest. Later in the year Morse and Vail set up a similar demonstration across a canal in Washington DC with the same result
1843	Dr William Montgomerie introduces gutta percha to the Society for the Encouragement of Arts, Manufacture and Commerce in London
1843	Cooke opens the first commercial telegraph system between Paddington and Slough in the UK
1845	Morse opens his first commercial telegraph system between Washington DC and Baltimore in the US
1845	Michael Faraday suggests the use of gutta percha as an insulation for electrical wires
1845	On 16 June, Jacob Brett registers the General Oceanic Telegraph Company
1845	Henry Bewley develops a machine for producing gutta percha tubes
1845	The Gutta Percha Company is incorporated in the UK
1846	Charles Samuel West lays a 1 mile rubber insulated cable across Portsmouth Harbor
1848	Friedrich Clemens Gerke develops a telegraphy code that was initially used between Hamburg and Cuxhaven in Germany. This code became the basis of *International Morse Code*
1849	C.V. Walker tests the first gutta percha insulated cable in Folkston Harbor
	This is generally recognized as the start of the submarine cable industry and the beginning of the Telegraph Era
1850	The first international submarine telegraph cable is laid between Dover and Calais by the *Goliath*. It is a copper wire covered by gutta percha and weighed down with lead weights

Table (*cont.*)

1851	The first commercially successful submarine telegraph cable between Dover and Calais is laid by *Blazer*. This system is the first to use cable armoring
1853	The first telegraph cable is laid across the North Sea from Dover to Ostend
1853	The first telegraph cable is laid across the Irish Sea from Donaghadee to Port Patrick
1853	Dr Julius Willem Gintl devises the first practical method of duplex signaling
1853	The *Monarch I* is the first vessel to be permanently converted as a cable maintenance ship
1854	Glass, Elliot & Company is formed in Greenwich (now Alcatel-Lucent) to manufacture submarine cables
1854	Cyrus W. Field meets Frederick Gisborne and then forms the New York, Newfoundland and London Telegraph Company
1854	Charles Bourseul publishes a paper on the use of electricity for transmitting and receiving speech
1854	Antonio Santi Giuseppe Meucci invents the first successful system for the transmission of speech by electricity
1856	The Atlantic Telegraph Company is registered in the UK
1857	Manufacture of the first Atlantic Telegraph cable completed at Glass, Elliot and Company in Greenwich and at R.S. Newall in Liverpool
1857	Laying of the first Atlantic Telegraph fails after 330 nm of cable is laid from Valentina in Ireland
1858	*Agamemnon* and *Nigeria* successfully lay the first Atlantic Telegraph Cable
1858	After some 400 messages are sent the Atlantic Telegraph fails
1859	The British Government and the Atlantic Telegraph Company set up a joint commission to investigate the failure of the Atlantic Telegraph
1860	Johann Philippe Reis makes a device to electrically transmit speech. This is the first device to be called a *telephone*
1861	The Government Report on the Atlantic Telegraph is published. One of the outcomes is that Professor William Thomson (Lord Kelvin) sets up and chairs a committee to establish electrical standards
1864	The Telegraph Construction & Maintenance Company is formed merging Glass, Elliot and Company and the Gutta Percha Company
1865	The *Great Eastern* attempts to lay a new Atlantic Cable from Ireland but it breaks 600 nm from the Newfoundland landing
1865	The International Telegraph Convention is signed in Paris with 20 founding member countries, and the International Telegraph Union is set up to facilitate amendments to the Convention
1865	International Morse Code is standardized at the International Telegraphy Congress in Paris

Table (*cont.*)

1865	The concept of a repeater housing is patented in the UK
1866	The *Great Eastern* successfully installs a new cable across the Atlantic and recovers and completes the 1865 cable
1868	Benjamin Disraeli's government introduces the Telegraph Act allowing the British Government to take control of all national telegraph companies
1868	John Pender stands down as Chairman of Telcon in favor of Daniel Gooch
1869	The *Hibernia* becomes the first vessel to be permanently converted as a cable laying ship
1870	John Pender's companies complete and open a telegraph between London and Bombay
1870	R.S. Newall & Co cease submarine cable manufacture
1871	Antonio Meucci sets up an electrical voice communication system in his house in Staten Island. He files a patent caveat with the US Patent Office
1871	Lord Rayleigh publishes his theory on light scatter (Rayleigh Scatter)
1871	Pender Companies extend telegraph to Penang and Singapore
1872	The *HC Oersted* is launched, the first purpose built cable maintenance ship
1872	Pender Companies open telegraph services between the UK and Australia
1872	Four Pender telegraph companies are merged to form the Eastern Telegraph Co Ltd
1872	The first submarine telegraph cables go into operation in Japan for the Great Northern China and Japan Extension Company set up by Carl Fredrick Tietgen
1873	Three Pender telegraph companies merge to form the Eastern Extension, Australasia and China Telegraph Co Ltd
1873	The *Hooper* is launched; it is the first purpose built cable laying ship
1873	Frederick Guthrie discovers thermionic emission
1874	Antonio Meucci fails to renew his patent caveat for the telephone
1875	The International Telegraph Conference meets in St Petersburg and agrees on the first set of Telegraph Standards
1876	J.W. Brown and George Allan invent an electro-magnetic relay that doubles the transmission rate over submarine cables
1876	On 14 February Elisha Grey submits a patent caveat and Alexander Graham Bell submits a patent application to the US Patents Office, for the Telephone. On 7 March Patent 174,465 is granted to Bell
1876	Hungarian Tivada Puskas invents the telephone switchboard
1877	The Bell Telephone Company is established
1878	The world's first telephone exchange opens in New Haven, Connecticut

Table (*cont.*)

1879	Brass tape is introduced into cable design to protect the gutta percha against Toredo worms
1879	The first telephone exchange in Europe opens in London
1880	British Courts declare that the telephone is a telegraph within the meaning of the Telegraph Acts of 1869 and 1873
1880	The coaxial cable design is patented by Oliver Heaviside
1883	Western Electric (US) establishes Western Electric Ltd in the UK
1883	The first telephone exchange is installed in Rheims, France
1883	John W. Mackay and James Gordon Bennett found the Commercial Cable Company
1884	The Convention for the Protection of Submarine Telegraph Cables is adopted by an international convention in Paris
1884	The International Meridian Conference held in Washington DC confirms Greenwich as the 0° meridian
1885	American Telephone and Telegraph Corporation (AT&T) is formed as a subsidiary of the Bell Telephone Company
1885	Oliver Heaviside demonstrates the use of inductance to compensate for cable capacitance
1885	The International Telegraph Union begins to develop international legislation governing telephony
1886	Western Union starts a price war on trans-Atlantic telegraph traffic
1887	Heinrich Hertz confirms the existence of electro-magnetic (radio) waves
1887	All Pender's companies adopt the collective name of Eastern Associated Companies
1888	The Convention for the Protection of Submarine Telegraph Cables comes into force (40 signatories to date)
1888	Anglo America Telegraph Co, the Commercial Cable Company and Western Union agree a common rate for trans-Atlantic telegraph traffic
1890	First telephone service in Japan opens between Tokyo and Yokohama
1890	Domestic telegraph companies are nationalized in the UK under the General Post Office
1891	The first telephone cable is laid between the UK and France by *Monarch II*
1891	La société Générale des Téléphone (now Alcatel Lucent) opens a cable manufacturing factory in Calais
1892	La société Générale des Téléphone is acquired by La société Industrielle des Téléphone
1894	The French Atlantic Telegraph Company goes into liquidation

Table (*cont.*)

1895	The Simplex Electrical Company is incorporated in the US
1895	Willoughby Smith & W.P. Granville are granted a UK patent for a submarine telegraph cable with a copper, helically wrapped return conductor
1896	The Imperial Pacific Cable Committee meets in London for the first time
1897	The first reduced capacitance cable is laid between the UK mainland and the Isle of Wight
1898	A reduced capacitance cable is laid across the Irish Sea from Nevin, North Wales to Newcastle in County Wicklow
1898	La Compagnie Genérale d'Electricité is incorporated in France
1898	Western Electric Ltd acquires its factory in North Woolwich
1899	Marconi sends radio signals across the channel from North Foreland to Boulogne
1899	The Bell Telephone Co (American Bell) is absorbed by AT&T
1900	The Simplex Electrical Company completes two small submarine cable projects in Boston Harbor and Lake Michigan
1900	The Telegraph Construction and Maintenance Company is awarded a contract for a telegraph cable across the Pacific from New Zealand and Queensland, Australia to Vancouver, Canada
1901	Marconi demonstrates wireless communication across the Atlantic from Cornwall to Newfoundland
1902	The *Colonia* lays a telegraph cable from Fanning Island to Bamfield, British Colombia; at 3458 nm it is the longest telegraph cable ever laid
1902	Volcanic eruptions in St Vincent and Martinique linked to offshore telegraph cable breaks
1902	The British Colonial Government discuss nationalization of the British submarine telegraph cable companies while attending the coronation of Edward VII in London
1904	John Ambrose Flemming invents and patents the thermionic diode
1904	The first International Convention on Radio Telegraph is signed at a conference in Berlin
1907	Lee De Forest invents the thermionic triode
1910	The Western Electric Co Ltd is incorporated in the UK
1914	The Eastern, Eastern Extension, Western and nine other International Telegraph Companies are amalgamated to form the Eastern and Associated Telegraph Companies
1915	First submarine cable is manufactured in Japan, it is rubber insulated
1915	Skandinaviske Kabel og Gummifabrikker A/S is established in Norway
1919	Marconi is granted first license for commercial radio telegraph

Table (*cont.*)

1919	Walter H. Schottky invents the thermionic tetrode
1921	Three coaxial cables with gutta percha dielectric are laid between Florida and Cuba
1922	The Furukawa Electric Co manufactures the first gutta percha insulated submarine cable at its Yokohama factory
1922	Marconi Scientific Instrument Company broadcasts wireless telephony for the first time
1923	The first trans-Atlantic, one-way wireless telephone call is made from Rocky Point, New York to STC New Southgate
1924	The International Telephone Consultative Committee (CCIF) is established
1925	AT&T creates Bell Telephone Laboratories (Bell Labs)
1925	The International Telegraph Consultative Committee (CCIT) is established
1925	Western Electric Limited is purchased by ITT and becomes Standard Telephones and Cables Ltd (STC)
1927	The first trans-Atlantic wireless telephony service goes into operation
1927	The International Radio Consultative Committee is established
1928	The last trans-Atlantic Telegraph cable is installed from the Azores to Newfoundland
1928	AT&T approaches the British Post Office with the idea of developing a trans-Atlantic Telephone cable
1928	Bernard D.H. Tellegen invents the thermionic pentode
1928	The phenomena of light scattering in liquids is discovered by Chandrasekhara Venkata Raman and Kariamanickam Srinivasa Krishnan and independently by Grigory Landsberg and Leonid Mandelshtam in crystals. It becomes known as the Raman Effect
1929	The Eastern Associate Telegraph Companies are merged with Marconi Wireless Telephone Co Ltd, The Pacific Cable, Imperial Atlantic Cables and the West Indies Cable & Wireless Systems to form Imperial and International Communications Ltd
1929	The Newfoundland (Grand Banks) earthquake creates a mudslide that takes out a series of the trans-Atlantic telegraph cables
1930	Bell Laboratories coaxial submarine cable design is manufactured by Norddeutsche Seekabelwerke and laid between Havana and Key West
1932	At a conference in Madrid it is agreed that the 1865 Telegraph Convention and the 1906 Radio Telegraph Convention should be combined to form the International Telecommunications Convention. It is also agreed to change the name of the International Telegraph Union to the International Telecommunications Union
1933	Imperial Chemical Industries (ICI) discovers polyethylene

Table (*cont.*)

1934	The International Telecommunications Union (ITU) comes into effect with headquarters in Bern, Switzerland
1934	Skandinaviske Kabel og Gummifabrikker A/S is purchased by ITT and the name is changed to Standard Telefon og Kabelfabfikk A/S
1934	Imperial and International Communications Ltd becomes Cable & Wireless Ltd
1935	The Nippon Submarine Cable Co Ltd (NSCC) is established as a joint venture between Fujikura Ltd, Furukawa and Sumitomo
1935	Submarine Cables Ltd is formed by merging the Telegraph Construction and Maintenance Company with Siemens Brothers submarine cables division, the only submarine cable manufacturing company in the UK
1935	The Nihon Kaiteidensen Co Ltd is established, it later becomes the Ocean Cable Co Ltd
1936	A 300 km submarine coaxial cable is laid between the Australian states of Victoria and Tasmania
1938	Alec Reeves patents Pulse Code Modulation
1938	La Société Industrielle des Téléphone merges with La Compagnie Genérale d'Electricité and submarine cable manufacture is carried out by a new company, Câble de Lyon
1938	A plow developed by Western Union Telegraph Company is used to bury shallow water cables
1941	Nippon Submarine Cables Ltd builds a new cable factory in Yokohama
1941	Bell Laboratories completes a flexible steel tube design for a repeater housing
1943	The *Iris* recovers and inserts the first submerged amplifier in a telephone coaxial cable, laid between Anglesey and Port Erin
1947	Cable & Wireless Ltd is nationalized by the British Government
1947	Submarine Cables Ltd manufactures and lays 1.7 inch, air spaced, coaxial, submarine cable between the UK and Holland
1947	Bell Laboratories invent the germanium semi-conductor transistor
1947	The General Assembly of the newly formed United Nations approves and agreement to recognize the ITU as a specialist agency of the United Nations
1948	Câble de Lyon rebuilds the Calais cable factory destroyed in WWII
1948	A trial system containing 20 nm of Simplex cable and six underwater amplifiers is laid in the Bahamas
1948	Dr John Shive invents the phototransistor at Bell Labs. His work is not published until 1950
1949	The agreement between the ITU and the United Nations comes into force

Table (*cont.*)

1950	The Western Union Telegraph Company develops the first telegraph submerged booster to repeat signals. It is given the name *repeater*, a term which has been used ever since. It was installed in the 1881 'American Telegraph' cable
	This is generally considered as the end of the Telegraph Era and the start of the Telephone Era
1950	Twin coaxial cables are laid between Havana and Key West; each contains three uni-directional flexible repeaters. The cable is manufactured by Simplex Wire & Cable and the repeaters are designed by Bell Labs and made by Western Electric
1950	At the centenary of the Submarine Cable Industry some 469,500 nm of cable has been made and laid, 82 per cent by the Telegraph Construction and Maintenance Company
1950	Richard Hamming invents the first forward error correction code
1952	An Englishman, Geoffrey Dummer presents the concept of Integrated Circuits at a conference in Washington DC
1952	Using data from the telegraph cable breaks caused by the Grand Banks earthquake, Heezen and Ewing establish the presence of submarine landslides and turbidity currents in deep water
1952	Western Union develop a trial lightweight telegraph cable and deploy it in deep water
1953	Simplex opens a new cable factory in Newington, New Jersey
1954	Submarine Cables Ltd opens a new cable factory at Erith on the River Thames
1954	Gordon Teal of Texas Instruments demonstrates the first silicon transistor
1956	STC opens a new cable factory in Southampton, UK
1956	The first trans-Atlantic Telephone cable TAT-1 is commissioned. The repeaters are the Bell labs flexible, uni-directional design and are manufactured in the US. 7739 km of cable is manufactured by Submarine Cables Ltd at Erith and 616 km at Newington by Simplex
1956	The CCIF and the CCIT are merged to form the International Telephone and Telegraph Consultative Committee (CCITT)
1957	Russia launchs satellite Sputnik 1
1957	La Compagnie Industrielle des Téléphones factory in Calais commences manufacture of submarine coaxial cables
1958	The International Cable Protection Committee (ICPC) is formed in the UK
1958	Jack Kilby of Texas Instruments demonstrates the first working Integrated Circuit
1958	The first international cable damage committee is set up with attendees from Canada, Denmark, France, Italy, Norway, Sweden, UK, US and West Germany
1958	The Convention on the Territorial Sea and Contiguous Zone is adopted by the United Nations at its conference in Geneva

Table (*cont.*)

1958	The Convention on the High Seas is adopted by the United Nations at its conference in Geneva
1958	The Convention on the Continental Shelf is adopted by the United Nations at its conference in Geneva
1958	The US launch satellite Explorer 1
1959	The CCIR sets up a study group to consider 'Space Communications'
1960	Repeater parachutes are successfully trialed in Loch Fyne
1960	Theodore Maiman demonstrates the first (ruby) laser
1961	CANTAT 1 goes into operation. This is the first system to include lightweight cable and rigid repeater housings in deep water
1962	The first coherent light emission from a semiconductor laser is demonstrated in the US by research teams at General Electric and IBM
1962	Time Assignment Speech Interpolation (TASI) is developed by Bell Labs
1962	The United Nations Convention on the High Seas comes into force
1962	Communications Satellite Act sets up COMSAT in the US
1962	Communication satellites Telstar and Relay are launched
1963	The CCIR holds an Extraordinary Administrative Conference in Geneva to allocate frequencies for Space Communications
1963	AT&T Bell Labs begin the development of Sea Plow 1
1963	La Compangnie Industrielle des Telephones becomes La Compagnie industrielle des Telecommunications
1963	COMPAC the first trans-Pacific telephone cable is completed
1963	The first successful geosynchronous satellite Syncom II is launched
1964	Nippon Submarine Cables Ltd merges with the newly formed Ocean Cable Company to form a single company to manufacture submarine cables for the Pacific
1964	The United Nations Convention on the Continental Shelf comes into force
1964	The United Nations Convention of the Territorial Sea and Contiguous Zone comes into force
1964	Trans-Pacific Cable (TPC) 1 is installed between the US and Japan
1964	INTELSAT is formed
1964	The US Navy sets up the first Transit Satellite Navigation System NAVSAT
1964	Stewart Miller of Bell Labs deduces the first way to probe glass as a long distant transmission medium
1965	The Atlantic Cable Maintenance Agreement (ACMA) is formed, the first submarine cable maintenance consortium in the world
1965	INTELSAT 1 (early bird) is launched

Table (*cont.*)

1966	Drs Charles Kao and George Hockham propose glass fiber as a transmission medium for telecommunications
1968	Stronger 1.47 inch coaxial cable is introduced and repeater parachutes are abandoned
1969	SAT-1 is installed by STC connecting South Africa to the UK. This is the last system to use thermionic valves in the repeaters and provides 360 × 3 kHz voice channels
1969	Repeater manufacture commences in Japan at Fujitsu Ltd and the Nippon Electric Company (NEC)
1970	STC purchases Submarine Cables Ltd to form STC Submarine Systems
1970	Alcatel CIT is formed absorbing La Compagnie industrielle des Telecommunications
1970	STC supplies UK-Spain 1 The first 5 MHz, transistor amplified system providing 480 × 4 KHz voice channels
1970	Alcatel Submarcom is set up to market Alcatel submarine cable systems
1970	Bell Labs develop the first practical semiconductor laser
1971	STC supplies Penbal 1 between Barcelona and Mallorca. It is the first 14 Mhz system providing 1380 × 4 kHz voice channels
1971	The Post Office Corporation installs the first linear cable engine (14 wheel pair) on *CS Alert*. This was a joint development with Dowty Boulton Paul Ltd
1971	The first Japanese manufactured 36 Mhz system is installed between Kure and Marsuyama
1972	The Convention on the Prevention of Marine Pollution by Dumping of Wastes and Other Matter is adopted by the United Nations at its Conference in Stockholm
1972	The Convention on the International Regulations for Preventing Collisions at Sea (COLREGS) is adopted by the International Maritime Organization
1972	R.H. Stolen and E.P. Ippen demonstrate Raman amplification in optical fibers
1973	The US Department of Defense starts work on the satellite Global Positioning System GPS
1973	An improved Dowty Boulton Paul linear cable engine (18 wheel pair) is installed on *CS Mercury*
1974	Simplex Wire & Cable becomes a wholly owned subsidiary of Tyco International Ltd
1975	Corning Laboratories produce optical fiber with a loss of 4 dB/km
1975	Bell Labs sets up the first field trial of fiber optic telecommunication between offices in Atlanta, Georgia
1975	The United Nations Convention on the Prevention of Marine Pollution by the Dumping of Wastes and Other Matter comes into force
1976	STC installs its first 45 MHz system between Rome and Palermo; it provides 1800 × 4kHz voice channels plus a two-way color television channel

Table (*cont.*)

1976	TAT-6 is installed from France to the US, providing 4000 × 3 kHz voice circuits
1976	KDDI-designed cable plow is used for the first time on the East China Sea cable
1977	A 2.4 km fiber optic system is installed between two Bell offices in Chicago
1977	The first commercial fiber optic system is installed over the 4 km between Hitchin and Stevenage in the UK
1977	STC installs PENCAN 3 between the Spanish mainland and Grand Canaria. It was designed to provide 69 supergroups or 5520 × 3 kHz voice channels, but was equalized to carry 71 supergroups or 5680 × 3 kHz voice channels
1977	COLREGS comes into force
1977	STC closes its Erith and North Woolwich sites and ceases cable manufacture at Greenwich
1878	The experimental Block-I GPS satellite is launched
1979	A tidal wave hits the French Riviera, caused by a submarine landslide and turbidity currents, that takes out the Barcelona-Genoa and Genoa-Sassari systems at a point where they are 10 nm apart on the seabed
1980	STC install the first experimental submarine fiber optic system (containing an optical repeater) in Loch Fyne
1980	50 km of OCC supplied optical cable is deployed as a trial between Inatori and Kawazu
1981	The telecommunications arm of the Post Office Corporation becomes British Telecom in preparation for privatization
1981	Two experimental systems containing Fujitsu and NEC optical repeaters are deployed
1981	Cable & Wireless Ltd is floated on the London Stock Exchange as Cable & Wireless plc
1982	The United Nations Convention on the Law of the Sea (UNCLOS) is adopted at its conference in Montego Bay, Jamaica. The Convention introduces the concept of 200 nm Exclusive Economic Zones (EEZ)
1982	STC is partially floated on the London Stock Exchange. STC Submarine Systems becomes a division of STC plc
1982	Waves and currents forced by Hurricane Iwa trigger submarine landslides and turbidity currents that damage six cables off Oahu, Hawaii
1983	Cable & Wireless Marine becomes a wholly owned subsidiary of Cable & Wireless plc
1983	After a Russian fighter shoots down a civilian airline President Regan announces that GPS will be made available for civilian use. Almost immediately cableships are fitted with GPS receivers
1984	The FCC breaks up AT&T in the US giving birth to six 'Baby Bell' companies
1985	Dr S.B. Poole and David Payne develop the Erbium Doped Optical Amplifier at Southampton University

Table (*cont.*)

1985	A new plow designed by British Telecom in collaboration with Soil Marine Dynamics (SMD) is successfully sea trialed
1986	AT&T SSI deploy its first experimental optical system 'Optican' between Grand Canaria and Tenerife
1986	AT&T SSI report 'shark bite' damage to Optican and begin experiments to establish why this occurred
1986	STC installs its last 14 MHz system between India and the UAE. It goes into service in 1987 and provides 1380 × 4 kHz voice channels. It is the last submarine telephone cable system to be installed
	The three events that follow below are generally acknowledged as the start of the Optical Regenerator Era
1986	The first commercial, repeatered, optical submarine system deployed anywhere in the world is a Japanese domestic system owned by NTT. FS-400M is 300 km long and is laid by *Kuroshio Maru* between Hachinohe and Tomakomai, connecting the islands of Honshu and Hokkaido. The cable is supplied by OCC and contains six single mode fibers, the system contains seven NEC Repeaters; operating at 1310 nm, 400 Mb/s; 445,837 mB and a Line Code 10B1C RZ
1986	Alcatel install their first optical system from mainland France to Corsica
1986	First international commercial fiber optic system, UK-Belgium No 5 goes into service. It is manufactured by STC and comprises 112 km of 3 × fiber pairs armored cable, three repeaters, operating at 1310 nm and 280 Mbit/s. It is installed by *CS Alert* and buried using the SMD plow
1987	BT Marine is established as a wholly owned subsidiary of British Telecom
1987	Alcatel purchases Standard Telefon og Kabelfabfikk A/S and renames it Alcatel STK AS
1988	TAT-8 the first trans-Atlantic fiber optic system goes into operation between France, the UK and the US. It is manufactured by Alcatel, AT&T Submarine Systems Inc and STC Submarine Systems. This is the first system to contain a Branching Unit
1988	STC develops *lightweight screened* cable (LWS) for electrical safety and to address the concern of possible 'shark bite'
1989	PTAT-1 the first privately owned trans-Atlantic cable since the Telegraph Era goes into service. It is supplied by STC and laid by Cable & Wireless Marine. It operates at a line rate of 420 Mbit/s and provides 18,000 voice channels
1989	EMOS is installed by Alcatel operating at 1310 nm and 280 Mbit/s
1989	TPC-3, the first trans-Pacific fiber optic system goes into service. The system is supplied in separate segments by AT&T Submarine Networks and NEC/Fujitsu using OCC cable. The system operates at 1,310 nm and 280 Mbit/s
1989	STC opens a submarine cable factory in Portland Oregon
1989	The Universal Joint Consortium (UJC) is formed by Alcatel, AT&T SSI and BT Marine

Table (*cont.*)

1990	The US Department of Defense degrades the accuracy of GPS for non-military use by offsetting the clock signal by a random amount. This technique is called Selective Availability or SA
1990	STC and NEC supply NPC the first trans-Pacific privately owned cable system. It operates at 1,310 nm and 420 Mbit/s
1990	TAT-9 goes into operation; this is the first trans-Atlantic system to operate at the longer 1,550 nm wavelength and 565 Mbit/s line rate, providing 80,000 × 64 kbit/s voice channels. It contains the first and only submerged multiplexing Branching Units
1991	Câble de Lyon becomes Alcatel Câble and Alcatel STK AS becomes Alcatel Kabel Norge
1991	British Telecom changes its name to BT
1991	A Japanese weather buoy dragging its anchor, causes a fault to NPC, in 5,800 m of water. It is deepest anchor fault in submarine cable history
1991	Northern Telecom purchases ITT's holding in STC and becomes the majority share holder
1992	At a conference in Geneva the ITU is reformed into three sectors. These being: Telecommunication Standardization (ITU-T); Radio Communication (ITU-R); and Telecommunication Development (ITU-D)
1992	TPC-4 is the first trans-Pacific system to operate at 1,550 nm
1992	Alcatel Tasman Cable Company opens a new submarine cable factory in Botany Bay, New South Wales. The company is established to manufacture the cable for the Tasman 2 system
1994	At an ITU conference in Kyoto the World Telecommunications Policy Forum (WTPF) is established
1994	UNCLOS comes into force
1994	Alcatel Alstom purchases STC Submarine Systems and forms Alcatel Submarine Networks
1995	Alcatel Submarine Networks installs Roja between Spain, the UK, Belgium and the Netherlands. It is a repeaterless system, operating at 2.5 Gbit/s and is the first system to deploy a Remotely Pumped Optical Amplifier (ROPA)
1995	Cable & Wireless Marine and BT Marine merge to form Cable & Wireless Marine Ltd
1995	The first purpose built 'Stern Working Only' cableship *CS Innovator* is launched by C & W Marine
	The three events that follow below are generally acknowledged as the start of the Optical Amplifier Era
1996	The first optically amplified, fiber optic, submarine system in the world is a domestic system, installed between Kagoshima in Kyushu and Okinawa

Table (*cont.*)

1996	The first international, optically amplified, fiber optic submarine system is TPC-5. The system supply is split between AT&T Submarine Systems Inc and NEC/Fujitsu with OCC cable
1995–96	TAT-12 & 13 systems are installed to form a trans-Atlantic ring. The systems are optically amplified and operate at a line rate of 5 Gbit/s. The systems are jointly supplied by AT&T SSI and Alcatel Submarine Networks
1996	Alcatel Submarine Networks closes its Southampton cable factory bringing to an end 146 years of submarine cable manufacture in the UK
1996	The US Coast Guard begins transmitting Differential GPS signals (DGPS)
1997	Tyco purchases AT&T SSI for US$850M
1998	10 Gbit/s line rates with Wave Division Multiplexing (WDM) of up to 16 wavelengths become standard submarine cable system offerings
1998	KDD SCS wins supplies contract for TAT-14. This is the first TAT system not to be supplied by European or US suppliers
1999	Global Crossing purchases Cable & Wireless Marine Ltd and renames it Global Marine Systems Ltd
1999	The Venus submarine cable observatory is deployed using the retired TPC-2 coaxial cable
1999	The Ocean Cable Co Ltd is renamed the OCC Corporation
2000	The 150th Anniversary of the submarine cable industry is marked by the publishing of a history of the industry '*From Elektron to 'e' Commerce*'. The industry is on a crest of a wave with annual, global cable manufacturing capability of 340,000 km and a world cableship fleet that exceeds 80 purpose built or converted vessels
2000	Dense Wave Division Multiplexing ($64\lambda \times 10$ Gbit/s) offers 640 Gbit/s per fiber pair with repeatered systems being offered with up to eight fiber pairs. FEC and Raman Amplification stretches the spacing between repeaters and repeaterless spans to >400 km
2000	Due to the success of DGPS, President Clinton issues an executive order to turn off the GPS Selective Availability signal permanently
2000	The Universal Joint Consortium Agreement is re-negotiated to allow the supply of piece parts by all members. The members were Alcatel Submarine Networks, Global Marine Systems Ltd, KDD SCS, Pirelli SA and Tyco Telecoms
2000	Alcatel sells off Kabel Norge which becomes Nexans
2001	TAT-14 goes into service, it is a ring system with a design capacity of 1.87 Tbit/s, $16\lambda \times 10$ Gbit/s on 2 fiber pairs
2001	The bubble bursts and the industry goes from boom to bust
2001	Alcatel Submarine Networks closes its Portland, Oregon cable factory
2003	Alcatel Submarine Networks closes its Port Botany cable factory

Table (*cont.*)

2003	Alcatel Submarine Networks supplies Apollo a trans-Atlantic ring system. The design total capacity is 6.4 Tbit/s with 3.2 Tbit/s capability on each side of the ring. The system is jointly owned by Cable & Wireless and Alcatel Submarine Networks
2003	An earthquake creates submarine landslides and turbidity currents that break at least six cables off Algeria
2006	Western Union announced the sending of the last ever telegram after 150 years of continuous service
2006	Hengchun subsea earthquake, off Taiwan occurs with dramatic collapse of the internet in the region
2006	Cable & Wireless splits into Cable & Wireless International and Cable & Wireless Europe, Asia & USA
2006	Submarine cables support 99 per cent of all international telecommunications traffic. Only 1 per cent is carried by satellite
2006	Alcatel merges with Lucent Technologies to form Alcatel-Lucent
2007	Depredations by vessels on the high seas in the South China Sea take 98 km and 79 km from two cable systems, requiring over three months to repair
2008	The BP Gulf of Mexico system goes into operation. It is the first repeatered, backbone and spur system to support offshore oil and gas platforms. The system is supplied by Tyco Telecoms and includes OADM Branching Unit technology
2008	NEC Corporation and Sumitomo Electric Industries Ltd take control of OCC
2008	The MARS submarine cable scientific observatory goes into operation
2008	Breaks to cables off of Suez, in the Persian Gulf and Malaysia disrupt internet services to India and the Middle East
2008	Huawei Submarine Networks Co is formed as a joint venture between Global Marine Systems Ltd and Huawei Technologies Co
2009	The Neptune submarine cable scientific observatory goes into operation
2009	Sir Charles Kao is a joint winner of the Nobel Prize in physics for pioneering research into fiber optic cables
2009	Flood waters from super Typhoon Morakot, form submarine mud flows that rapidly descend to 4000 m depth breaking at least six fiber optic cables between Taiwan and the Philippines
2010	The Secretary-General of the United Nations addresses the subject of the world's submarine cable networks in his report
2010	The ICPC changes its constitution to allow governments and companies that are key players in the submarine cable industry to become members
2010	Tyco Telecommunications becomes TE SubCom
2010	Cable & Wireless demerges C&W International becomes Cable & Wireless Worldwide and C&W Europe, Asia and USA becomes Cable & Wireless Communications

Table (*cont.*)

2010	UN Omnibus Resolution on Oceans and Laws of the Sea recognizes the importance of submarine cables to the global economy
2010	SEA-ME-WE 4 is cut in three places off of Palermo, Italy
2011	The United Nations Omnibus Resolution on Oceans and Laws of the Sea includes for the first time the word 'repair'
2011	The Tohoku earthquake severely disrupts cables around Japan, including APCN 2, EAC, Japan–US and PC-1 and damages cable landing stations in Japan
2011	The Mitsubishi Electric Corporation signs a contract to upgrade TAT 14 with 40 Gbit/s DWDM technology. The upgrade was scheduled to be completed by Q4 2012
2012	Vodafone purchases Cable & Wireless Worldwide
2012	Marine Protected Areas (MPAs) now occupy 3.2 per cent of the world's oceans
2012	First direct fiber optic cable link between China and Taiwan established

Telegraph Era (1850–1950) 100 years

Telephone Era (1950–1986) 36 years

Optical Regenerator Era (1986–1996) 10 years

Optical Amplifier Era (1996–date) 16 years

APPENDIX TWO

MAJOR SUBMARINE SYSTEM SUPPLIERS (1850–2012)

Prepared by Stewart Ash

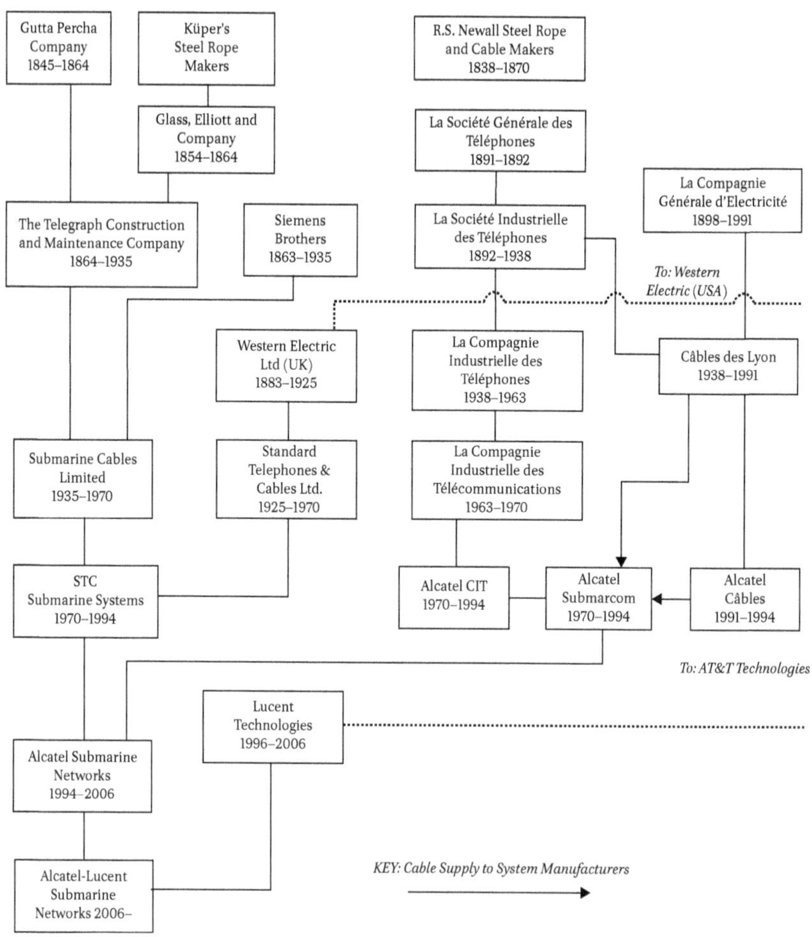

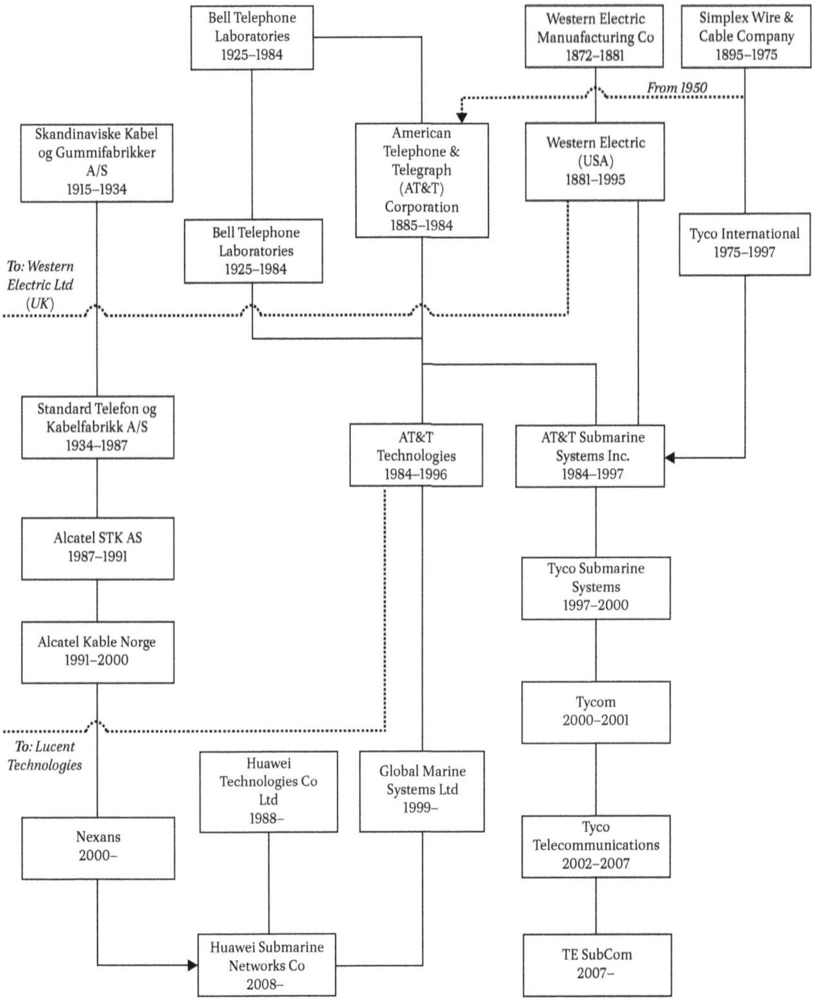
Bell Telephone Laboratories 1925–1984
Western Electric Manuafacturing Co 1872–1881
Simplex Wire & Cable Company 1895–1975
From 1950
Skandinaviske Kabel og Gummifabrikker A/S 1915–1934
American Telephone & Telegraph (AT&T) Corporation 1885–1984
Western Electric (USA) 1881–1995
Bell Telephone Laboratories 1925–1984
Tyco International 1975–1997
To: Western Electric Ltd (UK)
Standard Telefon og Kabelfabrikk A/S 1934–1987
AT&T Technologies 1984–1996
AT&T Submarine Systems Inc. 1984–1997
Alcatel STK AS 1987–1991
Tyco Submarine Systems 1997–2000
Alcatel Kable Norge 1991–2000
Tycom 2000–2001
To: Lucent Technologies
Huawei Technologies Co Ltd 1988–
Global Marine Systems Ltd 1999–
Nexans 2000–
Tyco Telecommunications 2002–2007
Huawei Submarine Networks Co 2008–
TE SubCom 2007–

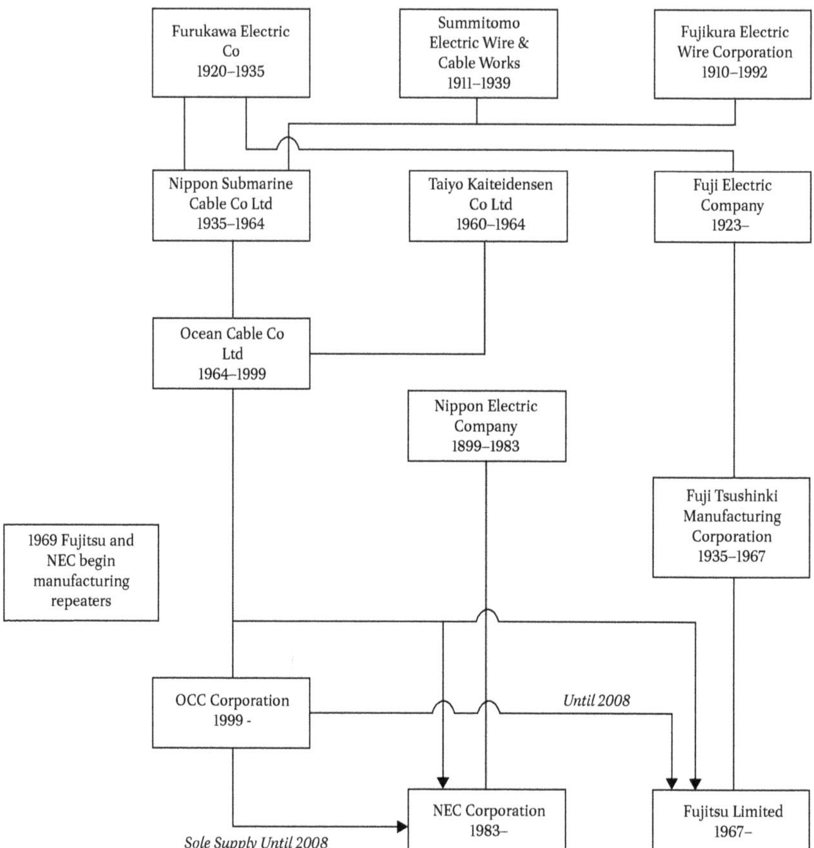
Furukawa Electric Co 1920–1935
Summitomo Electric Wire & Cable Works 1911–1939
Fujikura Electric Wire Corporation 1910–1992
Nippon Submarine Cable Co Ltd 1935–1964
Taiyo Kaiteidensen Co Ltd 1960–1964
Fuji Electric Company 1923–
Ocean Cable Co Ltd 1964–1999
Nippon Electric Company 1899–1983
Fuji Tsushinki Manufacturing Corporation 1935–1967
1969 Fujitsu and NEC begin manufacturing repeaters
OCC Corporation 1999 -
Until 2008
NEC Corporation 1983–
Fujitsu Limited 1967–
Sole Supply Until 2008

APPENDIX THREE

EXCERPTS OF MOST RELEVANT TREATY PROVISIONS

1982 UNITED NATIONS CONVENTION ON THE LAW OF THE SEA, 10 DECEMBER 1982, 1833 U.N.T.S. 3 (ENTERED INTO FORCE 16 NOVEMBER 1994)

PREAMBLE

The States Parties to this Convention,

PROMPTED by the desire to settle, in a spirit of mutual understanding and cooperation, all issues relating to the law of the sea and aware of the historic significance of this Convention as an important contribution to the maintenance of peace, justice and progress for all peoples of the world,

NOTING that developments since the United Nations Conferences on the Law of the Sea held at Geneva in 1958 and 1960 have accentuated the need for a new and generally acceptable Convention on the law of the sea,

CONSCIOUS that the problems of ocean space are closely interrelated and need to be considered as a whole,

RECOGNIZING the desirability of establishing through this Convention, with due regard for the sovereignty of all States, a legal order for the seas and oceans which will facilitate international communication, and will promote the peaceful uses of the seas and oceans, the equitable and efficient utilization of their resources, the conservation of their living resources, and the study, protection and preservation of the marine environment,

BEARING IN MIND that the achievement of these goals will contribute to the realization of a just and equitable international economic order which takes into account the interests and needs of mankind as a whole and, in particular, the special interests and needs of developing countries, whether coastal or land-locked,

DESIRING by this Convention to develop the principles embodied in resolution 2749 (XXV) of 17 December 1970 in which the General Assembly of the United Nations solemnly declared *inter alia* that the area of the seabed and ocean floor and the subsoil thereof, beyond the limits of national jurisdiction, as well as its resources, are the common heritage of mankind, the exploration and exploitation of which shall be carried out for the benefit of mankind as a whole, irrespective of the geographical location of States,

BELIEVING that the codification and progressive development of the law of the sea achieved in this Convention will contribute to the strengthening of peace, security, cooperation and friendly relations among all nations in conformity with the principles of justice and equal rights and will promote the economic and social advancement of all peoples of the world, in accordance with the Purposes and Principles of the United Nations as set forth in the Charter,

AFFIRMING that matters not regulated by this Convention continue to be governed by the rules and principles of general international law,

HAVE AGREED AS FOLLOWS:

ARTICLE 2: LEGAL STATUS OF THE TERRITORIAL SEA, OF THE AIR SPACE OVER THE TERRITORIAL SEA AND OF ITS BED AND SUBSOIL

1. The sovereignty of a coastal State extends, beyond its land territory and internal waters and, in the case of an archipelagic State, its archipelagic waters, to an adjacent belt of sea, described as the territorial sea.
2. This sovereignty extends to the air space over the territorial sea as well as to its bed and subsoil.
3. The sovereignty over the territorial sea is exercised subject to this Convention and to other rules of international law.

ARTICLE 3: BREADTH OF THE TERRITORIAL SEA

Every State has the right to establish the breadth of its territorial sea up to a limit not exceeding 12 nautical miles, measured from baselines determined in accordance with this Convention.

ARTICLE 19: MEANING OF INNOCENT PASSAGE

1. Passage is innocent so long as it is not prejudicial to the peace, good order or security of the coastal State. Such passage shall take place in conformity with this Convention and with other rules of international law.

2. Passage of a foreign ship shall be considered to be prejudicial to the peace, good order or security of the coastal State if in the territorial sea it engages in any of the following activities:

   (a) any threat or use of force against the sovereignty, territorial integrity or political independence of the coastal State, or in any other manner in violation of the principles of international law embodied in the Charter of the United Nations;
   (b) any exercise or practice with weapons of any kind;
   (c) any act aimed at collecting information to the prejudice of the defence or security of the coastal State;
   (d) any act of propaganda aimed at affecting the defence or security of the coastal State;
   (e) the launching, landing or taking on board of any aircraft;
   (f) the launching, landing or taking on board of any military device;
   (g) the loading or unloading of any commodity, currency or person contrary to the customs, fiscal, immigration or sanitary laws and regulations of the coastal State;
   (h) any act of wilful and serious pollution contrary to this Convention;
   (i) any fishing activities;
   (j) the carrying out of research or survey activities;
   (k) any act aimed at interfering with any systems of communication or any other facilities or installations of the coastal State;
   (l) any other activity not having a direct bearing on passage.

ARTICLE 21: LAWS AND REGULATIONS OF THE COASTAL STATE RELATING TO INNOCENT PASSAGE

1. The coastal State may adopt laws and regulations, in conformity with the provisions of this Convention and other rules of international law, relating to innocent passage through the territorial sea, in respect of all or any of the following:

   (a) the safety of navigation and the regulation of maritime traffic;
   (b) the protection of navigational aids and facilities and other facilities or installations;
   (c) the protection of cables and pipelines;
   (d) the conservation of the living resources of the sea;
   (e) the prevention of infringement of the fisheries laws and regulations of the coastal State;
   (f) the preservation of the environment of the coastal State and the prevention, reduction and control of pollution thereof;

(g) marine scientific research and hydrographic surveys;
(h) the prevention of infringement of the customs, fiscal, immigration or sanitary laws and regulations of the coastal State.

2. Such laws and regulations shall not apply to the design, construction, manning or equipment of foreign ships unless they are giving effect to generally accepted international rules or standards.
3. The coastal State shall give due publicity to all such laws and regulations.
4. Foreign ships exercising the right of innocent passage through the territorial sea shall comply with all such laws and regulations and all generally accepted international regulations relating to the prevention of collisions at sea.

ARTICLE 40: RESEARCH AND SURVEY ACTIVITIES

During transit passage, foreign ships, including marine scientific research and hydrographic survey ships, may not carry out any research or survey activities without the prior authorization of the States bordering straits.

ARTICLE 49: LEGAL STATUS OF ARCHIPELAGIC WATERS, OF THE AIR SPACE OVER ARCHIPELAGIC WATERS AND OF THEIR BED AND SUBSOIL

1. The sovereignty of an archipelagic State extends to the waters enclosed by the archipelagic baselines drawn in accordance with article 47, described as archipelagic waters, regardless of their depth or distance from the coast.
2. This sovereignty extends to the air space over the archipelagic waters, as well as to their bed and subsoil, and the resources contained therein.
3. This sovereignty is exercised subject to this Part.
4. The regime of archipelagic sea lanes passage established in this Part shall not in other respects affect the status of the archipelagic waters, including the sea lanes, or the exercise by the archipelagic State of its sovereignty over such waters and their air space, bed and subsoil, and the resources contained therein.

ARTICLE 51: EXISTING AGREEMENTS, TRADITIONAL FISHING RIGHTS AND EXISTING SUBMARINE CABLES

1. Without prejudice to article 49, an archipelagic State shall respect existing agreements with other States and shall recognize traditional fishing rights and other legitimate activities of the immediately adjacent neighbouring States in certain areas falling within archipelagic waters. The terms and conditions for the exercise of such rights and activities, including the nature, the extent and the areas to which they apply, shall, at the request of any of the States

concerned, be regulated by bilateral agreements between them. Such rights shall not be transferred to or shared with third States or their nationals.

2. An archipelagic State shall respect existing submarine cables laid by other States and passing through its waters without making a landfall. An archipelagic State shall permit the maintenance and replacement of such cables upon receiving due notice of their location and the intention to repair or replace them.

ARTICLE 54: DUTIES OF SHIPS AND AIRCRAFT DURING THEIR PASSAGE, RESEARCH AND SURVEY ACTIVITIES, DUTIES OF THE ARCHIPELAGIC STATE AND LAWS AND REGULATIONS OF THE ARCHIPELAGIC STATE RELATING TO ARCHIPELAGIC SEA LANES PASSAGE

Articles 39, 40, 42 and 44 apply *mutatis mutandis* to archipelagic sea lanes passage.

ARTICLE 55: SPECIFIC LEGAL REGIME OF THE EXCLUSIVE ECONOMIC ZONE

The exclusive economic zone is an area beyond and adjacent to the territorial sea, subject to the specific legal regime established in this Part, under which the rights and jurisdiction of the coastal State and the rights and freedoms of other States are governed by the relevant provisions of this Convention.

ARTICLE 56: RIGHTS, JURISDICTION AND DUTIES OF THE COASTAL STATE IN THE EXCLUSIVE ECONOMIC ZONE

1. In the exclusive economic zone, the coastal State has:
   (a) sovereign rights for the purpose of exploring and exploiting, conserving and managing the natural resources, whether living or non-living, of the waters superjacent to the seabed and of the seabed and its subsoil, and with regard to other activities for the economic exploitation and exploration of the zone, such as the production of energy from the water, currents and winds;
   (b) jurisdiction as provided for in the relevant provisions of this Convention with regard to:
      (i) the establishment and use of artificial islands, installations and structures;
      (ii) marine scientific research;
      (iii) the protection and preservation of the marine environment;
   (c) other rights and duties provided for in this Convention.

2. In exercising its rights and performing its duties under this Convention in the exclusive economic zone, the coastal State shall have due regard to the rights and duties of other States and shall act in a manner compatible with the provisions of this Convention.
3. The rights set out in this article with respect to the seabed and subsoil shall be exercised in accordance with Part VI.

ARTICLE 58: RIGHTS AND DUTIES OF OTHER STATES IN THE EXCLUSIVE ECONOMIC ZONE

1. In the exclusive economic zone, all States, whether coastal or land-locked, enjoy, subject to the relevant provisions of this Convention, the freedoms referred to in article 87 of navigation and overflight and of the laying of submarine cables and pipelines, and other internationally lawful uses of the sea related to these freedoms, such as those associated with the operation of ships, aircraft and submarine cables and pipelines, and compatible with the other provisions of this Convention.
2. Articles 88 to 115 and other pertinent rules of international law apply to the exclusive economic zone in so far as they are not incompatible with this Part.
3. In exercising their rights and performing their duties under this Convention in the exclusive economic zone, States shall have due regard to the rights and duties of the coastal State and shall comply with the laws and regulations adopted by the coastal State in accordance with the provisions of this Convention and other rules of international law in so far as they are not incompatible with this Part.

ARTICLE 74: DELIMITATION OF THE EXCLUSIVE ECONOMIC ZONE BETWEEN STATES WITH OPPOSITE OR ADJACENT COASTS

1. The delimitation of the exclusive economic zone between States with opposite or adjacent coasts shall be effected by agreement on the basis of international law, as referred to in Article 38 of the Statute of the International Court of Justice, in order to achieve an equitable solution.
2. If no agreement can be reached within a reasonable period of time, the States concerned shall resort to the procedures provided for in Part XV.
3. Pending agreement as provided for in paragraph 1, the States concerned, in a spirit of understanding and cooperation, shall make every effort to enter into provisional arrangements of a practical nature and, during this transitional period, not to jeopardize or hamper the reaching of the final agreement. Such arrangements shall be without prejudice to the final delimitation.

4. Where there is an agreement in force between the States concerned, questions relating to the delimitation of the exclusive economic zone shall be determined in accordance with the provisions of that agreement.

ARTICLE 77: RIGHTS OF THE COASTAL STATE OVER THE CONTINENTAL SHELF

1. The coastal State exercises over the continental shelf sovereign rights for the purpose of exploring it and exploiting its natural resources.
2. The rights referred to in paragraph 1 are exclusive in the sense that if the coastal State does not explore the continental shelf or exploit its natural resources, no one may undertake these activities without the express consent of the coastal State.
3. The rights of the coastal State over the continental shelf do not depend on occupation, effective or notional, or on any express proclamation.
4. The natural resources referred to in this Part consist of the mineral and other non-living resources of the seabed and subsoil together with living organisms belonging to sedentary species, that is to say, organisms which, at the harvestable stage, either are immobile on or under the seabed or are unable to move except in constant physical contact with the seabed or the subsoil.

ARTICLE 78: LEGAL STATUS OF THE SUPERJACENT WATERS AND AIR SPACE AND THE RIGHTS AND FREEDOMS OF OTHER STATES

1. The rights of the coastal State over the continental shelf do not affect the legal status of the superjacent waters or of the air space above those waters.
2. The exercise of the rights of the coastal State over the continental shelf must not infringe or result in any unjustifiable interference with navigation and other rights and freedoms of other States as provided for in this Convention.

ARTICLE 79: SUBMARINE CABLES AND PIPELINES ON THE CONTINENTAL SHELF

1. All States are entitled to lay submarine cables and pipelines on the continental shelf, in accordance with the provisions of this article.
2. Subject to its right to take reasonable measures for the exploration of the continental shelf, the exploitation of its natural resources and the prevention, reduction and control of pollution from pipelines, the coastal State may not impede the laying or maintenance of such cables or pipelines.
3. The delineation of the course for the laying of such pipelines on the continental shelf is subject to the consent of the coastal State.

4. Nothing in this Part affects the right of the coastal State to establish conditions for cables or pipelines entering its territory or territorial sea, or its jurisdiction over cables and pipelines constructed or used in connection with the exploration of its continental shelf or exploitation of its resources or the operations of artificial islands, installations and structures under its jurisdiction.
5. When laying submarine cables or pipelines, States shall have due regard to cables or pipelines already in position. In particular, possibilities of repairing existing cables or pipelines shall not be prejudiced.

ARTICLE 80: ARTIFICIAL ISLANDS, INSTALLATIONS AND STRUCTURES ON THE CONTINENTAL SHELF

Article 60 applies mutatis mutandis to artificial islands, installations and structures on the continental shelf.

ARTICLE 83: DELIMITATION OF THE CONTINENTAL SHELF BETWEEN STATES WITH OPPOSITE OR ADJACENT COASTS

1. The delimitation of the continental shelf between States with opposite or adjacent coasts shall be effected by agreement on the basis of international law, as referred to in Article 38 of the Statute of the International Court of Justice, in order to achieve an equitable solution.
2. If no agreement can be reached within a reasonable period of time, the States concerned shall resort to the procedures provided for in Part XV.
3. Pending agreement as provided for in paragraph 1, the States concerned, in a spirit of understanding and cooperation, shall make every effort to enter into provisional arrangements of a practical nature and, during this transitional period, not to jeopardize or hamper the reaching of the final agreement. Such arrangements shall be without prejudice to the final delimitation.
4. Where there is an agreement in force between the States concerned, questions relating to the delimitation of the continental shelf shall be determined in accordance with the provisions of that agreement.

ARTICLE 87: FREEDOM OF THE HIGH SEAS

1. The high seas are open to all States, whether coastal or land-locked. Freedom of the high seas is exercised under the conditions laid down by this Convention and by other rules of international law. It comprises, *inter alia*, both for coastal and land-locked States:

   (a) freedom of navigation;
   (b) freedom of overflight;
   (c) freedom to lay submarine cables and pipelines, subject to Part VI;

(d) freedom to construct artificial islands and other installations permitted under international law, subject to Part VI;
(e) freedom of fishing, subject to the conditions laid down in section 2;
(f) freedom of scientific research, subject to Parts VI and XIII.

2. These freedoms shall be exercised by all States with due regard for the interests of other States in their exercise of the freedom of the high seas, and also with due regard for the rights under this Convention with respect to activities in the Area.

ARTICLE 89: INVALIDITY OF CLAIMS OF SOVEREIGNTY OVER THE HIGH SEAS

No State may validly purport to subject any part of the high seas to its sovereignty.

ARTICLE 101: DEFINITION OF PIRACY

Piracy consists of any of the following acts:

(a) any illegal acts of violence or detention, or any act of depredation, committed for private ends by the crew or the passengers of a private ship or a private aircraft, and directed:

  (i) on the high seas, against another ship or aircraft, or against persons or property on board such ship or aircraft;
  (ii) against a ship, aircraft, persons or property in a place outside the jurisdiction of any State;

(b) any act of voluntary participation in the operation of a ship or of an aircraft with knowledge of facts making it a pirate ship or aircraft;
(c) any act of inciting or of intentionally facilitating an act described in subparagraph (a) or (b).

ARTICLE 105: SEIZURE OF A PIRATE SHIP OR AIRCRAFT

On the high seas, or in any other place outside the jurisdiction of any State, every State may seize a pirate ship or aircraft, or a ship or aircraft taken by piracy and under the control of pirates, and arrest the persons and seize the property on board. The courts of the State which carried out the seizure may decide upon the penalties to be imposed, and may also determine the action to be taken with regard to the ships, aircraft or property, subject to the rights of third parties acting in good faith.

ARTICLE 110: RIGHT OF VISIT

1. Except where acts of interference derive from powers conferred by treaty, a warship which encounters on the high seas a foreign ship, other than a ship entitled to complete immunity in accordance with articles 95 and 96, is not justified in boarding it unless there is reasonable ground for suspecting that:
   (a) the ship is engaged in piracy;
   (b) the ship is engaged in the slave trade;
   (c) the ship is engaged in unauthorized broadcasting and the flag State of the warship has jurisdiction under article 109;
   (d) the ship is without nationality; or
   (e) though flying a foreign flag or refusing to show its flag, the ship is, in reality, of the same nationality as the warship.
2. In the cases provided for in paragraph 1, the warship may proceed to verify the ship's right to fly its flag. To this end, it may send a boat under the command of an officer to the suspected ship. If suspicion remains after the documents have been checked, it may proceed to a further examination on board the ship, which must be carried out with all possible consideration.
3. If the suspicions prove to be unfounded, and provided that the ship boarded has not committed any act justifying them, it shall be compensated for any loss or damage that may have been sustained.
4. These provisions apply *mutatis mutandis* to military aircraft.
5. These provisions also apply to any other duly authorized ships or aircraft clearly marked and identifiable as being on government service.

ARTICLE 112: RIGHT TO LAY SUBMARINE CABLES AND PIPELINES

1. All States are entitled to lay submarine cables and pipelines on the bed of the high seas beyond the continental shelf.
2. Article 79, paragraph 5, applies to such cables and pipelines.

ARTICLE 113: BREAKING OR INJURY OF A SUBMARINE CABLE OR PIPELINE

Every State shall adopt the laws and regulations necessary to provide that the breaking or injury by a ship flying its flag or by a person subject to its jurisdiction of a submarine cable beneath the high seas done wilfully or through culpable negligence, in such a manner as to be liable to interrupt or obstruct telegraphic or telephonic communications, and similarly the breaking or injury of a submarine pipeline or high-voltage power cable, shall be a punishable offence. This provision shall apply also to conduct calculated or likely to result in such breaking or injury. However, it shall not apply to any break or injury caused by persons who acted merely with the legitimate object of saving their lives or their ships, after having taken all necessary precautions to avoid such break or injury.

ARTICLE 114: BREAKING OR INJURY BY OWNERS OF A SUBMARINE CABLE OR PIPELINE OF ANOTHER SUBMARINE CABLE OR PIPELINE

Every State shall adopt the laws and regulations necessary to provide that, if persons subject to its jurisdiction who are the owners of a submarine cable or pipeline beneath the high seas, in laying or repairing that cable or pipeline, cause a break in or injury to another cable or pipeline, they shall bear the cost of the repairs.

ARTICLE 115: INDEMNITY FOR LOSS INCURRED IN AVOIDING INJURY TO A SUBMARINE CABLE OR PIPELINE

Every State shall adopt the laws and regulations necessary to ensure that the owners of ships who can prove that they have sacrificed an anchor, a net or any other fishing gear, in order to avoid injuring a submarine cable or pipeline, shall be indemnified by the owner of the cable or pipeline, provided that the owner of the ship has taken all reasonable precautionary measures beforehand.

ARTICLE 245: MARINE SCIENTIFIC RESEARCH IN THE TERRITORIAL SEA

Coastal States, in the exercise of their sovereignty, have the exclusive right to regulate, authorize and conduct marine scientific research in their territorial sea. Marine scientific research therein shall be conducted only with the express consent of and under the conditions set forth by the coastal State.

ARTICLE 246: MARINE SCIENTIFIC RESEARCH IN THE EXCLUSIVE ECONOMIC ZONE AND ON THE CONTINENTAL SHELF

1. Coastal States, in the exercise of their jurisdiction, have the right to regulate, authorize and conduct marine scientific research in their exclusive economic zone and on their continental shelf in accordance with the relevant provisions of this Convention.
2. Marine scientific research in the exclusive economic zone and on the continental shelf shall be conducted with the consent of the coastal State.
3. Coastal States shall, in normal circumstances, grant their consent for marine scientific research projects by other States or competent international organizations in their exclusive economic zone or on their continental shelf to be carried out in accordance with this Convention exclusively for peaceful purposes and in order to increase scientific knowledge of the marine environment

for the benefit of all mankind. To this end, coastal States shall establish rules and procedures ensuring that such consent will not be delayed or denied unreasonably.

4. For the purposes of applying paragraph 3, normal circumstances may exist in spite of the absence of diplomatic relations between the coastal State and the researching State.
5. Coastal States may however in their discretion withhold their consent to the conduct of a marine scientific research project of another State or competent international organization in the exclusive economic zone or on the continental shelf of the coastal State if that project:

   (a) is of direct significance for the exploration and exploitation of natural resources, whether living or non-living;
   (b) involves drilling into the continental shelf, the use of explosives or the introduction of harmful substances into the marine environment;
   (c) involves the construction, operation or use of artificial islands, installations and structures referred to in articles 60 and 80;
   (d) contains information communicated pursuant to article 248 regarding the nature and objectives of the project which is inaccurate or if the researching State or competent international organization has outstanding obligations to the coastal State from a prior research project.

6. Notwithstanding the provisions of paragraph 5, coastal States may not exercise their discretion to withhold consent under subparagraph (a) of that paragraph in respect of marine scientific research projects to be undertaken in accordance with the provisions of this Part on the continental shelf, beyond 200 nautical miles from the baselines from which the breadth of the territorial sea is measured, outside those specific areas which coastal States may at any time publicly designate as areas in which exploitation or detailed exploratory operations focused on those areas are occurring or will occur within a reasonable period of time. Coastal States shall give reasonable notice of the designation of such areas, as well as any modifications thereto, but shall not be obliged to give details of the operations therein.
7. The provisions of paragraph 6 are without prejudice to the rights of coastal States over the continental shelf as established in article 77.
8. Marine scientific research activities referred to in this article shall not unjustifiably interfere with activities undertaken by coastal States in the exercise of their sovereign rights and jurisdiction provided for in this Convention.

ARTICLE 286: APPLICATION OF PROCEDURES UNDER THIS SECTION

Subject to section 3, any dispute concerning the interpretation or application of this Convention shall, where no settlement has been reached by recourse to section 1, be submitted at the request of any party to the dispute to the court or tribunal having jurisdiction under this section.

ARTICLE 297: LIMITATIONS ON APPLICABILITY OF SECTION 2

1. Disputes concerning the interpretation or application of this Convention with regard to the exercise by a coastal State of its sovereign rights or jurisdiction provided for in this Convention shall be subject to the procedures provided for in section 2 in the following cases:

   (a) when it is alleged that a coastal State has acted in contravention of the provisions of this Convention in regard to the freedoms and rights of navigation, overflight or the laying of submarine cables and pipelines, or in regard to other internationally lawful uses of the sea specified in article 58;
   (b) when it is alleged that a State in exercising the aforementioned freedoms, rights or uses has acted in contravention of this Convention or of laws or regulations adopted by the coastal State in conformity with this Convention and other rules of international law not incompatible with this Convention; or
   (c) when it is alleged that a coastal State has acted in contravention of specified international rules and standards for the protection and preservation of the marine environment which are applicable to the coastal State and which have been established by this Convention or through a competent international organization or diplomatic conference in accordance with this Convention.

2.

   (a) Disputes concerning the interpretation or application of the provisions of this Convention with regard to marine scientific research shall be settled in accordance with section 2, except that the coastal State shall not be obliged to accept the submission to such settlement of any dispute arising out of:

      (i) the exercise by the coastal State of a right or discretion in accordance with article 246; or
      (ii) a decision by the coastal State to order suspension or cessation of a research project in accordance with article 253.

   (b) A dispute arising from an allegation by the researching State that with respect to a specific project the coastal State is not exercising its rights under articles 246 and 253 in a manner compatible with this Convention shall be submitted, at the request of either party, to conciliation under

Annex V, section 2, provided that the conciliation commission shall not call in question the exercise by the coastal State of its discretion to designate specific areas as referred to in article 246, paragraph 6, or of its discretion to withhold consent in accordance with article 246, paragraph 5.

3.

(a) Disputes concerning the interpretation or application of the provisions of this Convention with regard to fisheries shall be settled in accordance with section 2, except that the coastal State shall not be obliged to accept the submission to such settlement of any dispute relating to its sovereign rights with respect to the living resources in the exclusive economic zone or their exercise, including its discretionary powers for determining the allowable catch, its harvesting capacity, the allocation of surpluses to other States and the terms and conditions established in its conservation and management laws and regulations.

(b) Where no settlement has been reached by recourse to section 1 of this Part, a dispute shall be submitted to conciliation under Annex V, section 2, at the request of any party to the dispute, when it is alleged that:

(i) a coastal State has manifestly failed to comply with its obligations to ensure through proper conservation and management measures that the maintenance of the living resources in the exclusive economic zone is not seriously endangered;

(ii) a coastal State has arbitrarily refused to determine, at the request of another State, the allowable catch and its capacity to harvest living resources with respect to stocks which that other State is interested in fishing; or

(iii) a coastal State has arbitrarily refused to allocate to any State, under articles 62, 69 and 70 and under the terms and conditions established by the coastal State consistent with this Convention, the whole or part of the surplus it has declared to exist.

(c) In no case shall the conciliation commission substitute its discretion for that of the coastal State.

(d) The report of the conciliation commission shall be communicated to the appropriate international organizations.

(e) In negotiating agreements pursuant to articles 69 and 70, States Parties, unless they otherwise agree, shall include a clause on measures which they shall take in order to minimize the possibility of a disagreement concerning the interpretation or application of the agreement, and on how they should proceed if a disagreement nevertheless arises.

1958 CONVENTION ON THE HIGH SEAS, 29 APRIL 1958, 450 U.N.T.S. 11 (ENTERED INTO FORCE 30 SEPTEMBER 1962)

ARTICLE 26

1. All States shall be entitled to lay submarine cables and pipelines on the bed of the high seas.
2. Subject to its right to take reasonable measures for the exploration of the continental shelf and the exploitation of its natural resources, the coastal State may not impede the laying or maintenance of such cables or pipelines.
3. When laying such cables or pipelines the State in question shall pay due regard to cables or pipelines already in position on the seabed. In particular, possibilities of repairing existing cables or pipelines shall not be prejudiced.

ARTICLE 27

Every State shall take the necessary legislative measures to provide that the breaking or injury by a ship flying its flag or by a person subject to its jurisdiction of a submarine cable beneath the high seas done willfully or through culpable negligence, in such a manner as to be liable to interrupt or obstruct telegraphic or telephonic communications, and similarly the breaking or injury of a submarine pipeline or high-voltage power cable shall be a punishable offence. This provision shall not apply to any break or injury caused by persons who acted merely with the legitimate object of saving their lives or their ships, after having taken all necessary precautions to avoid such break or injury.

ARTICLE 28

Every State shall take the necessary legislative measures to provide that, if persons subject to its jurisdiction who are the owners of a cable or pipeline beneath the high seas, in laying or repairing that cable or pipeline, cause a break in or injury to another cable or pipeline, they shall bear the cost of the repairs.

ARTICLE 29

Every State shall take the necessary legislative measures to ensure that the owners of ships who can prove that they have sacrificed an anchor, a net or any other fishing gear, in order to avoid injuring a submarine cable or pipeline, shall be indemnified by the owner of the cable or pipeline, provided that the owner of the ship has taken all reasonable precautionary measures beforehand.

ARTICLE 30

The provisions of this Convention shall not affect conventions or other international agreements already in force, as between States Parties to them.

1958 CONVENTION ON THE CONTINENTAL SHELF, 29 APRIL 1958, 499 U.N.T.S. 311 (ENTERED INTO FORCE 10 JUNE 1964)

ARTICLE 4

Subject to its right to take reasonable measures for the exploration of the continental shelf and the exploitation of its natural resources, the coastal State may not impede the laying or maintenance of submarine cables or pipe lines on the continental shelf.

1972 CONVENTION ON THE INTERNATIONAL REGULATIONS FOR PREVENTING COLLISIONS AT SEA, 20 OCTOBER 1972, 1050 U.N.T.S. 16 (ENTERED INTO FORCE 15 JULY 1977)

RULE 3: GENERAL DEFINITIONS

For the purpose of these Rules, except where the context otherwise requires:

(a) The word "vessel" includes every description of water craft, including non-displacement craft and seaplanes, used or capable of being used as a means of transportation on water.
(b) The term "power-driven vessel" means any vessel propelled by machinery.
(c) The term "sailing vessel" means any vessel under sail provided that propelling machinery, if fitted, is not being used.
(d) The term "vessel engaged in fishing" means any vessel fishing with nets, lines, trawls or other fishing apparatus which restrict manoeuvrability, but does not include a vessel fishing with trolling lines or other fishing apparatus which do not restrict manoeuvrability.
(e) The word "seaplane" includes any aircraft designed to manoeuvre on the water.
(f) The term "vessel not under command" means a vessel which through some exceptional circumstance is unable to manoeuvre as required by these Rules and is therefore unable to keep out of the way of another vessel.
(g) The term "vessel restricted in her ability to manoeuvre" means a vessel which from the nature of her work is restricted in her ability to manoeuvre as required by these Rules and is therefore unable to keep out of the way of another vessel.

The following vessels shall be regarded as vessels restricted in their ability to manoeuvre:

(i) a vessel engaged in laying, servicing or picking up a navigation mark, submarine cable or pipeline;
(ii) a vessel engaged in dredging, surveying or underwater operations;
(iii) a vessel engaged in replenishment or transferring persons, provisions or cargo while underway;
(iv) a vessel engaged in the launching or recovery of aircraft;
(v) a vessel engaged in minesweeping operations;
(vi) a vessel engaged in a towing operation such as renders her unable to deviate from her course.

(h) The term "vessel constrained by her draught" means a power-driven vessel which because of her draught in relation to the available depth of water is severely restricted in her ability to deviate from the course she is following.
(i) The word "underway" means that a vessel is not at anchor, or made fast to the shore, or aground.
(j) The words "length" and "breath" of a vessel mean her length overall and greatest breadth.
(k) Vessels shall be deemed to be in sight of one another only when one can be observed visually from the other.
(l) The term "restricted visibility" means any condition in which visibility is restricted by fog, mist, falling snow, heavy rainstorms, sandstorms or any other similar causes.

RULE 18: RESPONSIBILITIES BETWEEN VESSELS

Except where Rules 9, 10 and 13 otherwise require:

(a) A power-driven vessel underway shall keep out of the way of:

(i) a vessel not under command;
(ii) a vessel restricted in her ability to manoeuvre;
(iii) a vessel engaged in fishing;
(iv) a sailing vessel.

(b) A sailing vessel underway shall keep out of the way of:

(i) a vessel not under command;
(ii) a vessel restricted in her ability to manoeuvre;
(iii) a vessel engaged in fishing.

(c) A vessel engaged in fishing when underway shall, so far as possible, keep out of the way:

(i) a vessel not under command;
(ii) a vessel restricted in her ability to manoeuvre.

(d)

(i) Any vessel other than a vessel not under command or a vessel restricted in her ability to manoeuvre shall, if the circumstances of the case admit, avoid impeding the safe passage of a vessel constrained by her draught, exhibiting the signals in Rule 28.
(ii) A vessel constrained by her draught shall navigate with particular caution having full regard to her special condition.

(e) A seaplane on the water shall, in general, keep well clear of all vessels and avoid impeding their navigation. In circumstances, however, where risk of collision exists, she shall comply with the Rules of this Part.

RULE 27: VESSELS NOT UNDER COMMAND OR RESTRICTED IN THEIR ABILITY TO MANOEUVRE

(a) A vessel not under command shall exhibit:

(i) two all-round red lights in a vertical line where they can best be seen;
(ii) two balls or similar shapes in a vertical line where they can best be seen;
(iii) when making way through the water, in addition to the lights prescribed in this paragraph, sidelights and a sternlight.

(b) A vessel restricted in her ability to manoeuvre, except a vessel engaged in minesweeping operations, shall exhibit:

(i) three all-round lights in a vertical line where they can best be seen. The highest and lowest of these lights shall be red and the middle light shall be white;
(ii) three shapes in a vertical line where they can best be seen. The highest and lowest of these shapes shall be balls and the middle one a diamond;
(iii) when making way through the water, masthead lights, sidelights and a sternlight, in addition to the lights prescribed in sub-paragraph (i);
(iv) when at anchor, in addition to the lights or shapes prescribed in sub-paragraphs (i) and (ii), the light, lights or shape prescribed in Rule 30.

(c) A vessel engaged in a towing operation such as renders her unable to deviate from her course shall, in addition to the lights or shapes prescribed in sub-paragraph (b)(i) and (ii) of this Rule, exhibit the lights or shape prescribed in Rule 24(a).

(d) A vessel engaged in dredging or underwater operations, when restricted in her ability to manoeuvre, shall exhibit the lights and shapes prescribed in paragraph (b) of this Rule and shall in addition, when an obstruction exists, exhibit:

(i) two all-round red lights or two balls in a vertical line to indicate the side on which the obstruction exists;
(ii) two all-round green lights or two diamonds in a vertical line to indicate the side on which another vessel may pass;
(iii) when making way through the water, in addition to the lights prescribed in this paragraph, masthead lights, sidelights and a sternlight;
(iv) a vessel to which this paragraph applies when at anchor shall exhibit the lights or shapes prescribed in sub-paragraphs (i) and (ii) instead of the lights or shape prescribed in Rule 30.

(e) Whenever the size of a vessel engaged in diving operations makes it impracticable to exhibit the shapes prescribed in paragraph (d) of this Rule, a rigid replica of the International Code flag "A" not less than 1 metre in height shall be exhibited. Measures shall be taken to ensure all-round visibility.
(f) A vessel engaged in minesweeping operations shall, in addition to the lights prescribed for a power-driven vessel in Rule 23, exhibit three all-round green lights or three balls. One of these lights or shapes shall be exhibited at or near the foremast head and one at each end of the fore yard. These lights or shapes indicate that it is dangerous for another vessel to approach closer than 1,000 metres astern or 500 metres on either side of the minesweeper.
(g) Vessels of less than 7 metres in length shall not be required to exhibit the lights prescribed in this Rule.
(h) The signals prescribed in this Rule are not signals of vessels in distress and requiring assistance. Such signals are contained in Annex IV to these Regulations.

1884 CONVENTION FOR THE PROTECTION OF SUBMARINE TELEGRAPH CABLES, 14 MARCH 1884, TS 380 (ENTERED INTO FORCE 1 MAY 1888)

ARTICLE I

The present Convention applies outside territorial waters to all legally established submarine cables landed on the territories, colonies or possessions of one or more of the High Contracting Parties.

ARTICLE II

It is a punishable offence to break or injure a submarine cable, wilfully or by culpable negligence, in such manner as might interrupt or obstruct telegraphic communication, either wholly or partially, such punishment being without prejudice to any civil action for damages.

This provision does not apply to cases where those who break or injure a cable do so with the lawful object of saving their lives or their ship, after they have taken every necessary precaution to avoid so breaking or injuring the cable.

ARTICLE III

The High Contracting Parties undertake that, on granting a concession for landing a submarine cable, they will insist, so far as possible, upon proper measures of safety being taken, both as regards the track of the cable and its dimensions.

ARTICLE IV

The owner of a cable who, on laying or repairing his own cable, breaks or injures another cable, must bear the cost of repairing the breakage or injury, without prejudice to the application, if need by, of Article II of the present Convention.

ARTICLE V

Vessels engaged in laying or repairing submarine cables shall conform to the regulations as to signals which have been, or may be, adopted by mutual agreement among the High Contracting Parties, with the view of preventing collisions at sea.

When a ship engaged in repairing a cable exhibits the said signals, other vessels which see them, or are able to see them, shall withdraw to or keep beyond a distance of one nautical mile at least from the ship in question, so as not to interfere with her operations.

Fishing gear and nets shall be kept at the same distance.

Nevertheless, fishing vessels which see, or are able to see, a telegraph-ship exhibiting the said signals, shall be allowed a period of 24 hours at most within which to obey the notice so given, during which time they shall not be interfered with in any way.

The operations of the telegraph-ships shall be completed as quickly as possible.

ARTICLE VI

Vessels which see, or are able to see, the buoys showing the position of a cable when the latter is being laid, is out of order, or is broken, shall keep beyond a distance of one-quarter of a nautical mile at least from the said buoys.

Fishing nets and gear shall be kept at the same distance.

ARTICLE VII

Owners of ships or vessels who can prove that they have sacrificed an anchor, a net, or other fishing gear in order to avoid injuring a submarine cable, shall receive compensation from the owner of the cable.

In order to establish a claim to such compensation, a statement, supported by the evidence of the crew, should, whenever possible, be drawn up immediately after the occurrence; and the master must, within 24 hours after his return to or next putting into port, make a declaration to the proper authorities.

The latter shall communicate the information to the consular authorities of the country to which the owner of the cable belongs.

ARTICLE VIII

The tribunals competent to take cognizance of infractions of the present Convention are those of the country to which the vessel on board of which the offence was committed belongs.

It is, moreover, understood that, in cases where the provisions in the previous paragraph cannot apply, offences against the present Convention will be dealt with in each of the Contracting States in accordance, so far as the subjects and citizens of those States respectively are concerned, with the general rules of criminal jurisdiction prescribed by the laws of that particular State, or by international treaties.

ARTICLE IX

Prosecutions for infractions provided against by Articles II, V and VI of the present Convention shall be instituted by the State, or in its name.

ARTICLE X

Offences against the present Convention may be verified by all means of proof allowed by the legislation of the country of the court. When the officers commanding the ships of war, or ships specially commissioned for the purpose by one of the High Contracting Parties, have reason to believe that an infraction of the measures provided for in the present Convention has been committed by a vessel other than a vessel of war, they may demand from the captain or master the production of the official documents proving the nationality of the said vessel. The fact of such document having been exhibited shall then be endorsed upon it immediately. Further, formal statements of the facts may be prepared by the said officers, whatever may be the nationality of the vessel incriminated. These formal statements shall be drawn up in the form and in the language used in the country to which the officer making them belongs; they may be considered, in the country where they are adduced, as evidence in accordance with the laws of that country. The accused and the witnesses shall have the right to add, or to have added thereto, in their own language, any explanations they may consider useful. These declarations shall be duly signed.

ARTICLE XI

The proceedings and trial in cases of infraction of the provisions of the present Convention shall always take place as summarily as the laws and regulations in force will permit.

ARTICLE XII

The High Contracting Parties engage to take or to propose to their respective legislatures the necessary measures for insuring the execution of the present Convention, and especially for punishing, by either fine or imprisonment, or both, those who contravene the provisions of Articles II, V and VI.

ARTICLE XIII

The High Contracting Parties will communicate to each other laws already made, or which may hereafter be made, in their respective countries, relating to the object of the present Convention.

ARTICLE XIV

States which have not signed the present Convention may adhere to it on making a request to that effect. This adhesion shall be notified through the diplomatic channel to the Government of the French Republic, and by the latter to the other Signatory Powers.

ARTICLE XV

It is understood that the stipulations of the present Convention do not in any way restrict the freedom of action of belligerents.

ARTICLE XVI

The present Convention shall be brought into force on a day to be agreed upon by the High Contracting Powers.[1]

---

[1] The Convention entered into force for Queensland, South Australia and Victoria, and generally, 1 May 1888 pursuant to Final Protocol of 7 July 1887.

It shall remain in force for five years from that day, and unless any of the High Contracting Parties have announced, 12 months before the expiration of the said period of five years, its intention to terminate its operation, it shall continue in force for a period of one year, and so on from year to year.

If one of the Signatory Powers denounces the Convention, such denunciation shall have effect only as regards that Power.

DECLARATION, EXPLANATORY OF ARTICLES II AND IV, OF THE PLENIPOTENTIARIES OF THE SIGNATORY GOVERNMENTS OF THE CONVENTION FOR THE PROTECTION OF SUBMARINE TELEGRAPH CABLES OF 14 MARCH 1884

The undersigned plenipotentiaries of the Signatory Governments of the Convention of 14 March 1884 for the protection of submarine cables, having recognised the expediency of stating precisely the meaning of the terms of Articles II and IV of the said Convention, have agreed upon the following Declaration by common consent:

Certain doubts having been raised as to the meaning of the word "wilfully" used in Article II of the Convention of 14 March 1884, it is understood that the provision in respect of penal responsibility contained in the said Article does not apply to cases of breakage or injury caused accidentally or of necessity in the repair of a cable, when all precautions have been taken to avoid such breakage or injury.

It is equally understood that Article IV of the Convention had no other object and is to have no other effect than to empower the competent tribunals of each country to decide in conformity with their laws and according to the circumstances, the question of the civil responsibility of the owner of a cable, who, in laying or repairing his own cable, breaks or injures another cable, as well as the consequences of such responsibility if it is recognized as existing.

2001 CONVENTION ON THE PROTECTION OF THE UNDERWATER CULTURAL HERITAGE, 2 NOVEMBER 2001, 41 I.L.M. 37 (2002) (ENTERED INTO FORCE 2 JANUARY 2009)

ARTICLE 1—DEFINITIONS

For the purposes of this Convention:

1. (a) "Underwater cultural heritage" means all traces of human existence having a cultural, historical or archaeological character which have been partially or totally under water, periodically or continuously, for at least 100 years such as:

(i) sites, structures, buildings, artefacts and human remains, together with their archaeological and natural context;
(ii) vessels, aircraft, other vehicles or any part thereof, their cargo or other contents, together with their archaeological and natural context; and
(iii) objects of prehistoric character.

(b) Pipelines and cables placed on the seabed shall not be considered as underwater cultural heritage.
(c) Installations other than pipelines and cables, placed on the seabed and still in use, shall not be considered as underwater cultural heritage.

2. (a) "States Parties" means States which have consented to be bound by this Convention and for which this Convention is in force.
(b) This Convention applies *mutatis mutandis* to those territories referred to in Article 26, paragraph 2(b), which become Parties to this Convention in accordance with the conditions set out in that paragraph, and to that extent "States Parties" refers to those territories.

3. "UNESCO" means the United Nations Educational, Scientific and Cultural Organization.
4. "Director-General" means the Director-General of UNESCO.
5. "Area" means the seabed and ocean floor and subsoil thereof, beyond the limits of national jurisdiction.
6. "Activities directed at underwater cultural heritage" means activities having underwater cultural heritage as their primary object and which may, directly or indirectly, physically disturb or otherwise damage underwater cultural heritage.
7. "Activities incidentally affecting underwater cultural heritage" means activities which, despite not having underwater cultural heritage as their primary object or one of their objects, may physically disturb or otherwise damage underwater cultural heritage.
8. "State vessels and aircraft" means warships, and other vessels or aircraft that were owned or operated by a State and used, at the time of sinking, only for government non-commercial purposes, that are identified as such and that meet the definition of underwater cultural heritage.
9. "Rules" means the Rules concerning activities directed at underwater cultural heritage, as referred to in Article 33 of this Convention.

INDEX

www.ingramcontent.com/pod-product-compliance
Lightning Source LLC
LaVergne TN
LVHW020631300926
849161LV00004B/665